THE FACTS ON FILE
SPACE AND ASTRONOMY
HANDBOOK

JOSEPH A. ANGELO, JR.
Adjunct Professor, College of Engineering, Florida Tech

Checkmark Books®
An imprint of Facts On File, Inc.

The Facts On File Space and Astronomy Handbook
Copyright © 2002 by Joseph A. Angelo, Jr.

Checkmark Books
An imprint of Facts On File, Inc.
132 West 31st Street
New York NY 10001

Library of Congress Cataloging-in-Publication Data
Angelo, Joseph A.
   The Facts on File space and astronomy handbook/Joseph A. Angelo, Jr.
      p. cm.
   Includes bibliographical references and index.
   ISBN 0-8160-4542-9 (hc: acid-free paper)   ISBN 0-8160-4960-2 (pbk)
   1. Astronomy—Handbooks, manuals, etc. 2. Space sciences—Handbooks, manuals, etc.
I. Facts on File Inc. II. Title.
QB43.3 .A44 2002
520—dc21   2001054323

Checkmark Books are available at special discounts when purchased in bulk quantities for businesses, associations, institutions, or sales promotions. Please call our Special Sales Department in New York at 212/967-8800 or 800/322-8755.

You can find Facts On File on the World Wide Web at
http://www.factsonfile.com

Cover design by Cathy Rincon
Illustrations by Sholto S. Ainslie, © Facts On File

Printed in the United States of America

VB Hermitage 10 9 8 7 6 5 4 3 2 1

This book is printed on acid-free paper.

R0402263048

To my sons, Joseph and James,
who grew up in the red glare of the rockets from Cape Canaveral
as the United States reached for the stars.

CONTENTS

ACKNOWLEDGMENTS

This book could not have been prepared without the generous support and assistance of the National Aeronautics and Space Administration (NASA), especially the Media Services Office at the Lyndon B. Johnson Space Center. Special thanks also go to the staff at the Evans Library of Florida Tech, whose sustained, high-quality assistance proved essential in the successful development of this project.

Finally, without the patient and loving support of my wife, Joan, this book would never have evolved from chaotic piles of manuscript into the draft version that the editorial team at Facts On File, particularly editor Frank K. Darmstadt, transformed into a pleasing final product.

INTRODUCTION

An understanding of astronomy and space exploration is the basis for discovering the universe and how it works. Our daily lives, exciting new materials, and the information-rich, space-age civilization we now enjoy have been developed only through scientific research into the principles that underpin the physical world. However, obtaining a full view of any branch of science may be difficult without resorting to a range of books. Dictionaries of terms, encyclopedias of facts, biographical dictionaries, chronologies of scientific events—all of these collections of facts usually encompass a range of scientific subjects. THE FACTS ON FILE HANDBOOK LIBRARY covers four major scientific areas— CHEMISTRY, PHYSICS, EARTH SCIENCE (including astronomy and space exploration), and BIOLOGY.

THE FACTS ON FILE SPACE AND ASTRONOMY HANDBOOK contains four sections—a glossary of terms, biographies of personalities, a chronology of events, and essential charts and tables. It also contains an extensive index.

GLOSSARY
The specialized words used in any science subject mean that students need a glossary in order to understand the phenomena and processes involved. THE FACTS ON FILE SPACE AND ASTRONOMY HANDBOOK glossary contains more than 1,200 entries, often accompanied by labeled illustrations and diagrams to help clarify the meanings.

BIOGRAPHIES
The giants of astronomy and space—Copernicus, Galileo, Newton, and Goddard—are widely known, but hundreds of other dedicated scientists contributed to scientific knowledge. THE FACTS ON FILE SPACE AND ASTRONOMY HANDBOOK contains biographies of more than 400 people. Many of their achievements may have gone unnoticed. However, their discoveries have pushed forward the world's understanding of space and astronomy.

CHRONOLOGY
Scientific discoveries often have no immediate impact. Nevertheless, their effects can influence lives more than wars, political changes, and

world rulers. THE FACTS ON FILE SPACE AND ASTRONOMY HANDBOOK covers nearly 8,000 years of events in the history of discoveries in astronomy and space exploration.

CHARTS & TABLES

Basic information on any subject can be hard to find, and books tend to be descriptive. THE FACTS ON FILE SPACE AND ASTRONOMY HANDBOOK puts together key charts and tables for easy reference. Scientific discoveries mean that any compilation of facts can never be comprehensive. Nevertheless, this assembly of current information about space and astronomy offers an important resource for today's students. In past centuries, scientists were curious about a wide range of sciences. Today, with disciplines so specialized and independent, students of one subject rarely learn much about others or how the subjects relate. THE FACTS ON FILE HANDBOOKS enable students to compare knowledge in biology, chemistry, earth science, and physics; to put each subject into context; and to understand the close connections between all the sciences.

SECTION ONE
GLOSSARY

Abell cluster A rich (high-density) CLUSTER OF GALAXIES as characterized by the American astronomer George Abell (1927–83). In 1958, Abell produced a catalog describing over 2,700 of these high-density galactic clusters using PALOMAR OBSERVATORY photographic data.

aberration of starligh The tiny apparent displacement of the position of a STAR from its true position due to a combination of the finite VELOCITY of LIGHT (symbol c), about 300,000 km/s, and the motion of an observer across the path of the incident starlight. For example, an astronomer on EARTH's surface has a velocity of about 30 km/s—the average speed of Earth in its ORBIT around the SUN. This motion causes an annual aberration of starlight.

ablation The removal of surface material from a body by vaporization, melting, sublimation, or other erosive processes. Ablation is a special form of heat transfer called *mass transfer cooling.* Aerospace engineers use this sacrificial phenomenon to provide thermal protection to the underlying structure of a REENTRY VEHICLE, PLANETARY PROBE, or AEROSPACE VEHICLE during high-speed movement through a planetary atmosphere.

ablative cooling Temperature reduction achieved by vaporization or melting of special, sacrificial surface materials.

abort To cut short or cancel an operation with a ROCKET, SPACECRAFT, or AEROSPACE VEHICLE, especially because of equipment failure. NASA's SPACE SHUTTLE system has two types of abort modes during the ascent phase of a flight: the intact abort and the contingency abort. An intact abort is designed to achieve a safe return of the ASTRONAUT crew and ORBITER vehicle to a planned landing site. A contingency abort involves a ditching operation in which the crew is saved, but the orbiter vehicle is damaged or destroyed.

absolute magnitude (M) The measure of the brightness (or APPARENT MAGNITUDE) that a STAR would have if it were hypothetically located at a reference distance of 10 PARSECs (10 pc), about 32.6 LIGHT-YEARS, from the SUN.

absolute temperature The temperature value relative to ABSOLUTE ZERO, which corresponds to 0 K or –273.15°C (after ANDERS CELSIUS). In the international system (SI) of units, absolute temperatures are expressed in degrees kelvin (K), a unit named in honor of the Scottish physicist BARON WILLIAM THOMSON KELVIN.

absolute zero The temperature at which molecular motion vanishes and an object has no thermal energy (or heat). Absolute zero is the lowest possible temperature.

absorption line The gap, dip, or dark-line feature in a stellar SPECTRUM occurring at a specific WAVELENGTH. It is caused by the absorption of the radiation emitted from a STAR's hotter interior regions by an absorbing substance in its relatively cooler outer regions. Analysis of absorption lines lets astronomers determine the chemical composition of stars.

absorption spectrum The collection of dark lines superimposed upon a continuous SPECTRUM that occurs when RADIATION from a hot source passes through a cooler medium, allowing some of that radiant energy to get absorbed at selected WAVELENGTHS.

abundance of elements (*in the universe*) Stellar spectra provide an estimate of the cosmic abundance of ELEMENTS as a percentage of the total MASS of the UNIVERSE. The 10 most common elements are HYDROGEN (H) at 73.5 percent of the total mass, HELIUM (He) at 24.9 percent, oxygen (O) at 0.7 percent, carbon (C) at 0.3 percent, iron (Fe) at 0.15 percent, neon (Ne) at 0.12 percent, nitrogen (N) at 0.10 percent, silicon (Si) at 0.07 percent, magnesium (Mg) at 0.05 percent, and sulfur (S) at 0.04 percent.

accelerated life tests The series of test procedures for a SPACECRAFT or AEROSPACE system that approximate in a relatively short period of time the deteriorating effects and possible failures that might be encountered under normal, long-term space mission conditions.

acceleration (*a*) The rate at which the velocity of an object changes with time. Acceleration is a VECTOR quantity and has the physical dimensions of length per unit time to the second power (for example, meters per second per second, or m/s^2).

acceleration of gravity The local ACCELERATION due to GRAVITY on or near the surface of a PLANET. On EARTH, the acceleration due to gravity (*g*) of a FREE-FALLing object has the standard value of $9.80665\ m/s^2$ by international agreement. According to legend, GALILEO GALILEI simultaneously dropped a large and small cannonball from the top of the Tower of Pisa to investigate the acceleration of gravity. As he anticipated, each object fell to the ground in exactly the same amount of time (neglecting air resistance)—despite the difference in their MASSes. Galileo's pioneering work helped SIR ISAAC NEWTON unlock the secrets of motion of the mechanical UNIVERSE.

accelerometer An instrument that measures ACCELERATION or gravitational FORCES capable of imparting acceleration. It is frequently used on SPACE VEHICLES to assist in guidance and navigation and on PLANETARY PROBES to support scientific data collection.

accretion The gradual accumulation of small PARTICLES of gas and dust into larger material bodies, mostly due to the influence of GRAVITY. For example, in the early stages of stellar formation, matter begins to collect or accrete into a NEBULA (a giant interstellar cloud of gas and dust). Eventually, STARs are born in this nebula. When a particular star forms, small quantities of residual matter may collect into one or more PLANETs that orbit the new star.

accretion disk The whirling disk of inflowing (or infalling) material from a normal stellar companion that develops around a massive COMPACT BODY, such as a NEUTRON STAR or a BLACK HOLE. The conservation of ANGULAR MOMENTUM shapes this disk, which is often accompanied by a pair of very high-speed material jets that depart in opposite directions perpendicular to the plane of the disk.

Achilles The first ASTEROID of the TROJAN GROUP discovered. This 115 km diameter MINOR PLANET was found by MAXIMILLIAN WOLF in 1906 and is also called Asteroid 588.

acquisition The process of locating the ORBIT of a SATELLITE or the TRAJECTORY of a SPACE PROBE so that mission control personnel can track the object and collect its TELEMETRY data.

acronym A word formed from the first letters of a name, such as HST—which means the *HUBBLE* SPACE TELESCOPE. It is also a word formed by combining the initial parts of a series of words, such as lidar—which means light detection and ranging. Acronyms are frequently used in space technology and astronomy.

active galactic nucleus (AGN) The central region of a distant (active) GALAXY that appears to be a pointlike source of intense X-RAY or GAMMA RAY emissions. Astrophysicists speculate that the AGN is caused by the presence of a centrally located, super-heavy BLACK HOLE accreting nearby matter.

active galaxies Collectively, those unusual celestial objects, including QUASARs, BL LAC OBJECTS, and SEYFERT GALAXIES, that have extremely energetic central regions, called ACTIVE GALACTIC NUCLEI (AGN). These emit enormous amounts of ELECTROMAGNETIC RADIATION, ranging from RADIO WAVES to X RAYS and GAMMA RAYS.

Companion Star (Normal)

Material Inflow

Jet

Accretion Disk

Jet

Neutron Star or Black Hole

Accretion disk

active remote sensing A REMOTE-SENSING technique in which the sensor supplies its own source of ELECTROMAGNETIC RADIATION to illuminate a target. A SYNTHETIC APERTURE RADAR (SAR) system is an example.

active satellite A SATELLITE that transmits a signal, in contrast to a passive (dormant) satellite.

active Sun The name scientists give to the collection of dynamic SOLAR phenomena, including SUNSPOTS, SOLAR FLARES, and PROMINENCES, associated with intense variations in the SUN's magnetic activity. *Compare with* QUIET SUN.

adapter skirt A flange or extension on a LAUNCH VEHICLE stage or SPACECRAFT section that provides a means of fitting on another stage or section.

adaptive optics Optical systems, such as TELESCOPES, that are modified to compensate for distortions, usually through the use of a component mirror whose shape can be easily changed and controlled. In ground-based observational astronomy, adaptive optics helps eliminate the twinkling of STARS caused by variations and distortions in EARTH's intervening ATMOSPHERE.

adiabatic A process or phenomenon that takes place without gain or loss of thermal ENERGY (heat).

aero- A prefix that means of or pertaining to the AIR, the ATMOSPHERE, aircraft, or flight through a PLANET's atmosphere.

aeroassist The use of the thin, upper regions of a planet's ATMOSPHERE to provide the lift or drag needed to maneuver a SPACECRAFT. Near a PLANET with a SENSIBLE ATMOSPHERE, aeroassist allows a spacecraft to change direction or to slow down without expending PROPELLANT from the CONTROL ROCKET.

aerobraking The use of a specially designed SPACECRAFT structure to deflect rarefied (very low-density) airflow around a spacecraft, thereby supporting AEROASSIST maneuvers in the vicinity of a PLANET. Such maneuvers reduce the spacecraft's need to perform the large propulsive burns when making orbital changes near a planet. In 1993, NASA's *MAGELLAN* MISSION became the first planetary exploration system to use aerobraking as a means of changing its ORBIT around the target planet (VENUS).

aerodynamic force The lift *(L)* or drag *(D)* exerted by a moving gas upon a body completely immersed in it. Lift acts in a direction normal to the flight path, while drag acts in a direction parallel and opposite to the flight path. *See also* AIRFOIL.

aerodynamic heating Frictional surface heating experienced by an AEROSPACE VEHICLE or space system as it enters the upper regions of a planetary ATMOSPHERE at very high velocities. Special thermal protection is needed to prevent structural damage or destruction. NASA's SPACE SHUTTLE ORBITER vehicle, for example, uses thermal protection tiles to survive the intense aerodynamic heating environment that occurs during REENTRY and landing. *See also* ABLATIVE COOLING.

aerodynamic skip An atmospheric entry ABORT caused by entering a PLANET's ATMOSPHERE at too shallow an angle. Much like a stone skipping across the surface of a pond, this condition results in a TRAJECTORY back out into space rather than downward toward the planet's surface.

aerodynamic vehicle A craft that has lifting and control surfaces to provide stability, control, and maneuverability while flying through a PLANET's ATMOSPHERE.

aeropause A region of indeterminate limits in a PLANET's upper ATMOSPHERE, considered as a boundary between the denser (sensible) portion of the atmosphere and OUTER SPACE.

aerosol A very small dust particle or droplet of liquid (other than water or ice) in a PLANET's ATMOSPHERE, ranging in size from about 0.001 MICROMETER (μm) to larger than 100 micrometers (μm) in radius. Terrestrial aerosols include smoke, dust, haze, and fumes.

aerospace A term, derived from *aero*nautics and *space,* meaning of or pertaining to EARTH's atmospheric envelope and OUTER SPACE beyond it. NASA's SPACE SHUTTLE ORBITER vehicle is called an AEROSPACE VEHICLE because it operates both in the ATMOSPHERE and in outer space.

aerospace ground equipment (AGE) All the support and test equipment needed on EARTH's surface to make an AEROSPACE system or SPACECRAFT function properly during its intended space mission.

aerospace medicine The branch of medical science that deals with the effects of flight upon the human body. The treatment of SPACE SICKNESS (space adaptation syndrome) falls within this field.

aerospace vehicle A vehicle capable of operating both within EARTH's SENSIBLE (measurable) ATMOSPHERE and in OUTER SPACE. The SPACE SHUTTLE ORBITER vehicle is an example.

aerospike nozzle A rocket NOZZLE design that allows combustion to occur around the periphery of a spike (or center plug). The THRUST-producing, hot-exhaust flow is then shaped and adjusted by the ambient (atmospheric) pressure.

aerozine A liquid ROCKET fuel consisting of a mixture of hydrazine (N_2H_4) and unsymmetrical dimethylhydrazine (UDMH), which has the chemical formula $(CH_3)_2NNH_2$.

afterbody Any companion body (usually JETTISONed, expended hardware) that trails a SPACECRAFT following LAUNCH and contributes to the SPACE (ORBITAL) DEBRIS problem. It is also any expended portion of a LAUNCH VEHICLE or ROCKET that enters EARTH'S ATMOSPHERE unprotected behind a returning NOSE CONE or SPACE CAPSULE that is protected against the AERODYNAMIC HEATING. Finally, it is any unprotected, discarded portion of a SPACE PROBE or SPACECRAFT that trails behind the protected probe or LANDER SPACECRAFT as either enters a PLANET'S atmosphere to accomplish the mission.

Agena A versatile, UPPER-STAGE ROCKET that supported numerous American military and civilian space missions in the 1960s and 1970s. One special feature of this LIQUID PROPELLANT system was its in-space engine restart capability.

age of the Moon The elapsed time, usually expressed in days, since the last new MOON. *See also* PHASES OF THE MOON.

agglutinate A common type of particle found on the MOON, consisting of small rock, mineral, and glass fragments impact bonded together with glass.

air The overall mixture of gases that make up EARTH'S ATMOSPHERE, primarily nitrogen (N_2) at 78 percent (by volume), oxygen (O_2) at 21 percent, argon (Ar) at 0.9 percent, and carbon dioxide (CO_2) at 0.03 percent. Sometimes aerospace engineers use this word for the breathable gaseous mixture found inside the crew compartment of a SPACE VEHICLE or in the PRESSURIZED HABITABLE ENVIRONMENT of a SPACE STATION.

airfoil A wing designed to provide AERODYNAMIC FORCE when it moves through the AIR (on EARTH) or through the SENSIBLE ATMOSPHERE of a PLANET (such as MARS or VENUS) or of TITAN, the largest MOON of SATURN.

air launch The process of launching a GUIDED MISSILE or ROCKET from an aircraft while it is in flight.

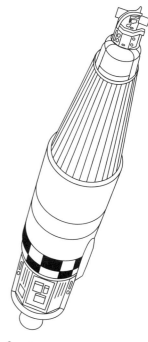

Agena

airlock　A small chamber with airtight doors that can be pressurized and depressurized. The airlock serves as a passageway for crew members and equipment between places at different pressure levels—for example, between a SPACECRAFT's pressurized crew cabin and OUTER SPACE.

albedo　The ratio of the amount of ELECTROMAGNETIC RADIATION (such as visible LIGHT) reflected by a surface to the total amount of electromagnetic radiation incident upon the surface. The albedo is usually expressed as a percentage. For example, the planetary albedo of EARTH is about 30 percent. This means that approximately 30 percent of the total SOLAR RADIATION falling upon Earth is reflected back to OUTER SPACE.

algorithm　A special mathematical procedure or rule for solving a particular type of problem.

alien life-form (ALF)　A general, though at present hypothetical, expression for EXTRATERRESTRIAL life, especially life that exhibits some degree of intelligence.

Almagest　The Arabic name (meaning "the greatest") for the collection of ancient Greek astronomical and mathematical knowledge written by PTOLEMY in about 150 C.E. and translated by Arab astronomers about 820 C.E. This compendium included the 48 ancient Greek CONSTELLATIONS upon which today's astronomers base the modern system of constellations.

Alpha Centauri　The closest STAR system, about 4.3 LIGHT-YEARS away. It is actually a triple-star system, with two stars orbiting around each other and a third star, called Proxima Centauri, revolving around the pair at some distance.

alphanumeric (*alphabet* plus *numeric*)　Including letters and numerical digits, for example, the term, JEN75WX11.

alpha particle (α particle)　A positively charged atomic PARTICLE emitted by certain radioactive NUCLIDES. It consists of two NEUTRONS and two PROTONS bound together and is identical to the NUCLEUS of a helium 4 ($^4_2$He) ATOM. Alpha particles are the least penetrating of the three common types of nuclear IONIZING RADIATION (alpha particle, BETA PARTICLE, and GAMMA RAY).

altazimuth mounting　A TELESCOPE mounting that has one AXIS pointing to the ZENITH.

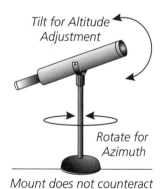

Tilt for Altitude Adjustment

Rotate for Azimuth

Mount does not counteract Earth's rotation.

Altazimuth mounting

altimeter An instrument for measuring the height (ALTITUDE) above a PLANET's surface; generally reported relative to a common planetary reference point, such as sea level on EARTH.

altitude (1) *(astronomy)* The angle between an observer's horizon and a target CELESTIAL BODY. The altitude is 0° if the object is on the horizon and 90° if the object is at ZENITH (directly overhead). (2) *(spacecraft)* In SPACE VEHICLE navigation, the height above the mean surface of the reference celestial body. Note that the *distance* of a space vehicle or SPACECRAFT from the reference celestial body is taken as the distance from the center of the object.

Amalthea The small (270 km × 150 km diameter), irregularly shaped, inner MOON of JUPITER, discovered as the fifth Jovian moon in 1892 by EDWARD EMERSON BARNARD.

ambient conditions *(planetary)* The environmental conditions, such as atmospheric pressure or temperature, that surround an AEROSPACE VEHICLE or PLANETARY PROBE. For example, a planetary probe on the surface of VENUS must function in an inferno-like environment where the ambient temperature is about 480°C (753 K).

Amor group A collection of NEAR-EARTH ASTEROIDS that cross the ORBIT of MARS but do not cross the orbit of EARTH. This ASTEROID group acquired its name from the 1 km diameter Amor asteroid, discovered by EUGÈNE JOSEPH DELPORTE in 1932.

amorphotoi Term used by the early Greek astronomers to describe the spaces in the night sky populated by dim STARs between the prominent groups of stars making up the ANCIENT CONSTELLATIONS. It is Greek for "unformed."

amplitude Generally, the maximum value of the displacement of a wave or other periodic phenomenon from a reference (average) position. Specifically, it is the overall range of brightness (from maximum MAGNITUDE to minimum magnitude) of a VARIABLE STAR.

ancient astronaut theory The (unproven) hypothesis that EARTH was visited in the past by a race of intelligent, EXTRATERRESTRIAL beings who were exploring this portion of the MILKY WAY GALAXY.

ancient constellations The collection of approximately 50 CONSTELLATIONS drawn up by ancient astronomers and recorded by PTOLEMY, including such familiar constellations as the signs of the ZODIAC, Ursa Major (the Great Bear), Boötes (the Herdsman), and Orion (the Hunter). *See also* SECTION IV CHARTS & TABLES.

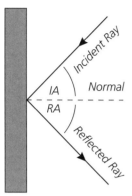

*Angle of Incidence (IA) =
Angle of Reflectance (RA)*

Angle of incidence

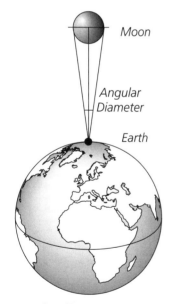

Angular diameter

Andromeda galaxy The Great Spiral Galaxy (or M31) in the CONSTELLATION of Andromeda, about 2.2 million LIGHT-YEARS away. It is the most distant object visible to the NAKED EYE and is the closest SPIRAL GALAXY to the MILKY WAY GALAXY.

angle The inclination of two intersecting lines to each other, measured by the arc of a circle intercepted between the two lines forming the angle. An acute angle is less than 90°; a right angle is precisely 90°; an obtuse angle is greater than 90° but less that 180°; and a straight angle is 180°.

angle of incidence The ANGLE at which a ray of LIGHT (or other type of ELECTROMAGNETIC RADIATION) impinges on a surface. This angle is usually measured between the direction of propagation and a perpendicular to the surface at the point of incidence.

angle of reflection The ANGLE at which a reflected ray of LIGHT (or other type of ELECTROMAGNETIC RADIATION) leaves a reflecting surface. This angle is usually measured between the direction of the outgoing ray and a perpendicular to the surface at the point of reflection. For a plane mirror, the angle of reflection equals the ANGLE OF INCIDENCE.

angstrom (Å) A unit of length used to indicate the WAVELENGTH of ELECTROMAGNETIC RADIATION in the visible, near-infrared, and near-ultraviolet portions of the SPECTRUM. Named after ANDERS JONAS ÅNGSTRÖM, 1 angstrom equals 0.1 nanometer (10^{-10} m).

angular acceleration (α) The time rate of change of ANGULAR VELOCITY (ω).

angular diameter The angle formed by the lines projected from a common point to the opposite sides of a body.

angular measure Units of angle generally expressed in terms of degrees (°), arc minutes ('), and arc seconds ("), where 1 degree of angle equals 60 arc minutes, and 1 arc minute equals 60 arc seconds.

angular momentum *(L)* A measure of an object's tendency to continue rotating at a particular rate around a certain AXIS. It is defined as the product of the ANGULAR VELOCITY (ω) of the object and its moment of INERTIA *(I)* about the axis of rotation.

angular velocity (ω) The change of angle per unit time; usually expressed in radians per second.

annihilation radiation Upon collision, the conversion of a PARTICLE and its corresponding antiparticle into pure electromagnetic ENERGY (called *annihilation radiation*). For example, when an ELECTRON (e⁻) and

POSITRON (e^+) collide, the minimum annihilation radiation released consists of a pair of GAMMA RAYS, each of approximately 0.511 million ELECTRON VOLTS (MeV) energy.

annual parallax (π) The PARALLAX of a STAR that results from the change in the position of a reference observing point during EARTH's annual REVOLUTION around the SUN. It is the maximum angular displacement of the star that occurs when the star-Sun-Earth ANGLE is 90° (as illustrated). Also called the HELIOCENTRIC parallax.

annular nozzle A NOZZLE with a ring-shaped (annular) throat formed by an outer wall and a center body wall.

anomalistic period The time interval between two successive PERIGEE passages of a SATELLITE in ORBIT about its PRIMARY BODY. For example, the term *anomalistic month* defines the mean time interval between successive passages of the MOON through its closest point to EARTH (perigee), about 27.555 days.

anomaly (1) *(astronomy)* The ANGLE used to define the position (at a particular time) of a celestial object, such as a PLANET or ARTIFICIAL SATELLITE in an elliptical ORBIT about its PRIMARY BODY. The *true anomaly* of a planet is the angle (in the direction of the planet's motion) between the point of closest approach (the PERIHELION), the focus (the SUN), and the planet's current orbital position. (2) *(space operations)* A deviation from the normal or anticipated result.

antenna A device used to detect, collect, or transmit RADIO WAVES. A RADIO TELESCOPE is a large receiving antenna. Many SPACECRAFT have both a DIRECTIONAL ANTENNA and an omnidirectional antenna to transmit (DOWNLINK) TELEMETRY and to receive (UPLINK) instructions.

antenna array A group of ANTENNAS coupled together into a system to obtain directional effects or to increase sensitivity. *See also* VERY LARGE ARRAY.

anthropic principle The controversial hypothesis in modern COSMOLOGY suggesting that the UNIVERSE evolved in just the right way after the BIG BANG event to allow for the emergence of human life.

antimatter Matter in which the ordinary nuclear PARTICLES (such as ELECTRONS, PROTONS, and NEUTRONS) are replaced by their corresponding antiparticles—POSITRONS, antiprotons, antineutrons, and so on. It is sometimes called *mirror matter*. Normal matter and antimatter mutually annihilate each other upon contact and are converted into pure ENERGY, called ANNIHILATION RADIATION.

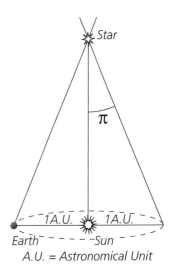

A.U. = Astronomical Unit

Annual parallax

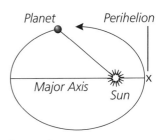

Anomaly

antisatellite (ASAT) spacecraft A SPACECRAFT designed to destroy other SATELLITEs in space. An ASAT spacecraft could be deployed in space disguised as a peaceful satellite that quietly lurks as a secret hunter/killer satellite, awaiting instructions to track and attack its prey.

antislosh baffle A device installed in the PROPELLANT tank of a liquid-fuel ROCKET to dampen unwanted liquid motion, or sloshing, during flight.

apastron The point in an body's ORBIT around a STAR at which it is at a maximum distance from the star. *Compare with* PERIASTRON.

aperture The opening in front of a TELESCOPE, camera, or other optical instrument through which light passes.

aperture synthesis A resolution-improving technique in RADIO ASTRONOMY that uses a variable-aperture radio INTERFEROMETER to mimic the full-dish size of a huge RADIO TELESCOPE.

apex The direction in the sky toward which the SUN and its system of PLANETs appear to be moving relative to the local STARs. Also called the *solar apex,* it is located in the CONSTELLATION of Hercules.

aphelion The point in an object's ORBIT around the SUN that is most distant from the Sun. *Compare with* PERIHELION.

Aphrodite Terra A large, fractured highland region near the EQUATOR of VENUS.

apogee The point in the ORBIT of a SATELLITE that is farthest from EARTH. The term applies both to the orbit of the MOON as well as to the orbits of ARTIFICIAL SATELLITEs around Earth. At apogee, the orbital VELOCITY of a satellite is at a minimum. *Compare with* PERIGEE.

apogee motor A SOLID-PROPELLANT ROCKET motor that is attached to a SPACECRAFT and fired when the deployed spacecraft is at the APOGEE of an initial (relatively low-ALTITUDE) PARKING ORBIT around EARTH. This firing establishes a new ORBIT farther from Earth or permits the spacecraft to achieve ESCAPE VELOCITY.

Apollo group A collection of NEAR-EARTH ASTEROIDs that have PERIHELION distances of 1.017 ASTRONOMICAL UNITS (AU) or less, taking them across the ORBIT of EARTH around the SUN. This group acquired its name from the first ASTEROID to be discovered, Apollo, in 1932 by KARL REINMUTH.

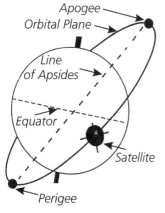

Apogee (Courtesy of NASA)

Apollo Project (Courtesy of NASA)

Apollo Project The American effort in the 1960s and early 1970s to place ASTRONAUTS successfully onto the surface of the MOON and return them safely to EARTH. The project was launched in May 1961 by President JOHN F. KENNEDY in response to a growing space technology challenge from the former Soviet Union. Managed by NASA, the *Apollo 8* mission sent the first three humans to the vicinity of the Moon in December 1968. The *Apollo 11* mission involved the first human landing on another world (20 July 1969). *Apollo 17,* the last lunar landing mission under this project, took place in December 1972. The project is often considered one of the greatest technical accomplishments in all human history. *See also* SECTION IV CHARTS & TABLES.

Apollo-Soyuz Test Project (ASTP) The joint United States-former Soviet Union space mission (July 1975), centering on the RENDEZVOUS and DOCKING of the *APOLLO 18* spacecraft (three-ASTRONAUT crew) and the *SOYUZ 19* SPACECRAFT (two-COSMONAUT crew).

apolune That point in an ORBIT around the MOON of a SPACECRAFT launched from the LUNAR surface that is farthest from the Moon. *Compare with* PERILUNE.

apparent In astronomy, observed. True values are reduced from apparent (observed) values by eliminating those factors, such as refraction and flight time, that can affect the observation.

apparent diameter The observed diameter (but not necessarily the actual diameter) of a CELESTIAL BODY. It is usually expressed in degrees, minutes, and seconds of arc. *See also* ANGULAR DIAMETER.

apparent magnitude *(m)* The brightness of a STAR (or other CELESTIAL BODY) as measured by an observer on EARTH. Its value depends on the star's intrinsic brightness (LUMINOSITY), how far away it is, and how much of its light has been absorbed by the intervening INTERSTELLAR MEDIUM. *See also* ABSOLUTE MAGNITUDE; MAGNITUDE.

apparent motion The observed motion of a heavenly body across the CELESTIAL SPHERE, assuming that EARTH is at the center of the celestial sphere and is standing still (stationary).

approach The maneuvers of a SPACECRAFT or AEROSPACE VEHICLE from its normal orbital position (STATION-KEEPING position) toward another orbiting spacecraft for the purpose of conducting RENDEZVOUS and DOCKING operations.

archaeological astronomy Scientific investigation concerning the astronomical significance of ancient structures and sites, such as STONEHENGE in the United Kingdom.

arc-jet engine An electric ROCKET engine that heats a PROPELLANT gas by passing through it an electric arc.

Arecibo Observatory The world's largest radio/radar telescope with a 305 m diameter dish. It is located in a large, bowl-shaped natural depression in the tropical jungles of Puerto Rico. When it operates as a RADIO WAVE receiver, the giant RADIO TELESCOPE can listen for signals from celestial objects at the farthest reaches of the UNIVERSE.

Ariane Family of modern LAUNCH VEHICLEs developed by the French Space Agency (CENTRE NATIONAL D'ETUDES SPATIALES, or CNES) and the EUROPEAN SPACE AGENCY (ESA). The *Ariane 4* ROCKET, Europe's space workhorse, has carried many scientific and commercial PAYLOADS into ORBIT from the Guiana Space Center in Kourou, French Guiana. *See also* SECTION IV CHARTS & TABLES.

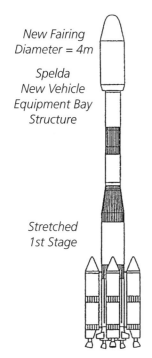

New Fairing Diameter = 4m

Spelda New Vehicle Equipment Bay Structure

Stretched 1st Stage

Ariane (Arianespace)

**Arecibo Observatory
(Courtesy of NASA)**

artificial gravity Simulated GRAVITY conditions established within a
SPACECRAFT, SPACE STATION, or SPACE SETTLEMENT. Rotating the
space system about an AXIS creates this condition since the
CENTRIFUGAL FORCE generated by the rotation produces effects
similar to the FORCE of gravity within the vehicle. This technique was
first suggested by KONSTANTIN EDUARDOVICH TSIOLKOVSKY at the
start of the 20th century.

artificial intelligence (AI) Information-processing functions (including
thinking and perceiving) performed by machines that imitate (to
some extent) the mental activities performed by the human brain.
Advances in AI will allow "very smart" ROBOT SPACECRAFT to
explore distant alien worlds with minimal human supervision.

artificial satellite A human-made object, such as a SPACECRAFT, placed into
ORBIT around EARTH or another CELESTIAL BODY. *SPUTNIK 1* was the
first artificial satellite to be placed into orbit around Earth.

ascending node That point in the ORBIT of a CELESTIAL BODY when it travels
from south to north across a reference plane, such as the equatorial

plane of the CELESTIAL SPHERE or the plane of the ECLIPTIC. Also called the *northbound node. Compare with* DESCENDING NODE.

asteroid A small, solid, rocky object that orbits the SUN but is independent of any major PLANET. Most asteroids (or minor planets) are found in the main ASTEROID BELT. The largest asteroid is Ceres, about 1,000 km in diameter and discovered in 1801 by GIUSEPPE PIAZZI. EARTH-CROSSING ASTEROIDS or NEAR-EARTH ASTEROIDS (NEAs) have ORBITS that take them near or across EARTH's orbit around the Sun and are divided into the ATEN, APOLLO, and AMOR GROUPS. *See also* TROJAN GROUP.

asteroid belt The region of OUTER SPACE between the ORBITS of MARS and JUPITER that contains the great majority of the ASTEROIDs. These minor planets or planetoids have ORBITAL PERIODS of between three and six years and travel around the SUN at distances of between 2.2 to 3.3 ASTRONOMICAL UNITS (AUs).

astro- A prefix that means STAR or (by extension) OUTER SPACE or CELESTIAL; for example, ASTRONAUT, ASTRONAUTICS, or ASTROPHYSICS.

astrobiology *(exobiology)* The search for and study of living organisms found on CELESTIAL BODIES beyond Earth.

astrobleme A geologic structure (often eroded) produced by the hypervelocity impact of a METEOROID, COMET, or ASTEROID.

astrochimps Nickname given to the primates used in the early U.S. space program to test SPACE CAPSULE and LAUNCH VEHICLE hardware prior to human missions.

astrodynamics The application of CELESTIAL MECHANICS, PROPULSION SYSTEM theory, and related fields of science and engineering to the problem of carefully planning and directing the TRAJECTORY of a SPACE VEHICLE.

astrolabe Instrument used by ancient astronomers to measure the ALTITUDE of a STAR.

astrology The attempt by many early astronomers to forecast future events on EARTH by observing and interpreting the relative positions of the FIXED STARS, the SUN, the PLANETs, and the MOON. Such mystical stargazing was a common activity in most ancient societies, was enthusiastically practiced in western Europe up through the 17th century, and still lingers today as daily horoscopes. At the dawn of the scientific revolution, GALILEO GALILEI taught a required

Astrochimp Ham (31 January 1961) (Courtesy of USAF)

university course on medical astrology. JOHANNES KEPLER earned a living as a court astrologer. The popular "science" of astrology is based on the unscientific hypothesis that the motion of CELESTIAL BODIES controls and influences human lives and terrestrial events. *See also* ZODIAC.

astrometric binary A BINARY (DOUBLE) STAR SYSTEM in which irregularities in the PROPER MOTION (wobbling) of a visible STAR imply the presence of an undetected companion.

astrometry Branch of ASTRONOMY that involves the very precise measurement of the motion and position of CELESTIAL BODIES.

astronaut Within the American space program, a person who travels in OUTER SPACE; a person who flies in an AEROSPACE VEHICLE to an ALTITUDE of more than 80 km (50 mi.). The word comes from a combination of two ancient Greek words that literally mean "STAR" *(astro)* and "sailor or traveler" *(naut). Compare with* COSMONAUT.

astronautics The branch of engineering science dealing with spaceflight and the design and operation of SPACE VEHICLES.

Astronomer Royal The honorary title created in 1675 by King Charles II and given to a prominent English astronomer. Up until 1971, the Astronomer Royal also served as the director of the ROYAL GREENWICH OBSERVATORY. JOHN FLAMSTEED was the first to hold this position, from 1675 to 1719.

astronomical unit (AU) A convenient unit of distance defined as the semimajor axis of EARTH'S ORBIT around the SUN. One AU, the average distance between Earth and the Sun, equals approximately 149.6×10^6 km or 499.01 light-seconds.

astronomy The branch of science that deals with CELESTIAL BODIES and studies their size, composition, position, origin, and dynamic behavior. *See also* ASTROPHYSICS; COSMOLOGY.

astrophotography The use of photographic techniques to create images of CELESTIAL BODIES. Astronomers are now replacing light-sensitive photographic emulsions with CHARGE-COUPLED DEVICES (CCDs) to create digital images in the visible, INFRARED, and ULTRAVIOLET portions of the ELECTROMAGNETIC SPECTRUM.

astrophysics The branch of physics that investigates the nature of STARS and star systems. It provides the theoretical principles enabling scientists to understand astronomical observations. By using space technology, astrophysicists now place sensitive REMOTE SENSING instruments above EARTH'S ATMOSPHERE and view the UNIVERSE in all portions of the ELECTROMAGNETIC SPECTRUM. High-energy astrophysics includes GAMMA RAY ASTRONOMY, COSMIC RAY ASTRONOMY, and X-RAY ASTRONOMY. *See also* COSMOLOGY.

Aten group A collection of NEAR-EARTH ASTEROIDS that cross the ORBIT of EARTH but whose average distances from the SUN lie inside Earth's orbit. This ASTEROID group acquired its name from the 0.9 km diameter asteroid Aten, discovered in 1976 by the American astronomer Eleanor Kay Helin (née Francis).

Atlas Family of versatile liquid-fuel ROCKET vehicles originally developed by General BERNARD SCHRIEVER of the United States Air Force in the late 1950s as the first operational American INTERCONTINENTAL BALLISTIC MISSILE (ICBM). Evolved and improved Atlas LAUNCH VEHICLES now serve many government and commercial space transportation needs. *See also* SECTION IV CHARTS & TABLES.

atmosphere (1) The gravitationally bound gaseous envelope that forms an outer region around a PLANET or other CELESTIAL BODY. (2) *(cabin)* The breathable environment inside a SPACE CAPSULE, AEROSPACE VEHICLE, SPACECRAFT, or SPACE STATION. (3) *(Earth's)* The life-sustaining gaseous envelope surrounding EARTH. Near sea level it contains the following composition of gases (by volume): nitrogen 78 percent, oxygen 21 percent, argon 0.9 percent, and carbon dioxide 0.03 percent. There are also lesser amounts of many other gases, including water vapor and human-generated chemical pollutants. Earth's electrically neutral atmosphere is composed of four primary layers: troposphere, stratosphere, mesosphere, and thermosphere. Life occurs in the troposhere, the lowest region that extends up to about 16 km ALTITUDE. It is also the place within which most of Earth's weather occurs. *See also* SECTION IV CHARTS & TABLES.

atmospheric pressure The pressure (FORCE per unit area) at any point in a PLANET'S ATMOSPHERE due solely to the WEIGHT of the atmospheric gases above that point.

atmospheric probe The special collection of scientific instruments (usually released by a MOTHER SPACECRAFT) for determining the pressure, composition, and temperature of a PLANET'S ATMOSPHERE at different ALTITUDES. An example is the probe released by NASA'S GALILEO PROJECT SPACECRAFT in December 1995. As it plunged into the Jovian atmosphere, the probe successfully transmitted its scientific data to the *Galileo* spacecraft (the mother spacecraft) for about 58 minutes.

atmospheric window A WAVELENGTH interval within which a PLANET'S ATMOSPHERE is transparent to (that is, easily transmits) ELECTROMAGNETIC RADIATION.

atom A tiny particle of matter (the smallest part of an element) indivisible by chemical means. It is the fundamental building block of the chemical ELEMENTS. The elements, such as HYDROGEN (H), HELIUM (He), carbon (C), iron (Fe), lead (Pb), and uranium (U), differ from each other because they consist of different types of atoms. According to (much simplified) modern atomic theory, an atom consists of a dense inner core (the NUCLEUS) that contains PROTONS and NEUTRONS, and a cloud of orbiting ELECTRONS. Atoms are electrically neutral, with the number of (positively charged) protons equal to the number of (negatively charged) electrons.

atomic clock A precise device for measuring or standardizing time that is based on periodic vibrations of certain ATOMS (cesium) or MOLECULES (ammonia). It is widely used in military and civilian SPACECRAFT, such as, for example, the GLOBAL POSITIONING SATELLITE (GPS) system.

atomic mass The mass of a neutral ATOM of a particular NUCLIDE usually expressed in ATOMIC MASS UNITS (AMU). *See also* MASS NUMBER.

atomic mass unit (amu) One-twelfth (1/12) the MASS of a neutral ATOM of the most abundant ISOTOPE of carbon, carbon 12.

atomic number *(Z)* The number of PROTONS in the NUCLEUS of an ATOM and also its positive charge.

atomic weight The MASS of an ATOM relative to other atoms. At present, the most abundant ISOTOPE of the ELEMENT carbon, namely carbon 12, is assigned an atomic weight of exactly 12. As a result, 1/12 the mass of a carbon 12 atom is called one ATOMIC MASS UNIT, which is approximately the mass of one PROTON or one NEUTRON. Also called relative atomic mass.

attenuation The decrease in intensity (strength) of an electromagnetic wave as it passes through a transmitting medium. This loss is due to absorption of the incident ELECTROMAGNETIC RADIATION (EMR) by the transmitting medium or to scattering of the EMR out of the path of the detector. Attenuation does not include the reduction in EMR wave strength due to geometric spreading as a consequence of the INVERSE SQUARE LAW.

attitude The position of an object as defined by the inclination of its AXES with respect to a frame of reference. It is the orientation of a SPACE VEHICLE (for example, a SPACECRAFT or AEROSPACE VEHICLE) that is

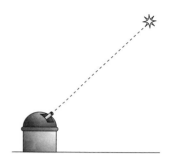

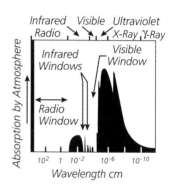

Atmospheric window

either in motion or at rest, as established by the relationship between the vehicle's axes and a reference line or plane. Attitude is often expressed in terms of PITCH, ROLL, and YAW.

attitude control system The onboard system of computers, low-THRUST ROCKETs (thrusters), and mechanical devices (such as a MOMENTUM wheel) used to keep a SPACECRAFT stabilized during flight and to point its instruments precisely in some desired direction. Stabilization is achieved by spinning the spacecraft or by using a three-axis active approach that maintains the spacecraft in a fixed, reference ATTITUDE by firing a selected combination of thrusters when necessary.

aurora The visible glow in a PLANET's upper ATMOSPHERE (IONOSPHERE) caused by the interaction of the planet's MAGNETOSPHERE and PARTICLES from the SUN (SOLAR WIND). On Earth, the aurora borealis (or northern lights) and the aurora australis (or southern lights) are visible manifestations of the magnetosphere's dynamic behavior. At high latitudes, disturbances in Earth's geomagnetic field accelerate trapped particles into the upper ATMOSPHERE, where they excite nitrogen MOLECULES (red emissions) and oxygen ATOMS (red and green emissions). Auroras also occur on JUPITER, SATURN, URANUS, and NEPTUNE.

auxiliary power unit (APU) A power unit carried on a SPACECRAFT or AEROSPACE VEHICLE that supplements the main source of electric power on the craft.

axis *(plural: axes)* Straight line about which a body rotates (axis of rotation) or along which its CENTER OF GRAVITY moves (axis of translation). Also, one of a set of reference lines for a coordinate system, such as the *x*-axis, *y*-axis, and *z*-axis in the CARTESIAN COORDINATE system.

azimuth The horizontal direction or bearing to a CELESTIAL BODY measured in degrees clockwise from north around a terrestrial observer's horizon. On EARTH, azimuth is 0° for an object that is due north, 90° for an object due east, 180° for an object due south, and 270° for an object due west. *See also* ALTITUDE (ASTRONOMY).

backout The process of undoing tasks that have already been completed during the COUNTDOWN of a LAUNCH VEHICLE, usually in reverse order.

Baikonur Cosmodrome The major LAUNCH SITE for the space program of the former Soviet Union and later the Russian Federation. The complex is located just east of the Aral Sea in Kazakhstan (now an independent republic). Also known as the Tyuratam Launch Site during the COLD WAR, the Soviets launched *SPUTNIK 1* (1957), the first ARTIFICIAL SATELLITE, and COSMONAUT YURI A. GAGARIN, the first human to fly in OUTER SPACE (1961), from this location.

Baily's beads An optical phenomenon that appears just before or immediately after totality in a SOLAR ECLIPSE, when sunlight bursts through gaps in the mountains on the MOON and a string of light beads appears along the lunar disk. FRANCIS BAILY first described this phenomenon in 1836.

ballistic missile A MISSILE that is propelled by ROCKET engines and guided only during the initial (THRUST-producing) phase of its flight. In the nonpowered and nonguided phase of its flight, it assumes a BALLISTIC TRAJECTORY similar to that of an artillery shell. After thrust termination, REENTRY VEHICLES (RVs) can be released. These RVs also follow free-falling (ballistic) trajectories toward their targets. *Compare with* GUIDED MISSILE.

ballistic missile defense (BMD) A proposed defense system designed to protect a territory from incoming BALLISTIC MISSILES and their WARHEAD-carrying REENTRY VEHICLES. A variety of BMD technologies have been suggested, including high-energy laser (HEL) weapons, high-performance interceptor missiles, and KINETIC ENERGY weapon (KEW) systems. However, the BMD problem is technically challenging and can be likened to stopping an incoming high-VELOCITY rifle bullet with another rifle bullet.

ballistic trajectory The path an object (that does not have lifting surfaces) follows while being acted upon by only the FORCE of GRAVITY and any resistive AERODYNAMIC FORCES of the medium through which it passes. A stone tossed into the air follows a ballistic trajectory. Similarly, after its propulsive unit stops operating, a ROCKET vehicle describes a ballistic trajectory.

band A range of (RADIO WAVE) FREQUENCIES. Alternatively, a closely spaced set of spectral lines that are associated with the ELECTROMAGNETIC RADIATION (EMR) characteristic of some particular atomic or molecular ENERGY levels.

bandwidth The number of HERTZ (cycles per second) between the upper and lower limits of a FREQUENCY band.

barbecue mode The slow roll of an orbiting AEROSPACE VEHICLE or SPACECRAFT to help equalize its external temperature and to promote a more favorable heat (thermal ENERGY) balance. This maneuver is performed during certain missions. In OUTER SPACE, SOLAR

RADIATION is intense on one side of a SPACE VEHICLE while the side opposite the SUN can become extremely cold.

Barnard's star A RED DWARF STAR approximately six LIGHT-YEARS from the SUN, making it the fourth-nearest star to the SOLAR SYSTEM. Discovered in 1916 by EDWARD EMERSON BARNARD, it has the largest PROPER MOTION (some 10.3 seconds of arc per year) of any known star.

barred spiral galaxy A type of SPIRAL GALAXY that has a bright bar of STARS across the central regions of the GALACTIC NUCLEUS.

barycenter The CENTER OF MASS of a system of masses at which point the total MASS of the system is assumed to be concentrated. In a system of two PARTICLES or two CELESTIAL BODIES (that is, a binary system), the barycenter is located somewhere on a straight line connecting the geometric center of each object but closer to the more massive object. For example, the barycenter for the EARTH-MOON system is located about 4,700 km from the center of Earth—a point actually inside Earth, which has a radius of about 6,400 km.

basin *(impact)* A large, shallow, lowland area in the crust of a TERRESTRIAL PLANET formed by the impact of an ASTEROID or COMET.

baud *(rate)* A unit of signaling speed. The baud rate is the number of electronic signal changes or data symbols that can be transmitted by a communications channel per second. It is named after J. M. Baudot (1845–1903), a French telegraph engineer.

beam A narrow, well-collimated stream of PARTICLES (such as ELECTRONS or PROTONS) or ELECTROMAGNETIC RADIATION (such as GAMMA RAY PHOTONS) that are traveling in a single direction.

beam rider A MISSILE guided to its target by a BEAM of ELECTROMAGNETIC RADIATION, such as a radar beam or a laser beam.

bell nozzle A NOZZLE with a circular opening for a throat and an axisymmetric contoured wall downstream of the throat that gives this type of nozzle a characteristic bell shape.

Belt of Orion The line of three bright STARS (Alnilam, Alnitak, and Mintaka) that form the Belt of Orion, a very conspicuous CONSTELLATION on the EQUATOR of the CELESTIAL SPHERE. It honors the great hunter in Greek mythology.

Bernal sphere A large, spherical SPACE SETTLEMENT first proposed by JOHN DESMOND BERNAL in 1929.

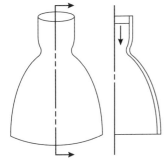

Bell nozzle (courtesy of NASA)

berthing The joining of two orbiting SPACECRAFT, using a MANIPULATOR or other mechanical device, to move one into contact (or very close proximity) with the other at a selected interface. For example, NASA ASTRONAUTS use the SPACE SHUTTLE'S REMOTE MANIPULATOR SYSTEM to berth a large FREE-FLYING SPACECRAFT (like the *HUBBLE SPACE TELESCOPE*) carefully onto a special support fixture located in the ORBITER'S PAYLOAD BAY during an on-orbit servicing and repair mission. *See also* DOCKING; RENDEZVOUS.

beta decay RADIOACTIVITY in which an atomic NUCLEUS spontaneously decays and emits two subatomic PARTICLES: a BETA PARTICLE (β) and a NEUTRINO (ν). In beta-minus (β^-) decay, a NEUTRON in the transforming (parent) nucleus becomes a PROTON, and a negative beta particle and an antineutrino are emitted. The resultant (daughter) nucleus has its ATOMIC NUMBER (Z) increased by one (thereby changing its chemical properties), while its total ATOMIC MASS (A) remains the same as that of the parent nucleus. In beta-plus (β^+) decay, a proton is converted into a neutron, and a positive beta particle (POSITRON) is emitted along with a neutrino. Here, the atomic number (Z) of the resultant (daughter) nucleus is decreased by one— a process that also changes its chemical properties.

beta particle *(β)* The negatively charged subatomic PARTICLE emitted from the atomic NUCLEUS during the process of BETA DECAY. It is identical to the ELECTRON. *See also* POSITRON.

big bang *(theory)* A contemporary theory in COSMOLOGY concerning the origin of the UNIVERSE. It suggests that a very large, ancient explosion, called the initial singularity, started space and time of the present universe, which has been expanding ever since. The big bang event is thought to have occurred about 15 to 20 billion years ago. Astrophysical observations, especially discovery of the COSMIC MICROWAVE BACKGROUND RADIATION in 1964, tend to support this theory.

big crunch Within the CLOSED UNIVERSE model of COSMOLOGY, the postulated end state that occurs after the present UNIVERSE expands to its maximum physical dimensions and then collapses in on itself under the influence of GRAVITATION, eventually reaching an infinitely dense end point, or SINGULARITY.

binary digit *(bit)* Only two possible values (or digits) are in the binary number system, namely 0 and 1. Binary notation is a common TELEMETRY (information)-encoding scheme that uses binary digits to represent numbers and symbols. For example, digital computers use a sequence of bits, such as an eight-bit-long byte (*binary* digi*t e*ight) to create a more complex unit of information.

binary (double) star system A pair of STARS that orbit around a common CENTER OF MASS and are bound together by their mutual GRAVITATION.

biogenic elements Those ELEMENTS generally considered by scientists (astrobiologists) as essential for all living systems, including HYDROGEN (H), carbon (C), nitrogen (N), oxygen (O), sulfur (S), and phosphorous (P). The availability of the chemical compound water (H_2O) is also considered necessary for life both here on EARTH and possibly elsewhere in the UNIVERSE. *See also* ASTROBIOLOGY.

biosphere The life zone of a planetary body; for example, that part of the EARTH SYSTEM inhabited by living organisms. On this PLANET, the biosphere includes portions of the ATMOSPHERE, the HYDROSPHERE, the CRYOSPHERE, and surface regions of the SOLID EARTH. *See also* ECOSPHERE; GLOBAL CHANGE.

biotelemetry The remote measurement of life functions. For example, data from biosensors attached to an ASTRONAUT or COSMONAUT are sent back to EARTH (as TELEMETRY) for the purposes of space crew health monitoring and evaluation by medical experts and mission managers.

bipropellant rocket A ROCKET that uses two unmixed (uncombined) liquid chemicals as its fuel and OXIDIZER, respectively. The two chemical PROPELLANTS flow separately into the rocket's COMBUSTION CHAMBER, where they are combined and combusted to produce high-temperature, THRUST-generating gases. The combustion gases then exit the rocket system through a suitably designed NOZZLE.

bird A popular aerospace industry expression (jargon) for a ROCKET, MISSILE, SATELLITE, or SPACECRAFT.

blackbody A perfect emitter and perfect absorber of ELECTROMAGNETIC RADIATION. According to PLANCK'S RADIATION LAW, the radiant energy emitted by a BLACKBODY is a function only of the ABSOLUTE TEMPERATURE of the emitting object.

black box A unit or subsystem (often involving an electronic device) of a SPACECRAFT or AEROSPACE VEHICLE that is considered only with respect to its input and output characteristics without any specification of its internal elements.

black dwarf The cold remains of a WHITE DWARF STAR that no longer emits visible radiation or a nonradiating ball of INTERSTELLAR gas that has contracted under GRAVITATION but contains too little MASS to initiate nuclear FUSION.

black hole An incredibly compact, gravitationally collapsed MASS from which nothing (light, matter, or any other kind of information) can escape. Astrophysicists believe that a black hole is the natural end product when a massive STAR dies and collapses beyond a certain critical dimension, called the SCHWARZSCHILD RADIUS. Once the massive star shrinks to this critical radius, its gravitational ESCAPE VELOCITY equals the SPEED OF LIGHT, and nothing can escape from it. Inside this radius, called the *event horizon,* lies an extremely dense point mass (SINGULARITY).

Black hole (distorting space-time continuum)

blastoff The moment a ROCKET or AEROSPACE VEHICLE rises from its LAUNCH PAD under full THRUST. *See also* LIFTOFF.

blazar A variable EXTRAGALACTIC object (possibly a high-speed jet from an ACTIVE GALACTIC NUCLEUS) that exhibits very dynamic, sometimes violent behavior. *See also* BL LAC (BL LACERTAE) OBJECT.

bl lac (bl lacertae) object A CLASS OF EXTRAGALACTIC objects thought to be the active centers of faint ELLIPTICAL GALAXIES that vary considerably in brightness over very short periods of time (typically hours, days, or weeks). Scientists further speculate that a very high-speed (relativistic) jet is emerging from such an object straight at an observer on EARTH.

blockhouse *(block house)* A reinforced-concrete structure, often built partially underground, that provides protection against blast, heat, and possibly an ABORT explosion during ROCKET launchings.

blue giant A massive, very high-LUMINOSITY STAR with a surface temperature of about 30,000 K that has exhausted all its HYDROGEN thermonuclear fuel and left the MAIN SEQUENCE.

blueshift When a celestial object (like a distant GALAXY) approaches an observer at high velocity, the ELECTROMAGNETIC RADIATION it emits in the visible portion of the SPECTRUM appears shifted toward the blue (higher frequency, shorter wavelength) region. *Compare with* REDSHIFT. *See also* DOPPLER SHIFT.

boiloff The loss of a CRYOGENIC PROPELLANT, such as LIQUID OXYGEN or LIQUID HYDROGEN, due to vaporization. This happens when the temperature of the cryogenic propellant rises slightly in the PROPELLANT tank of a ROCKET being prepared for LAUNCH. The longer a fully fueled rocket vehicle sits on its LAUNCH PAD, the more significant the problem of boiloff becomes.

bolide A brilliant METEOR, especially one that explodes into fragments near the end of its TRAJECTORY in EARTH'S ATMOSPHERE.

Boltzmann constant *(K)* The physical constant describing the relationship between ABSOLUTE TEMPERATURE and the KINETIC ENERGY of the ATOMS or MOLECULES in a perfect gas. It equals 1.380658×10^{-23} JOULES per KELVIN (J/K) and is named after LUDWIG BOLTZMANN.

Bond albedo The fraction of the total ELECTROMAGNETIC RADIATION (such as the total amount of light) falling upon a nonluminous spherical body that is reflected in all directions by that body. The Bond albedo is measured or calculated over all WAVELENGTHS and is named after GEORGE PHILLIPS BOND.

booster rocket A ROCKET motor, with either SOLID or LIQUID PROPELLANT, that assists the main propulsive system (called the sustainer engine) of a LAUNCH VEHICLE during some part of its flight.

brown dwarf A very low-LUMINOSITY, substellar (almost a star) CELESTIAL BODY that contains starlike material (that is, HYDROGEN and HELIUM) but has too low a MASS (typically 1 to 10 percent of a SOLAR MASS) to allow its core to initiate thermonuclear FUSION (hydrogen burning).

bulge of the Earth The extra extension of EARTH'S EQUATOR, caused by the CENTRIFUGAL FORCE of Earth's ROTATION, which slightly flattens the spherical shape of Earth. This bulge causes the planes of SATELLITE ORBITS inclined to the equator (but not POLAR ORBITS) to rotate slowly around Earth's AXIS.

burnout The moment in time or the point in a ROCKET'S TRAJECTORY when combustion of fuels in the engine is terminated. This usually occurs when all the PROPELLANTS are consumed.

bus The ROCKET-propelled final stage of an INTERCONTINENTAL BALLISTIC MISSILE (ICBM) that, after booster BURNOUT, places WARHEADS and (possibly) decoys onto BALLISTIC TRAJECTORIES toward their targets. This is also called the postboost vehicle (PBV).

Byurakan Astrophysical Observatory The observatory located at an ALTITUDE of 1.4 km on Mount Aragatz, near Yerevan, the capital of the Republic of Armenia. The facility was founded in 1946 by VIKTOR AMBARTSUMIAN. It played a major role in the ASTRONOMY activities of the former Soviet Union.

caldera A large volcanic depression, more or less circular in form and much larger than the included volcanic vents. A caldera may be formed by three basic geologic processes: explosion, collapse, or erosion. *See also* OLYMPUS MONS.

calendar A system of marking days of the YEAR, usually devised in a way to give each date a fixed place in the cycle of seasons. *See also* GREGORIAN CALENDAR; JULIAN CALENDAR.

calibration The process of translating the signals collected by a measuring instrument (such as a telescope) into something that is scientifically useful. The calibration procedure generally removes most of the errors caused by instabilities in the instrument or in the environment through which the signal has traveled.

Callisto The second largest MOON of JUPITER and the outermost of the four GALILEAN SATELLITES.

calorie (cal) A unit of thermal energy (heat) originally defined as the amount of ENERGY required to raise 1 g of water through 1°C. This energy unit (often called a *small calorie*) is related to the JOULE as follows: 1 cal = 4.1868 J. Scientists use the term *kilocalorie* (1,000 small calories) as one *big calorie* when describing the energy content of food.

Caloris basin A very large, ringed-impact BASIN (about 1,300 km across) on MERCURY.

canali The Italian word for *channels* used by GIOVANNI VIRGINIO SCHIAPARELLI in 1877 to describe natural surface features he observed on MARS. Subsequent pre–space age investigators, including PERCIVAL LOWELL, took the Italian word quite literally as meaning *canals* and sought additional evidence of an intelligent civilization on Mars. Since the 1960s, many SPACECRAFT have visited Mars, dispelling such popular speculations and revealing no evidence of any Martian canals constructed by intelligent beings.

canard A horizontal trim and control surface on an AERODYNAMIC VEHICLE.

cannibalize The process of taking functioning parts from a nonoperating SPACECRAFT or LAUNCH VEHICLE and installing these salvaged parts into another spacecraft or launch vehicle in order to make the latter operational.

Cape Canaveral The region on Florida's east-central coast from which the United States Air Force and NASA have launched more than 3,000 ROCKETs since 1950. Cape Canaveral Air Force Station (CCAFS) is the major East Coast LAUNCH SITE for the Department of Defense, while the adjacent NASA KENNEDY SPACE CENTER is the SPACEPORT for the fleet of SPACE SHUTTLE vehicles.

capital satellite A very important or very expensive SATELLITE, distinct from a decoy satellite or a scientific satellite of minimal national security significance. *See also* ANTISATELLITE (ASAT) SPACECRAFT.

captive firing The firing of a ROCKET PROPULSION SYSTEM at full or partial THRUST while the rocket is restrained in a test stand facility. Usually, engineers instrument the propulsion system to obtain test data that verify rocket design and demonstrate performance levels. Sometimes called a *holddown test.*

carbon cycle (1) *(astrophysics)* The chain of thermonuclear FUSION reactions thought to be the main energy-liberating mechanisms in STARS with interior temperatures much hotter (> 16 million K) than the SUN. In this cycle, HYDROGEN is converted to HELIUM. Large quantities of ENERGY are released, and the ISOTOPE carbon 12 serves as a nuclear reaction catalyst. It is also called the carbon-nitrogen (CN) cycle, the carbon-nitrogen-oxygen (CNO) cycle, or the BETHE-WEIZSÄCKER cycle. *See also* ASTROPHYSICS.
(2) *(Earth system)* The planetary BIOSPHERE cycle that consists of four central biochemical processes: photosynthesis, autotrophic respiration (carbon intake by green plants), aerobic oxidation, and anaerobic oxidation. Scientists believe excessive human activities involving the combustion of fossil fuels, the destruction of forests, and the conversion of wild lands to agriculture may now be causing undesirable perturbations in the planet's overall carbon cycle, thereby endangering important balances within EARTH's highly interconnected biosphere. *See also* EARTH SYSTEM; GLOBAL CHANGE; GREENHOUSE EFFECT.

cargo bay The unpressurized middle portion of NASA's SPACE SHUTTLE ORBITER vehicle. *See* PAYLOAD BAY.

cartesian coordinates A coordinate system, developed by the French mathematician René Descartes (1596–1650), in which locations of points in space are expressed by reference to three mutually perpendicular planes, called *coordinate planes.* The three planes intersect in straight lines called the *coordinate axes.* The distances and the AXES are usually marked *(x, y, z)* and the origin is the (zero) point at which the three axes intersect.

case *(rocket)* The structural envelope for the PROPELLANT in a SOLID-PROPELLANT ROCKET motor.

Cassegrain telescope A compound REFLECTING TELESCOPE in which a small, convex, secondary mirror reflects the convergent beam from the parabolic primary mirror through a hole in the primary mirror to an EYEPIECE in the back of the primary mirror. It was designed by the Frenchman GUILLAUME CASSEGRAIN in 1672.

Cassini mission The joint NASA-EUROPEAN SPACE AGENCY planetary exploration mission to SATURN launched from CAPE CANAVERAL on 15 October 1997. Starting in 2004, the *Cassini* SPACECRAFT will perform detailed studies of Saturn, its RINGS, and its MOONS. It will also deliver an ATMOSPHERIC PROBE, called *Huygens,* into the nitrogen-rich ATMOSPHERE of the largest Saturnian moon, TITAN. The spacecraft is named after GIOVANNI DOMENICO CASSINI and the Titan probe after CHRISTIAAN HUYGENS.

cavitation The formation of bubbles (vapor-filled cavities) in a flowing liquid. The formation of these cavities can adversely impact the performance of high-speed hydraulic machinery, such as the TURBOPUMP SYSTEM of a LIQUID-PROPELLANT ROCKET ENGINE.

celestial Of or pertaining to the heavens.

celestial body A heavenly body. It is any aggregation of matter in OUTER SPACE constituting a unit for study in ASTRONOMY, such as PLANETs, MOONS, COMETS, ASTEROIDS, STARS, NEBULAS, and GALAXIES.

celestial latitude *(β)* With respect to the CELESTIAL SPHERE, the angular distance of a CELESTIAL BODY from 0° to 90° north (considered positive) or south (considered negative) of the ECLIPTIC.

celestial longitude *(λ)* With respect to the CELESTIAL SPHERE, the angular distance of a CELESTIAL BODY from 0° to 360° measured eastward along the ECLIPTIC to the intersection of the body's circle of CELESTIAL LONGITUDE. The VERNAL EQUINOX is taken as 0°.

celestial mechanics The scientific study of the dynamic relationships among the CELESTIAL BODIES of the SOLAR SYSTEM. It analyzes the relative motions of objects under the influence of gravitational fields, such as the motion of a MOON or ARTIFICIAL SATELLITE in ORBIT around a PLANET.

celestial sphere To create a consistent coordinate system for the heavens, early astronomers developed the concept of a celestial sphere. It is an imaginary sphere of very large radius with EARTH as its center and on which all observable CELESTIAL BODIES are assumed projected. The rotational AXIS of Earth intersects the north and south poles of the celestial sphere. An extension of Earth's equatorial plane cuts the celestial sphere and forms a great circle, called the *celestial equator*. The direction to any STAR or other celestial body can then be plotted in two dimensions on the inside of this imaginary sphere by using the CELESTIAL LATITUDE and the CELESTIAL LONGITUDE.

Celsius temperature scale The widely used relative temperature scale, originally developed by ANDERS CELSIUS, in which the range between two reference points (ice at 0° and boiling water at 100°) is divided into 100 equal units or degrees.

Centaur *(rocket)* A powerful and versatile UPPER-STAGE ROCKET originally developed by the United States in the 1950s for use with the ATLAS LAUNCH VEHICLE. Engineered by KRAFFT A. EHRICKE, it was the first American rocket to use LIQUID HYDROGEN as its PROPELLANT. Centaur has supported many important military and scientific missions, including the CASSINI MISSION (launched on 15 October 1997) to SATURN.

Centaurs A group of unusual CELESTIAL OBJECTS residing in the outer SOLAR SYSTEM, such as CHIRON, that exhibit a dual ASTEROID/COMET nature. They are named after the centaurs in Greek mythology, who were half-human and half-horse beings.

center of gravity That point in a rigid body at which all the external FORCES appear to act.

center of mass The point at which the entire MASS of a body (or system of bodies) appears to be concentrated. For a body (or system of bodies) in a uniform gravitational field, the center of mass coincides with the CENTER OF GRAVITY. *See also* BARYCENTER.

central force A FORCE that, for the purposes of computation, can be assumed to be concentrated at one central point with its intensity at any other point being a function of the distance from the central point. For example, GRAVITATION is considered as a central force in CELESTIAL MECHANICS.

Centre National d'Etudes Spatiales (CNES) The public body responsible for all aspects of French space activity including LAUNCH VEHICLES and SPACECRAFT. CNES has four main centers: Headquarters (Paris), the Launch Division at Evry, the Toulouse Space Center, and the Guiana Space Center (LAUNCH SITE) in Kourou, French Guiana (South America). *See also* ARIANE; EUROPEAN SPACE AGENCY.

centrifugal force A reaction FORCE that is directed opposite to a CENTRIPETAL FORCE such that it points out along the radius of curvature away from the center of curvature.

centripetal force The central (inward-acting) FORCE on a body that causes it to move in a curved (circular) path. Consider a person carefully whirling a stone secured by a strong (but lightweight) string in a circular path at a constant speed. The string exerts a radial tug on the stone, which is called the centripetal force. Now as the stone keeps moving in a circle at constant speed, the stone also exerts a reaction force on the string, which is called the CENTRIFUGAL FORCE. It is equal in magnitude but opposite in direction to the centripetal force exerted by the string on the stone.

Cepheid variable A type of very bright, SUPERGIANT STAR that exhibits a regular pattern of changing its brightness as a function of time. The period of this pulsation pattern is directly related to the star's intrinsic brightness. So modern astronomers can use Cepheid variables to determine astronomical distances.

Ceres The first and largest (940 km diameter) ASTEROID to be found. It was discovered on 1 January 1801 by GIUSEPPE PIAZZI.

Cerro Tololo Inter-American Observatory (CTIO) An astronomical observatory at an ALTITUDE of about 2.2 km on Cerro Tololo Mountain in the Chilean Andes near La Serena. The main instrument is a 4 m REFLECTING TELESCOPE. The observatory is operated by the Association of Universities for Research in Astronomy (AURA).

Challenger accident NASA's SPACE SHUTTLE *Challenger* was launched from Complex 39-B at the KENNEDY SPACE CENTER on 28 January 1986 as part of the STS 51-L mission. At approximately 74 seconds into the flight, an explosion occurred that caused the loss of the AEROSPACE VEHICLE and its entire crew, including ASTRONAUTS FRANCIS R. SCOBEE, MICHAEL J. SMITH, ELLISON S. ONIZUKA, JUDITH A. RESNIK, RONALD E. MCNAIR, S. CHRISTA CORRIGAN MCAULIFFE, and GREGORY B. JARVIS.

Chandrasekhar limit In the 1920s, SUBRAHMANYAN CHANDRASEKHAR used RELATIVITY theory and QUANTUM MECHANICS to show that if the MASS of a DEGENERATE STAR is more than about 1.4 SOLAR MASSes (a maximum mass called the *Chandrasekhar limit*), it will not evolve into a WHITE DWARF star but, rather, it will continue to collapse under the influence of GRAVITY and become a NEUTRON STAR or a BLACK HOLE, or blow itself apart in a SUPERNOVA explosion.

Chandra X-Ray
Observatory (Courtesy of
NASA)

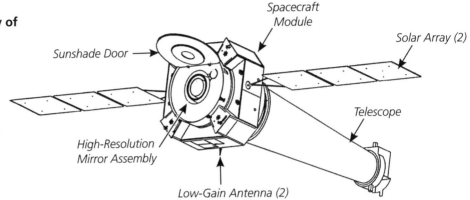

Sunshade Door

Spacecraft
Module

Solar Array (2)

High-Resolution
Mirror Assembly

Low-Gain Antenna (2)

Telescope

Chandra X-Ray Observatory (CXO) One of NASA's major orbiting
astronomical observatories launched in July 1999 and named after
SUBRAHMANYAN CHANDRASEKHAR. It was previously called the
Advanced X-Ray Astrophysics Facility (AXAF). This EARTH-
orbiting facility studies some of the most interesting and puzzling X-
RAY sources in the UNIVERSE, including emissions from ACTIVE
GALACTIC NUCLEI, exploding STARS, NEUTRON STARS, and matter
falling into BLACK HOLES.

chaos The branch of physics (mechanics) that studies unstable systems.

chaotic orbit The ORBIT of a CELESTIAL BODY that changes in a highly
unpredictable manner, usually when a small object, such as an
ASTEROID or COMET, passes close to a massive PLANET (like SATURN
or JUPITER) or the SUN. For example, the chaotic orbit of CHIRON is
influenced by both URANUS and Saturn.

chaotic terrain A planetary surface feature (first observed on MARS in 1969)
characterized by a jumbled, irregular assembly of fractures and
blocks of rock.

charge-coupled device (CCD) An electronic (solid-state) device containing
a regular array of sensor elements that are sensitive to various types
of ELECTROMAGNETIC RADIATION (for example, LIGHT) and emit
ELECTRONS when exposed to such radiation. The emitted electrons
are collected and the resulting charge analyzed. CCDs are used as the
light-detecting component in modern television cameras and
TELESCOPES.

charged particle An ION; an elementary PARTICLE that carries a positive or negative electric charge, such as an ELECTRON, a PROTON, or an ALPHA PARTICLE.

Charon The large (about 1,200 km diameter) MOON of PLUTO discovered in 1978 by the American astronomer James Walter Christy. It orbits at a mean distance of 20,000 km and keeps one face permanently toward the PLANET.

chaser spacecraft The SPACECRAFT or AEROSPACE VEHICLE that actively performs the key maneuvers during orbital RENDEZVOUS and DOCKING/BERTHING operations. The other SPACE VEHICLE serves as the target and remains essentially passive during the encounter.

chasm A canyon or deep linear feature on a PLANET's surface.

checkout The sequence of actions (such as functional, operational, and calibration tests) performed to determine the readiness of a SPACECRAFT or LAUNCH VEHICLE to carry out its intended mission.

chemical rocket A ROCKET that uses the combustion of a chemical fuel in either solid or liquid form to generate THRUST. The chemical fuel requires an OXIDIZER to support combustion.

chilldown Cooling all or part of a cryogenic (very cold) ROCKET engine system from ambient (room) temperature down to cryogenic temperature by circulating CRYOGENIC PROPELLANT (fluid) through the system prior to engine start.

Chiron An unusual CELESTIAL BODY in the outer SOLAR SYSTEM with a CHAOTIC ORBIT that lies almost entirely between the ORBITs of SATURN and URANUS. This massive ASTEROID-sized object has a diameter of about 200 km and is the first object placed into the CENTAURS group because it also has a detectable COMA, a feature characteristic of COMETs.

choked flow A flow condition in a duct or pipe such that the flow upstream of a certain critical section (like a NOZZLE or valve) cannot be increased by further reducing downstream pressure.

chromatic aberration A phenomenon that occurs in a refracting optical system because light of different WAVELENGTHS (colors) is refracted (bent) by a different amount. As a result, a simple lens will give red light a longer focal length than blue light.

chromosphere The reddish layer in the SUN's ATMOSPHERE located between the PHOTOSPHERE (the apparent solar surface) and the base of the CORONA. It is the source of solar PROMINENCES.

Chryse Planitia A large plain on MARS characterized by many ancient channels that could have once contained flowing surface water. It was the landing site for NASA's *VIKING 1* LANDER SPACECRAFT in July 1976.

chugging A form of combustion instability that occurs in a LIQUID-PROPELLANT ROCKET ENGINE. It is characterized by a pulsing operation at a fairly low FREQUENCY.

circumsolar space Around the SUN or HELIOCENTRIC (Sun-centered) space.

circumstellar Around a STAR, as opposed to *interstellar,* which means between the stars.

cislunar Of or pertaining to phenomena, projects, or activities happening in the region of OUTER SPACE between EARTH and the MOON. It comes from the Latin word *cis,* meaning "on this side" and *lunar,* which means "of or pertaining to the Moon." Therefore, it means "on this side of the Moon."

Clarke orbit A GEOSTATIONARY ORBIT, named after SIR ARTHUR C. CLARKE, who first proposed in 1945 the use of this special ORBIT around EARTH for COMMUNICATIONS SATELLITES.

clean room A controlled work environment for spacecraft and aerospace systems in which dust, temperature, and humidity are carefully controlled during the fabrication, assembly, and/or testing of critical components.

closed universe The model in COSMOLOGY that assumes the total MASS of the UNIVERSE is sufficiently large that one day the GALAXIES will slow down and stop expanding because of their mutual gravitational attraction. At that time, the universe will have reached its maximum size. Then GRAVITATION will make it slowly contract, ultimately collapsing to a single point of infinite density (sometimes called the BIG CRUNCH). Also called *bounded-universe model. Compare with* OPEN UNIVERSE.

cluster of galaxies An accumulation of GALAXIES that lie within a few million LIGHT-YEARS of each other and are bound by GRAVITATION. Galactic clusters can occur with just a few member galaxies (say 10 to 100), such as the LOCAL GROUP, or they can occur in great groupings involving thousands of galaxies.

cold-flow test The thorough testing of a LIQUID-PROPELLANT ROCKET ENGINE without actually firing (igniting) it. This type of test helps AEROSPACE engineers verify the performance and efficiency of a PROPULSION

SYSTEM since all aspects of PROPELLANT flow and conditioning, except combustion, are examined. Tank pressurization, propellant loading, and propellant flow into the COMBUSTION CHAMBER (without ignition) are usually included in a cold-flow test. *Compare with* HOT-FIRE TEST.

cold war The ideological conflict between the United States and the former Soviet Union from approximately 1946–89, involving rivalry, mistrust, and hostility just short of overt military action. The tearing down of the Berlin Wall in November 1989 is generally considered as the symbolic end of the cold war period.

collimator A device for focusing or confining a BEAM of PARTICLES or ELECTROMAGNETIC RADIATION, such as X-RAY PHOTONS.

color A quality of LIGHT that depends on its WAVELENGTH. The *spectral color* of emitted light corresponds to its place in the SPECTRUM of the rainbow. *Visual light* or *perceived color* is the quality of light emission as recognized by the human eye. Simply stated, the human eye contains three basic types of light-sensitive cells that respond in various combinations to incoming spectral colors. For example, the color brown occurs when the eye responds to a particular combination of blue, yellow, and red light. Violet light has the shortest wavelength, while red light has the longest wavelength. All the other colors have wavelengths that lie in between.

coma The gaseous envelope that surrounds the NUCLEUS of a COMET.

combustion chamber The part of a ROCKET engine in which the combustion of chemical PROPELLANTs takes place at high pressure. The combustion chamber and the diverging section of the NOZZLE make up a rocket's THRUST chamber. It is sometimes called the firing chamber or simply the chamber.

comet A dirty ice "rock" consisting of dust, frozen water, and gases that orbits the SUN. As a comet approaches the inner SOLAR SYSTEM from deep space, solar radiation causes its frozen materials to vaporize (sublime), creating a COMA and a long TAIL of dust and IONs. Scientists think these icy PLANETESIMALS are the remainders of the primordial material from which the OUTER PLANETS were formed billions of years ago. *See also* KUIPER BELT; OORT CLOUD.

Comet Halley *(1P/Halley)* The most famous PERIODIC COMET. Named after EDMOND HALLEY, who successfully predicted its 1758 return. Reported since 240 B.C.E., this COMET reaches PERIHELION approximately every 76 years. During its most recent inner SOLAR

SYSTEM appearance, an international fleet of five different SPACECRAFT, including the *GIOTTO* SPACECRAFT, performed scientific investigations that supported the dirty ice rock model of a comet's NUCLEUS.

command destruct An intentional action leading to the destruction of a ROCKET or MISSILE in flight. Whenever a malfunctioning vehicle's performance creates a safety hazard on or off the rocket test range, the range safety officer sends the command destruct signal to destroy it.

communications satellite A SATELLITE that relays or reflects electromagnetic signals between two (or more) communications stations. An *active communications satellite* receives, regulates, and retransmits electromagnetic signals between stations, while a *passive communications satellite* simply reflects signals between stations. In 1945, SIR ARTHUR C. CLARKE proposed placing communications satellites into GEOSTATIONARY ORBIT around EARTH. Numerous active communications satellites now maintain a global telecommunications infrastructure.

compact body A small, very dense CELESTIAL BODY that represents the end product of stellar evolution: a WHITE DWARF, a NEUTRON STAR, or a BLACK HOLE.

companion body A NOSE CONE, protective shroud, last-stage ROCKET, or PAYLOAD separation hardware that ORBITS EARTH along with an operational SATELLITE or SPACECRAFT. Companion bodies contribute significantly to a growing SPACE (ORBITAL) DEBRIS population in LOW EARTH ORBIT.

compressible flow Flow conditions in which DENSITY changes in the fluid cannot be neglected.

Compton Gamma Ray Observatory (CGRO) A major NASA orbiting astrophysical observatory dedicated to GAMMA RAY ASTRONOMY. The CGRO was placed into ORBIT around EARTH in April 1991. At the end of its useful scientific mission, flight controllers intentionally commanded the massive (16,300 kg) SPACECRAFT to perform a DE-ORBIT BURN. This caused it to reenter and safely crash in June 2000 in a remote region of the Pacific Ocean. The spacecraft was named in honor of ARTHUR HOLLY COMPTON.

Compton scattering *(Compton effect)* The scattering of energetic PHOTONS (either X-RAY or GAMMA RAY) by ELECTRONS. In this process, the electron gains ENERGY and recoils. The scattered photon changes direction while losing some of its some of its energy and increasing in WAVELENGTH. Although first observed in 1923 by ARTHUR HOLLY COMPTON, this phenomenon demonstrated that photons have MOMENTUM. It has proven very useful in gamma ray detection.

concave lens *(or mirror)* A LENS or mirror with an inward curvature.

conduction *(thermal)* The transport of heat (thermal ENERGY) through an object by means of a temperature difference from a region of higher temperature to a region of lower temperature. For solids and liquid metals, thermal conduction is accomplished by the migration of fast-moving electrons, while atomic and molecular collisions support thermal conduction in gases and other liquids. *Compare with* CONVECTION.

conic section A curve formed by the intersection of a plane and a right circular cone. Also called *conic*. The conic sections are the ELLIPSE, the *parabola,* and the *hyperbola*—all curves that describe the paths of bodies moving in space. The *circle* is simply an ellipse with an ECCENTRICITY of zero.

conjunction The alignment of two bodies in the SOLAR SYSTEM so that they have the same CELESTIAL LONGITUDE as seen from EARTH (that is, when they appear closest together in the sky). For example, a SUPERIOR PLANET forms a *superior conjunction* when the SUN lies between it and Earth. An INFERIOR PLANET (either VENUS or MERCURY) forms an *inferior conjunction* when it lies directly between the Sun and Earth and a *superior conjunction* when it lies directly behind the Sun.

conservation of angular momentum The principle of physics that states that absolute ANGULAR MOMENTUM is a property that cannot be created or destroyed but can only be transferred from one physical system to another through the action of a net torque on the system. As a consequence, the total angular momentum of an isolated system remains constant.

conservation of energy The principle of physics that states that the total ENERGY of an isolated system remains constant if no interconversion of MASS and energy takes place within the system. It is also called the first law of thermodynamics.

conservation of mass and energy From special RELATIVITY and ALBERT EINSTEIN's famous mass-energy equivalence formula ($E = \Delta mc^2$), this conservation principle states that for an isolated system, the sum of

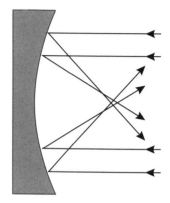

Concave lens

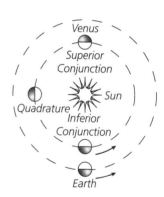

Conjunction

the MASS and ENERGY remains constant, although interconversion of mass and energy can occur within the system.

conservation of momentum The principle of physics that states that in the absence of external FORCES, absolute MOMENTUM is a property that cannot be created or destroyed. Consequently, the total momentum of an isolated system remains constant. *See also* NEWTON'S LAWS OF MOTION.

console A desklike array of controls, indicators, and video display devices for monitoring and controlling AEROSPACE operations, such as the CHECKOUT, countdown, and LAUNCH of a ROCKET. During the critical phases of a space mission, the console becomes the central place from which to issue commands to or at which to display information concerning an AEROSPACE VEHICLE, a deployed PAYLOAD, an EARTH-orbiting SPACECRAFT, or a PLANETARY PROBE. The mission control center generally contains clusters of consoles (each assigned to specific monitoring and control tasks). Depending on the nature and duration of a particular space mission, operators will remain at their consoles continuously or work there only intermittently.

constellation (1) *(aerospace)* A term used to describe collectively the number and orbital disposition of a set of SATELLITES, such as the constellation of GLOBAL POSITIONING SYSTEM satellites.
(2) *(astronomy)* An easily identifiable configuration of the brightest stars in a moderately small region of the night sky. Originally, constellations, such as Orion the Hunter and Ursa Major (the Great Bear), were named by early astronomers after heroes and creatures from various ancient cultures and mythologies.

continuously crewed spacecraft A SPACECRAFT that has accommodations for continuous habitation (human occupancy) during its mission. The INTERNATIONAL SPACE STATION (ISS) is an example. It is sometimes (though not preferably) called a *continuously manned spacecraft.*

continuously habitable zone (CHZ) The region around a STAR in which one or several PLANETs can maintain conditions appropriate for the emergence and sustained existence of life. One important characteristic of a planet in the CHZ is that its environmental conditions support the retention of significant amounts of liquid water on the planetary surface.

control rocket A low-THRUST ROCKET, such as a RETROROCKET or a VERNIER ENGINE, used to guide, to change the ATTITUDE of, or to make small corrections in the velocity of an AEROSPACE VEHICLE, SPACECRAFT, or EXPENDABLE LAUNCH VEHICLE.

convection *(thermal)* The transfer of heat (thermal ENERGY) brought about by the mass motion and mixing of a fluid. Because the DENSITY of a MASS (glob) of heated fluid is lower than the surrounding cooler fluid, it rises *(natural convection).* When cooled, the glob will sink as its density increases slightly. Pumps and fans promote *forced convection* when a mass of warmer fluid is driven through or across cooler surfaces or a mass of cooler fluid is driven across warmer surfaces.

converging-diverging (CD) nozzle A THRUST-producing flow device for expanding and accelerating hot exhaust gases from a ROCKET engine. A properly designed NOZZLE efficiently converts the thermal ENERGY of combustion into kinetic energy of the combustion product gases. In a supersonic converging-diverging nozzle, the hot gas upstream of the nozzle throat is at subsonic VELOCITY (that is, the MACH NUMBER $(M) < 1$), reaches sonic velocity (the speed of sound, for which $M = 1$) at the throat of the nozzle, and then expands to supersonic velocity $(M > 1)$ downstream of the nozzle throat region while flowing through the diverging section of the nozzle. *See also* DE LAVAL NOZZLE.

converging lens *(or mirror)* A LENS (or mirror) that refracts (or reflects) a parallel beam of light, making it converge at a point (called the principal focus). A converging mirror is a concave mirror, while a converging lens is generally a CONVEX LENS that is thicker in the middle than at its edges. *Compare with* DIVERGING LENS. *See also* CONCAVE LENS.

convex lens *(or mirror)* A LENS or mirror with an outward curvature. Compare with CONCAVE LENS.

cooperative target A three-axis, stabilized, orbiting object that has signaling devices to support RENDEZVOUS and DOCKING/capture operations by a CHASER SPACECRAFT.

co-orbital Sharing the same or very similar ORBIT. For example, during a RENDEZVOUS operation, the CHASER SPACECRAFT and its COOPERATIVE TARGET are said to be co-orbital.

coordinated universal time *See* UNIVERSAL TIME.

Copernican system The theory of planetary motions, proposed by NICHOLAS COPERNICUS, in which all PLANETs (including EARTH) move in *circular* ORBITs around the SUN, with the planets closer to the Sun moving faster. In this system, the hypothesis of which helped trigger the scientific revolution of the 16th and 17th centuries, Earth was viewed not as an immovable object at the center of the UNIVERSE (as in the GEOCENTRIC PTOLEMAIC SYSTEM) but rather as a planet orbiting the Sun between VENUS and MARS. Early in the 17th century, JOHANNES KEPLER showed that while Copernicus's HELIOCENTRIC hypothesis was correct, the PLANETS actually moved in (slightly) ELLIPTICAL ORBITS around the Sun.

Copernicus Observatory A scientific SPACECRAFT launched into ORBIT around EARTH by NASA on 21 August 1972. Also called the *Orbiting Astronomical Observatory-3 (OAO-3)* and named in honor of NICHOLAS COPERNICUS, this space-based observatory examined the UNIVERSE in the ULTRAVIOLET RADIATION portion of the ELECTROMAGNETIC (EM) SPECTRUM from 1972 to 1981. *See also* ULTRAVIOLET ASTRONOMY.

core (1) *(planetary)* The high-density, central region of a PLANET. (2) *(stellar)* The very high-temperature, central region of a STAR. For MAIN SEQUENCE STARS, FUSION processes within the core burn HYDROGEN. For stars that have left the main sequence, nuclear fusion processes in the core involve HELIUM and oxygen.

corona The outermost region of a STAR. The SUN's corona consists of low-density clouds of very hot gases (> 1 million K) and ionized materials.

coronal mass ejection (CME) A high-speed (10 to 1,000 km/s) ejection of matter from the SUN's CORONA. A CME travels through space disturbing the SOLAR WIND and giving rise to geomagnetic storms when the disturbance reaches EARTH.

cosmic Of or pertaining to the UNIVERSE, especially that part outside EARTH's ATMOSPHERE. This term frequently appears in the Russian (former Soviet Union) space program as the equivalent to *space* or ASTRO-, such as cosmic station (versus SPACE STATION) or COSMONAUT (versus ASTRONAUT).

Cosmic Background Explorer *(COBE)* A NASA SPACECRAFT placed into ORBIT around EARTH in November 1989. It successfully measured the SPECTRUM and intensity distribution of the COSMIC MICROWAVE BACKGROUND (CMB).

cosmic microwave background (CMB) The background of MICROWAVE RADIATION that permeates the UNIVERSE and has a BLACKBODY temperature of about 2.7 K. Sometimes called the *primal glow,* scientists believe it represents the remains of the ancient fireball in which the universe was created. *See also* BIG BANG.

cosmic ray astronomy The branch of high-energy ASTROPHYSICS that uses COSMIC RAYS to provide information on the origin of the chemical ELEMENTS through NUCLEOSYNTHESIS during stellar explosions.

cosmic rays Extremely energetic PARTICLES (usually bare atomic NUCLEI) that move through OUTER SPACE at speeds just below the SPEED OF LIGHT and bombard EARTH from all directions. Their existence was discovered in 1912 by VICTOR FRANCIS HESS. HYDROGEN nuclei (PROTONs) make up the highest proportion of the cosmic ray population (approximately 85 percent), but these particles range over the entire periodic table of ELEMENTs. *Galactic cosmic rays* are samples of material from outside the SOLAR SYSTEM and provide direct evidence of phenomena that occur as a result of explosive processes in STARs throughout the MILKY WAY GALAXY. *Solar cosmic rays* (mostly protons and ALPHA PARTICLES) are ejected from the SUN during SOLAR FLARE events.

cosmological principle The hypothesis that the expanding UNIVERSE is ISOTROPIC and homogeneous. In other words, there is no special location for observing the universe, and all observers anywhere in the universe would see the same recession of distant GALAXIES.

cosmology The study of the origin, evolution, and structure of the UNIVERSE. Contemporary cosmology centers around the BIG BANG hypothesis. This theory states that about 15 to 20 billion (10^9) years ago, the universe began in a great explosion and has been expanding ever since. In the OPEN UNIVERSE (or STEADY-STATE UNIVERSE) model, scientists postulate that the universe is infinite and will continue to expand forever. In the more widely accepted CLOSED UNIVERSE model, the total MASS of the universe is assumed sufficiently large to stop its expansion eventually and then start contracting by GRAVITATION, leading ultimately to a BIG CRUNCH. In the *flat universe model,* the expansion gradually comes to a halt. However, instead of collapsing, the universe achieves an equilibrium condition with expansion FORCEs precisely balancing the forces of gravitational contraction.

cosmonaut The title given by Russia (formerly the Soviet Union) to its space travelers. Equivalent to ASTRONAUT.

Cosmos spacecraft The general name given to a large number of Soviet and later Russian SPACECRAFT, ranging from MILITARY SATELLITEs to scientific platforms investigating near-EARTH space. *Cosmos 1* was launched in March 1962. Since then, well over 2,000 Cosmos satellites have been sent into outer space. Also called *Kosmos*.

countdown The step-by-step process that leads to the LAUNCH of a ROCKET or AEROSPACE VEHICLE. A countdown takes place in accordance with a specific schedule—with zero being the go or activate time.

Crab nebula The SUPERNOVA remnant of an exploding STAR observed in 1054 C.E. by Chinese astronomers. Called M1 in the MESSIER CATALOGUE, it is about 6,500 LIGHT-YEARS away in the CONSTELLATION Taurus and contains a PULSAR that flashes optically.

crater A bowl-shaped topographic depression with steep slopes on the surface of a PLANET or MOON. There are two general types: *impact crater* (formed by an ASTEROID, COMET, or METEOROID strike) and *eruptive* (formed when a VOLCANO erupts).

crew-tended spacecraft A SPACECRAFT that is visited and/or serviced by ASTRONAUTS but can provide only temporary accommodations for human habitation during its overall mission. It is sometimes referred to as a *man-tended spacecraft.* Compare with CONTINUOUSLY CREWED SPACECRAFT.

critical mass density (Ω) The mean DENSITY of matter within the UNIVERSE (considered as a whole) that cosmologists consider necessary if GRAVITATION is to halt its expansion eventually. If the universe does not contain sufficient MASS (that is, if $\Omega < 1$), then the universe will continue to expand forever. If $\Omega > 1$, then the universe has enough mass to stop its expansion eventually and to start an inward collapse under the influence of gravitation. If the critical mass density is just right (that is, $\Omega = 1$), then the universe is considered flat and a state of equilibrium will exist in which the outward FORCE of expansion is precisely balanced by the inward force of gravitation. *See also* COSMOLOGY.

critical point The highest temperature at which the liquid and vapor phases of a fluid can coexist. The temperature and pressure corresponding to this point (thermodynamic state) are called the fluid's *critical temperature* and *critical pressure,* respectively.

cruise phase For a scientific SPACECRAFT on an INTERPLANETARY mission, the part of the mission (usually months or even years in duration) following LAUNCH and prior to planetary ENCOUNTER.

crust The outermost solid layer of a PLANET or MOON.

cryogenic propellant A ROCKET fuel, oxidizer, or propulsion fluid that is liquid only at very low (cryogenic) temperatures. LIQUID HYDROGEN (LH_2) and LIQUID OXYGEN (LO_2 or LOX) are examples.

cryosphere The portion of EARTH's climate system consisting of the world's ice masses and snow deposits. *See also* EARTH SYSTEM.

cutoff The act of shutting off the PROPELLANT flow in a ROCKET or of stopping the combustion of the propellant. *Compare with* BURNOUT.

Cygnus X-1 The strong X-RAY source in the CONSTELLATION Cygnus that scientists believe comes from a BINARY (DOUBLE) STAR SYSTEM, consisting of a SUPERGIANT star orbiting and a BLACK HOLE companion. Gas drawn off the supergiant star emits X rays as it is intensely heated while falling into the black hole.

Dactyl A natural SATELLITE of the ASTEROID Ida, discovered in February 1994 when NASA scientists were reviewing GALILEO PROJECT SPACECRAFT data from a FLYBY encounter with the asteroid on 28 August 1993. This tiny MOON (about 1.2 km × 1.4 km × 1.6 km) is the first to be discovered.

dark matter Matter in the UNIVERSE that cannot be observed directly because it emits very little or no ELECTROMAGNETIC RADIATION. Scientists infer its existence through secondary phenomena such as gravitational effects and suggest that it may make up about 90 percent of the total MASS of the universe. Also called *missing mass*. *See* COSMOLOGY.

dark nebula A cloud of INTERSTELLAR dust and gas sufficiently dense and thick that the light from more distant STARS and CELESTIAL BODIES (behind it) is obscured. The HORSEHEAD NEBULA in the CONSTELLATION Orion is an example of a dark nebula.

de-boost A retrograde (opposite-direction) burn of one or more low-THRUST ROCKETs or an AEROBRAKING maneuver that lowers the ALTITUDE of an orbiting SPACECRAFT.

debris JETTISONed human-made materials, discarded LAUNCH VEHICLE components, and derelict or nonfunctioning SPACECRAFT in ORBIT around EARTH. *See also* SPACE (ORBITAL) DEBRIS.

decay (l) *(orbital)* The gradual lessening of both the APOGEE and PERIGEE of an orbiting object from its PRIMARY BODY. For example, the orbital decay process for ARTIFICIAL SATELLITEs and DEBRIS often results in their ultimate fiery plunge back into the denser regions of EARTH's ATMOSPHERE.

(2) *(radioactive)* The spontaneous transformation of one radionuclide into a different NUCLIDE or into a different ENERGY state of the same nuclide. This natural process results in a decrease (with time) of the number of original radioactive ATOMS in a sample and involves the emission from the NUCLEUS of ALPHA PARTICLES, BETA PARTICLES, or GAMMA RAYS. *See also* RADIOACTIVITY; RADIOISOTOPE.

declination *(δ)* For a CELESTIAL BODY viewed on the CELESTIAL SPHERE, the angular distance north (0° to 90° positive) or south (0° to 90° negative) of the celestial equator.

Deep Space Network (Digital image courtesy of NASA/JPL)

Deep Space Network (DSN) NASA's global network of ANTENNAS that serve as the RADIO WAVE communications link to distant INTERPLANETARY SPACECRAFT and PROBES, transmitting instructions to them and receiving data from them. Large radio antennas of the DSN's three Deep Space Communications Complexes (DSCCs) are located in Goldstone, California; near Madrid, Spain; and near Canberra, Australia—providing almost continuous contact with a spacecraft in deep space as EARTH rotates on its AXIS.

deep-space probe A SPACECRAFT designed for exploring deep space, especially to the vicinity of the MOON and beyond. This includes LUNAR PROBES, MARS probes, OUTER PLANET probes, SOLAR probes, and so on.

Defense Meteorological Satellite Program (DMSP) A highly successful family of WEATHER SATELLITES operating in POLAR ORBIT around EARTH that have provided important environmental data to serve American defense and civilian needs for more than two decades.

Defense Support Program (DSP) The family of MISSILE surveillance SATELLITES operated by the U.S. Air Force since the early 1970s. When placed into GEOSYNCHRONOUS ORBIT around EARTH, these military surveillance satellites can detect missile launches, space launches, and nuclear detonations occurring around the world.

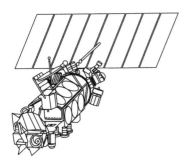

Defense Meteorological Satellite Program

degenerate star A STAR that has collapsed to a high-density condition, such as a WHITE DWARF or a NEUTRON STAR.

degrees of freedom (DOF) A mode of motion, either angular or linear, with respect to a coordinate system, independent of any other mode. A body in motion has six possible degrees of freedom, three linear (sometimes called *x-, y-,* and *z*-motion with reference to linear [axial] movements in the CARTESIAN COORDINATE system) and three angular (sometimes called PITCH, YAW, and ROLL with reference to angular movements).

Deimos The tiny, irregularly shaped (about 12 km average diameter) outer MOON of MARS, discovered by ASAPH HALL in 1877.

De Laval nozzle A flow device that efficiently converts the ENERGY content of a hot, high-pressure pressure gas into KINETIC ENERGY. Although originally developed by CARL GUSTAF PATRIK DE LAVAL for use in certain steam turbines, this versatile CONVERGING-DIVERGING (CD) NOZZLE is now used in practically all modern ROCKETS. The device constricts the outflow of the high-pressure (combustion) gas until it reaches the VELOCITY of sound (at the NOZZLE's throat) and then expands the exiting gas to very high velocities.

Delta *(launch vehicle)* A versatile family of American two- and three-stage LIQUID-PROPELLANT, EXPENDABLE LAUNCH VEHICLES (ELVs) that uses multiple strap-on BOOSTER ROCKETS in several configurations. The Delta ROCKET vehicle family has successfully launched more than 225 U.S. and foreign SATELLITES, earning it the nickname *space workhorse vehicle.*

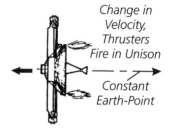

Change in Velocity, Thrusters Fire in Unison

Constant Earth-Point

Delta-V

delta-V *(V)* VELOCITY change; a numerical index of the maneuverability of a SPACECRAFT or ROCKET. This term often represents the maximum change in velocity that a space vehicle's PROPULSION SYSTEM can provide. It is typically described in terms of kilometers per second (km/s) or meters per second (m/s).

density *(ρ)* The MASS of a substance per unit volume at a specified temperature.

de-orbit burn A retrograde (opposite-direction) ROCKET engine firing by which a SPACE VEHICLE's VELOCITY is reduced to less than that required to remain in ORBIT around a CELESTIAL BODY.

descending node That point in the ORBIT of a CELESTIAL BODY when it travels from north to south across a reference plane, such as the equatorial plane of the CELESTIAL SPHERE or the plane of the ECLIPTIC. Also called the souththbound node. *Compare with* ASCENDING NODE.

destruct *(missile)* The deliberate action of destroying a MISSILE or ROCKET vehicle after it has been launched but before it has completed its course. Destruct commands are executed by the range safety officer when the missile or rocket veers off its intended (plotted) course or functions in a way so as to become a hazard. *See also* COMMAND DESTRUCT.

deuterium (D or $^2_1$H) A nonradioactive ISOTOPE of HYDROGEN whose NUCLEUS contains one NEUTRON and one PROTON. It is sometimes called *heavy hydrogen* because the deuterium nucleus is twice as heavy as that of ordinary hydrogen. *See also* TRITIUM.

Dewar flask A double-walled container with the interspace evacuated of gas (air) to prevent the contents from gaining or losing thermal ENERGY (heat). It is named for SIR JAMES DEWAR, and large, modern versions of this device are used to store CRYOGENIC PROPELLANTS.

diamonds The patterns of shock waves (pressure discontinuities) often visible in a ROCKET engine's exhaust. These patterns resemble a series of diamond shapes placed end to end.

diffraction The spreading out of ELECTROMAGNETIC RADIATION (such as LIGHT waves) as they pass by the edge of a body or through closely spaced parallel scratches in the surface of a diffraction grating. For example, when a ray of white light passes over a sharp, opaque edge (like that of a razor blade), it is broken up into its rainbow SPECTRUM of colors.

diffuse nebula *See* H II REGION.

Dirac cosmology An application of the large-numbers hypothesis within modern COSMOLOGY that tries to relate the fundamental physical constants found in subatomic physics to the age of the UNIVERSE and other large-scale COSMIC characteristics. It was suggested by PAUL ADRIAN MAURICE DIRAC. It is now not generally accepted but does influence the ANTHROPIC PRINCIPLE.

direct broadcast satellite (DBS) A class of COMMUNICATIONS SATELLITE, usually placed into GEOSTATIONARY ORBIT, that receives broadcast signals (such as television programs) from points of origin on EARTH and then amplifies, encodes, and retransmits these signals to individual end users scattered throughout some wide area or specific region. Many American households now receive hundreds of television channels directly from space by means of small (less than 0.5 m diameter), roof-top satellite dishes that are equipped to decode DBS transmissions.

direct conversion The conversion of thermal ENERGY (heat) or other forms of energy (such as sunlight) directly into electrical energy without intermediate conversion into mechanical work—that is, without the use of the moving components as found in a conventional electric generator system. The main approaches for converting heat directly into electricity include thermoelectric conversion, thermionic conversion, and MAGNETOHYDRODYNAMIC conversion. SOLAR ENERGY is directly converted into electrical energy by means of SOLAR CELLS (photovoltaic conversion). Batteries and FUEL CELLS directly convert chemical energy into electrical energy. *See also* RADIOISOTOPE THERMOELECTRIC GENERATOR.

directional antenna An ANTENNA that radiates or receives RADIO FREQUENCY (RF) signals more efficiently in some directions than in others. A collection of antennas arranged and selectively pointed for this purpose is called a directional antenna array.

direct readout The information technology capability that allows ground stations on EARTH to collect and interpret the data messages (TELEMETRY) being transmitted from SATELLITES.

disk (1) *(astronomy)* The visible surface of the SUN (or any other CELESTIAL BODY) seen in the sky or through a TELESCOPE. (2) *(of a galaxy)* The flattened, wheel-shaped region of STARS, gas, and dust that lies outside the central region (NUCLEUS) of a GALAXY.

diurnal Having a period of, occurirng in, or related to a day; daily.

diverging lens *(or mirror)* A LENS (or mirror) that refracts (or reflects) a parallel beam of light into a diverging beam. A diverging lens is generally a CONCAVE LENS, while a diverging mirror is a convex mirror. *Compare with* CONVERGING LENS; CONVEX LENS.

docking The act of physically joining two orbiting SPACECRAFT. This is usually accomplished by independently maneuvering one spacecraft (the CHASER SPACECRAFT) into contact with the other (the target SPACECRAFT) at a chosen physical interface. For spacecraft with human crews, a docking module assists in the process and often serves as a special passageway (AIRLOCK) that permits HATCHes to be opened and crew members to move from one spacecraft to the other without the use of a SPACESUIT and without losing cabin pressure.

doffing The act of removing wearing apparel or other apparatus, such as a SPACESUIT.

dogleg A directional turn made in a LAUNCH VEHICLE'S ascent TRAJECTORY to produce a more favorable ORBIT INCLINATION or to avoid passing over a populated (no-fly) region.

donning The act of putting on wearing apparel or other apparatus, such as a SPACESUIT.

Doppler shift The apparent change in the observed FREQUENCY and WAVELENGTH of a source due to the relative motion of the source and an observer. If the source is approaching the observer, the observed frequency is higher and the observed wavelength is shorter. This change to shorter wavelengths is often called the BLUESHIFT. If the

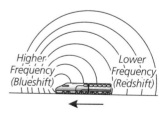

Doppler shift

source is moving away from the observer, the observed frequency will be lower and the wavelength will be longer. This change to longer wavelengths is called the REDSHIFT. It is named after CHRISTIAN JOHANN DOPPLER, who discovered this phenomenon in 1842 by observing sounds.

double-base propellant A solid ROCKET PROPELLANT using two unstable compounds, such as nitrocellulose and nitroglycerin. These unstable compounds contain enough chemically bonded OXIDIZER to sustain combustion.

downlink The TELEMETRY signal received at a ground station from a SPACECRAFT or SPACE PROBE.

downrange A location away from the LAUNCH SITE but along the intended flight path (TRAJECTORY) of a MISSILE or ROCKET flown from a rocket range. For example, the rocket vehicle tracking station on Ascension Island in the South Atlantic Ocean is far downrange from the launch sites at CAPE CANAVERAL Air Force Station in Florida.

drag *(D)* A retarding FORCE acting on a body in motion through a fluid parallel to (but opposite) the direction of motion of the body.

Drake equation A probabilistic expression, proposed by FRANK DONALD DRAKE in 1961, that is an interesting, though highly speculative, attempt to determine the number of advanced intelligent civilizations that might now exist in the MILKY WAY GALAXY and be communicating (via RADIO WAVES) across INTERSTELLAR distances. A basic assumption in Drake's formulation is the principle of mediocrity—namely that conditions in the SOLAR SYSTEM and even on EARTH are nothing particularly special but, rather, represent common conditions found elsewhere in the galaxy. *See also* SEARCH FOR EXTRATERRESTRIAL INTELLIGENCE.

drogue parachute A small parachute specifically used to pull a larger parachute out of stowage; a small parachute used to slow down a descending SPACE CAPSULE, AEROSPACE VEHICLE, or high-performance airplane.

dry emplacement A LAUNCH SITE that has no provision for water cooling of the pad during the LAUNCH of a ROCKET. *Compare with* WET EMPLACEMENT.

dwarf galaxy A small, often elliptical galaxy containing 1 million (10^6) to perhaps 1 billion (10^9) stars. The MAGELLANIC CLOUDS, humans' nearest galactic neighbors, are examples.

dwarf star Any STAR that is a MAIN-SEQUENCE star, according to the HERTZSPRUNG-RUSSELL (H-R) DIAGRAM. Most stars found in the GALAXY, including the SUN, are of this type and are from 0.1 to about 100 SOLAR MASSES in size. However, when astronomers use the term *dwarf star,* they are not referring to WHITE DWARFS, BROWN DWARFS, or BLACK DWARFS, which are CELESTIAL BODIES not in the collection of main-sequence stars.

Dyna-Soar *(Dynamic Soaring)* An early U.S. Air Force space project from 1958–63, involing a crewed boost-glide orbital vehicle that was to be sent into ORBIT by an EXPENDABLE LAUNCH VEHICLE, perform its military mission, and return to EARTH using wings to glide through the ATMOSPHERE during REENTRY (in a manner similar to NASA's SPACE SHUTTLE). The project was canceled in favor of the civilian (NASA) human spaceflight program, involving the MERCURY PROJECT, GEMINI PROJECT, and APOLLO PROJECT. Also called the *X-20 Project.*

early-warning satellite A military SPACECRAFT whose primary mission is the detection and notification of the LAUNCH of an enemy BALLISTIC MISSILE attack. This type of surveillance satellite uses sensitive INFRARED RADIATION sensors to detect the heat released when a MISSILE is launched. *See also* DEFENSE SUPPORT PROGRAM.

Earth The third PLANET from the SUN and the fifth largest in the SOLAR SYSTEM. When viewed from space, Earth is a beautiful world characterized by its distinctive blue waters, white clouds, and green vegetation. It circles the Sun in one YEAR (approximately 365.25 days) at an average distance of 149.6 million km and is the only planetary body currently known to possess life. *See also* SECTION IV CHARTS & TABLES.

Earth-based telescope A TELESCOPE that operates from the surface of EARTH, as opposed to a telescope placed onto an Earth-orbiting spacecraft, such as NASA's *HUBBLE* SPACE TELESCOPE.

Earth-crossing asteroid (ECA) An inner SOLAR SYSTEM ASTEROID whose orbital path takes it across EARTH's ORBIT around the SUN.

Earthlike planet A PLANET around another STAR (that is, an EXTRASOLAR PLANET) that orbits in a CONTINUOUSLY HABITABLE ZONE (CHZ) and maintains environmental conditions resembling EARTH. These conditions include a suitable ATMOSPHERE, a temperature range permitting the retention of liquid water on the planet's surface, and a sufficient quantity of radiant ENERGY striking the planet's surface from the parent star. Scientists in ASTROBIOLOGY hypothesize that under such conditions, the chemical evolution and the development of carbon-based life (as known on Earth) could also occur there.

Earth-observing spacecraft A SATELLITE in orbit around EARTH that has a specialized collection of sensors capable of monitoring important environmental variables. This is also called an environmental satellite or a green satellite. Data from such satellites help support EARTH SYSTEM science. *See also* LANDSAT; METEOROLOGICAL SATELLITE; *TERRA;* SPACECRAFT.

Earth radiation budget (ERB) The fundamental environmental phenomenon that influences EARTH's climate. Components include the incoming SOLAR RADIATION; the amount of solar radiation reflected back to space by clouds, other components in the ATMOSPHERE, and Earth's surface; and the long-WAVELENGTH (thermal) INFRARED RADIATION emitted by Earth's surface and stmosphere. The variation of Earth's radiation budget with LATITUDE represents the ultimate driving force for atmospheric and oceanic circulations and the resulting planetary climate.

Earth satellite An artificial (human-made) object placed into ORBIT around PLANET EARTH.

Earth's trapped radiation belts Two major belts (or zones) of very energetic atomic PARTICLE's (mainly ELECTRONS and PROTONS) that are trapped by EARTH's magnetic field hundreds of kilometers above the ATMOSPHERE. These are also called the Van Allen belts after JAMES ALFRED VAN ALLEN, who discovered them in 1958. *See also* MAGNETOSPHERE.

Earth system *(science)* The modern study of EARTH, facilitated by space-based observations, that treats the PLANET as an interactive, complex system. The four major components of the Earth system are the ATMOSPHERE, the HYDROSPHERE (which includes liquid water and ice), the BIOSPHERE (which includes all living things), and the SOLID EARTH (especially the planet's surface and soil).

eccentricity *(e)* A measure of the ovalness of an ORBIT. For example, when $e = 0$, the orbit is a circle; when $e = 0.9$, the orbit is a long, thin ELLIPSE.

eccentric orbit An ORBIT that deviates from a circle, thus forming an ELLIPSE.

eclipse (1) The reduction in visibility or the disappearance of a nonluminous CELESTIAL BODY when it passes into the shadow cast by another nonluminous body. The LUNAR eclipse is an example. It occurs when the MOON passes through the shadow cast by EARTH or when a

NATURAL SATELLITE (Moon) passes through the shadow cast by its PLANET.

(2) The apparent cutting off, totally or partially, of the light from a luminous body when a dark (nonluminous) body comes between it and an observer. The SOLAR eclipse is an example. It takes place when the Moon passes between the SUN and the Earth.

ecliptic *(plane)* The apparent annual path of the SUN among the STARS; the intersection of the plane of EARTH's ORBIT around the Sun with the CELESTIAL SPHERE. Because of the tilt in Earth's AXIS, the ecliptic is a great circle of the celestial sphere inclined at an angle of about 27.4° to the celestial EQUATOR.

ecosphere The CONTINUOUSLY HABITABLE ZONE (CHZ) around a MAIN-SEQUENCE STAR of a particular LUMINOSITY in which a PLANET could support environmental conditions favorable to the evolution and continued existence of life. For the chemical evolution of Earthlike, carbon-based, living organisms, global temperatures and atmospheric pressure conditions must allow the retention of liquid water on the planet's surface. A viable ECOSPHERE might lie between about 0.7 and 1.3 ASTRONOMICAL UNITS from a STAR like the SUN. However, if all the surface water has evaporated (the RUNAWAY GREENHOUSE effect) or has completely frozen (the ICE CATASTROPHE), then any Earthlike planet within this ecosphere cannot sustain life.

ejecta Any of a variety of rock fragments thrown out by an impact CRATER during its formation and subsequently deposited via BALLISTIC TRAJECTORIES onto the surrounding terrain. The deposits themselves are called ejecta blankets. It is also material thrown out of a VOLCANO during an explosive eruption.

electric propulsion A ROCKET engine that converts electric power into reactive THRUST by accelerating an ionized PROPELLANT (such as mercury, cesium, argon, or xenon) to a very high exhaust VELOCITY. There are three general types of electric rocket engine: electrothermal, electromagnetic, and electrostatic.

electromagnetic (EM) spectrum Comprises the entire range of WAVELENGTHS of ELECTROMAGNETIC RADIATION, from the most energetic, shortest-wavelength GAMMA RAYS to the longest-wavelength RADIO WAVES, and everything in between.

electromagnetic radiation (EMR) Radiation composed of oscillating electric and magnetic fields and propagated with the SPEED OF LIGHT. EMR includes (in order of decreasing ENERGY and increasing

WAVELENGTH) GAMMA RAYS, X RAYS, ULTRAVIOLET RADIATION, visible radiation (LIGHT), INFRARED RADIATION, radar waves, and RADIO WAVES.

electron (e) A stable elementary PARTICLE with a unit negative electrical charge (1.602×10^{-19} C) and a rest MASS $(m_0)_e$ of approximately 1/1,837 that of a proton (namely, 9.109×10^{-31} kg). Electrons surround the positively charged NUCLEUS and determine the chemical properties of the ATOM. Electrons detached from an atom are called free electrons. Positively charged electrons, or POSITRONS, also exist.

electron volt (eV) A unit of ENERGY equivalent to the energy gained by an ELECTRON when it moves through a potential difference of one volt. Larger multiple units of the electron volt are often encountered, such as keV for thousand (or kilo) electron volts (10^3 eV); MeV for million (or mega) electron volts (10^6 eV); and GeV for billion (or giga) electron volts (10^9 eV). One electron volt is also equivalent to 1.602×10^{-19} J.

element A chemical substance that cannot be divided or decomposed into simpler substances by chemical means. A substance whose ATOMS all have the same number of PROTONS (ATOMIC NUMBERS, Z) and ELECTRONS. There are 92 naturally occurring elements and over 15 human-made or transuranium elements, such as plutonium (atomic number 94).

ellipse A smooth, oval curve accurately fitted by the ORBIT of a SATELLITE around a much larger MASS. Specifically, a plane curve constituting the locus of all points the sum of whose distances from two fixed points (called focuses or foci) is constant; an elongated circle. The orbits of PLANETS, satellites, ASTEROIDS, and COMETS are ellipses; the center of attraction (that is, the PRIMARY BODY) is one focus.

elliptical galaxy A GALAXY with a smooth, elliptical shape without spiral arms and having little or no INTERSTELLAR gas and dust.

elliptical orbit A noncircular, Keplerian ORBIT. *See also* KEPLER'S LAWS.

emissivity *(e or ε)* The ratio of the radiant FLUX per unit area (sometimes called emittance, E) emitted by a body's surface at a specified WAVELENGTH *(λ)* and temperature *(T)* to the radiant flux per unit area emitted by a BLACKBODY radiator at the same temperature and under the same conditions. The greatest value for an emissivity is unity (1)—the emissivity value for a blackbody radiator—while the least value for an emissivity is zero (0).

empirical Derived from observation or experiment.

encounter The close FLYBY or RENDEZVOUS of a SPACECRAFT with a target body. The target of an encounter can be a natural CELESTIAL BODY (such as a PLANET, ASTEROID, or COMET) or a human-made object (such as another spacecraft).

endothermic reaction A reaction to which thermal ENERGY (heat) must be provided. *Compare with* EXOTHERMIC REACTION.

energy *(E)* The capacity to do WORK. Energy appears in many different forms, such as mechanical, thermal, electrical, chemical, and nuclear. According to the first law of thermodynamics, energy can neither be created nor destroyed but simply changes form (including mass-energy transformations).

energy satellite *See* SATELLITE POWER SYSTEM.

engine cutoff The specific time when a ROCKET engine is shut down during a flight.

entropy *(S)* A measure of the extent to which the ENERGY of a system is unavailable. As entropy increases, energy becomes less available to perform useful WORK. *See also* HEAT DEATH OF THE UNIVERSE.

environmental satellite *See* EARTH-OBSERVING SPACECRAFT.

ephemeris A collection of data about the predicted positions (or apparent positions) of CELESTIAL BODIES, including ARTIFICIAL SATELLITES, at various times in the future. *See also* KEPLERIAN ELEMENTS.

epicycle A small circle whose center moves along the circumference of a larger circle, called the deferent. Ancient astronomers, like PTOLEMY, used the epicycle in an attempt to explain the motions of CELESTIAL BODIES in their GEOCENTRIC (nonheliocentric) models of the SOLAR SYSTEM.

equator An imaginary circle around a CELESTIAL BODY that is everywhere equidistant (90°) from the poles of ROTATION. It defines the boundary between the Northern and Southern Hemisphere.

equatorial bulge The excess of a PLANET's equatorial diameter over its polar diameter. The increased size of the equatorial diameter is caused by CENTRIFUGAL FORCE associated with ROTATION about the polar AXIS.

equinox One of two points of intersection of the ECLIPTIC and the celestial equator that the SUN occupies when it appears to cross the celestial equator (that is, has a DECLINATION of 0°). In the Northern Hemisphere, the Sun appears to go from south to north at the vernal equinox, which occurs on or about 21 March. Similarly, the Sun

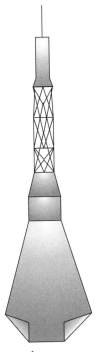

Escape rocket

Europa (Courtesy of NASA)

appears to travel from north to south at the autumnal equinox, which occurs on or about 23 September each year. The dates are reversed in the Southern Hemisphere.

ergometer A bicycle-like instrument used by ASTRONAUTS and COSMONAUTS for measuring muscular WORK and for exercising in place on extended orbital flights in MICROGRAVITY.

erosive burning An increased rate of burning (combustion) that occurs in certain SOLID-PROPELLANT ROCKETS as a result of the scouring influence of combustion gases moving at high speed across the burning surface.

escape rocket A small ROCKET engine, attached to the leading end of an escape tower, that can provide additional THRUST to the crew's SPACE CAPSULE so it can quickly separate from a malfunctioning or exploding EXPENDABLE LAUNCH VEHICLE during LIFTOFF.

escape velocity (V_e) The minimum VELOCITY that an object must acquire to overcome the gravitational attraction of a CELESTIAL BODY. The escape velocity for an object launched from the surface of EARTH is approximately 11.2 km/s, while the escape velocity from the surface of MARS is 5.0 km/s. *See also* SECTION IV CHARTS & TABLES.

Europa The smooth, ice-covered MOON of Jupiter, discovered by GALILEO GALILEI in 1610, and currently thought to have a liquid-water ocean beneath its frozen surface.

European Space Agency (ESA) An international organization that promotes the peaceful use of OUTER SPACE and cooperation among the European member states in space research and applications.

evening star Common name for the PLANET VENUS when it appears as a bright CELESTIAL BODY in the twilight sky just after sunset.

event horizon The point of no return for a BLACK HOLE; the distance from a black hole within which nothing can escape. Also called the SCHWARZSCHILD RADIUS.

evolved star A STAR near the end of its lifetime when most of its HYDROGEN fuel has been exhausted; a star that has left the main sequence. *See* MAIN-SEQUENCE STAR.

exhaust plume Hot gas ejected from the THRUST chamber of a ROCKET engine.

exoatmospheric Occurring outside EARTH's ATMOSPHERE; events and actions that take place at ALTITUDES above 100 km.

exobiology *See* ASTROBIOLOGY.

exosphere The outermost region of EARTH'S ATMOSPHERE.

exothermic reaction A chemical or physical reaction in which thermal ENERGY (heat) is released into the surroundings.

expanding universe Any model of the UNIVERSE in modern COSMOLOGY that has the distance between widely separated celestial objects (for example, distant GALAXIES) continuing to grow or expand with time.

expendable launch vehicle (ELV) A ground-launched ROCKET vehicle capable of placing a PAYLOAD into ORBIT around EARTH or on an Earth-escape TRAJECTORY whose various stages and supporting hardware are not designed or intended for recovery or reuse. A throwaway LAUNCH VEHICLE. *See also* SECTION IV CHARTS & TABLES.

experimental vehicle (X) A MISSILE, ROCKET, or AEROSPACE VEHICLE in the research, development, and testing portion of its technical life cycle; a vehicle not yet approved for operational use.

exploding galaxies Violent, very energetic explosions centered in certain GALACTIC NUCLEI where the total MASS of ejected material is comparable to the mass of some 5 million average-size, SUNLIKE STARS. Jets of gas 1,000 LIGHT-YEARS long are also typical.

Explorer I The first American SATELLITE to ORBIT around EARTH successfully. It was launched from CAPE CANAVERAL on 31 January 1958 by a Juno I four-stage ROCKET vehicle. This satellite involved a quickly assembled team from the U.S. Army (under the direction of WERNHER VON BRAUN) and Caltech's Jet Propulsion Laboratory (JPL). Dr. JAMES ALFRED VAN ALLEN (State University of Iowa) provided the satellite's instruments that discovered a portion of EARTH'S TRAPPED RADIATION BELTS, which were subsequently named after him.

Explorer spacecraft The large family of NASA scientific SPACECRAFT, starting in 1958, that have investigated astronomical and astrophysical phenomena, the properties and structure of EARTH'S MAGNETOSPHERE and ATMOSPHERE, and the PLANET'S precise shape and geophysical surface features.

external tank The large tank that contains the cryogenic PROPELLANTs for the three SPACE SHUTTLE main engines (SSMEs). This tank forms the structural backbone of NASA'S SPACE TRANSPORTATION SYSTEM flight vehicle.

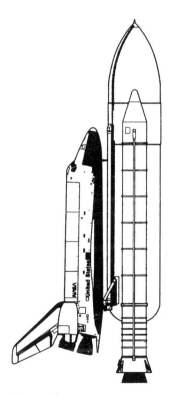

External tank

extragalactic Occurring, located, or originating beyond the MILKY WAY GALAXY—that is, more than 100,000 LIGHT-YEARS distant.

extragalactic astronomy A branch of ASTRONOMY that started about 1930 and deals with everything in the UNIVERSE outside of the MILKY WAY GALAXY.

extrasolar Occurring, located, or originating outside of the SOLAR SYSTEM.

extrasolar planet A PLANET around a STAR other than the SUN.

extraterrestrial (ET) Occurring, located, or originating beyond PLANET EARTH and its ATMOSPHERE.

extraterrestrial catastrophe theory The hypothesis that a large ASTEROID or COMET struck EARTH some 65 million years ago, causing global environmental consequences that annihilated over 90 percent of all animal species then living—including the dinosaurs.

extraterrestrial contamination The contamination of one world by life-forms, especially microorganisms, from another world. When using EARTH'S BIOSPHERE as the reference, planetary contamination is called forward contamination when an alien world is contaminated by contact with terrestrial organisms and back contamination when alien organisms are released into Earth's biosphere.

extraterrestrial life Life-forms that may have evolved independent of and now exist beyond the TERRESTRIAL BIOSPHERE.

extravehicular activity (EVA) Activites conducted by an ASTRONAUT or COSMONAUT in OUTER SPACE or on the surface of another PLANET (or MOON) outside of the protective environment of his/her AEROSPACE VEHICLE, SPACECRAFT, or LANDER. Astronauts and cosmonauts must put on SPACESUITS (which contain portable life-support systems) to perform EVA tasks.

extreme ultraviolet (EUV) radiation The region of the ELECTROMAGNETIC SPECTRUM that lies between the ULTRAVIOLET RADIATION and X-RAY regions. EUV PHOTONs have wavelengths between about 10 and 100 nm (or 100 and 1,000 ANGSTROMS).

Extreme Ultraviolet Explorer *(EUVE)* The 70th NASA EXPLORER SPACECRAFT. After being successfully launched from CAPE CANAVERAL in June 1992, this spacecraft went into orbit around Earth and provided astronomers with a survey of the (until then) relatively unexplored EXTREME ULTRAVIOLET (EUV) RADIATION portion of the ELECTROMAGNETIC SPECTRUM.

extremophile A hardy (TERRESTRIAL) microorganism that can exist under extreme environmental conditions, such as in frigid polar regions or boiling hot springs. Astrobiologists speculate that similar (EXTRATERRESTRIAL) microorganisms might exist elsewhere in this SOLAR SYSTEM, perhaps within subsurface biological niches on MARS or in a suspected liquid-water ocean beneath the frozen surface of EUROPA.

eyeballs in, eyeballs out Early American space program expression used to describing the acceleration-related sensations experienced by an ASTRONAUT at LIFTOFF or when RETROROCKETs fired. The experience at liftoff is eyeballs in (due to positive G FORCES on the human body when the LAUNCH VEHICLE accelerates). The experience when the retrorockets fire is eyeballs out (due to negative g forces on the human body as a SPACECRAFT decelerates).

eyepiece A magnifying LENS that helps an observer view the image produced by a TELESCOPE.

facula A bright region of the SUN's PHOTOSPHERE.

fallaway section A section of a ROCKET vehicle that is cast off and separates from the vehicle during flight, especially a section that falls back to EARTH.

farside The side of the MOON that never faces EARTH.

ferret satellite A military SPACECRAFT designed for the detection, location, recording, and analyzing of ELECTROMAGNETIC RADIATION (for example, enemy RADIO FREQUENCY [RF] transmissions).

ferry flight An in-the-ATMOSPHERE flight of NASA's SPACE SHUTTLE ORBITER vehicle while mated on top of a specially configured Boeing 747 shuttle carrier aircraft.

film cooling The cooling of body or surface, such as the inner surface of a ROCKET's COMBUSTION CHAMBER, by maintaining a thin fluid layer over the affected area.

field of view (FOV) The area or solid angle than can be viewed through or scanned by a REMOTE-SENSING (optical) instrument.

fire arrow An early gunpowder ROCKET attached to a large bamboo stick; developed by the Chinese about 1,000 years ago to confuse and startle enemy troops.

fireball *(meteor)* See BOLIDE.

Farside (courtesy of NASA)

fission   *(nuclear)* The process during which the NUCLEUS of certain heavy RADIOISOTOPES, such as uranium 235, captures a NEUTRON, becomes an unstable compound nucleus, and soon breaks apart. As the compound nucleus splits, or fissions, it forms two lighter nuclei (called *fission products*) and also releases a large amount of ENERGY (about 200 million ELECTRON VOLTS per reaction) plus additional neutrons and GAMMA RAYS.

fixed stars   A term used by early astronomers to distinguish between the apparently motionless background STARS and the wandering stars (PLANETS). Modern astronomers now use this term to describe stars that have no detectable PROPER MOTION.

flame bucket   A deep, cavelike metal construction built beneath a LAUNCH PAD. It is open at the top to receive the hot engine exhaust gases from the ROCKET positioned above it and has one to three sides open below. During THRUST buildup and the beginning of LAUNCH, water can be sprayed onto the flame bucket to keep it from melting.

flare   *(solar)* A bright eruption from the SUN's corona. An intense flare represents a major IONIZING RADIATION hazard to ASTRONAUTS traveling beyond EARTH's MAGNETOSPHERE through INTERPLANETARY space or while exploring the surface of the MOON or MARS.

flight crew   Personnel assigned to an AEROSPACE VEHICLE (like NASA's SPACE SHUTTLE), a SPACE STATION, or an INTERPLANETARY SPACECRAFT for a specific flight or mission. The Space Shuttle flight crew usually consists of ASTRONAUTS serving as the commander, the pilot, and one or several mission specialists.

flight test vehicle   A ROCKET, MISSILE, or AEROSPACE VEHICLE used for performing flight tests that demonstrate the capabilities of the vehicle itself or of specific equipment carried onboard.

flux   (Φ) The rate of transport or flow of some quantity per unit area; often used in reference to the flow of some form of ENERGY. A particle flux is defined as the number of PARTICLES passing through one square centimeter of a given target in one second, while a radiant flux represents the total POWER per unit area of some form of electromagnetic radiation—that is, WATTS per square centimeter.

flyby   An INTERPLANETARY or deep-space mission in which the FLYBY SPACECRAFT passes close to its target CELESTIAL BODY (for example, a distant PLANET, MOON, ASTEROID. or COMET) but does not impact the target or go into ORBIT around it.

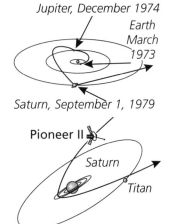

Jupiter, December 1974
Earth March 1973
Saturn, September 1, 1979
Pioneer II
Saturn
Titan

Flyby (*Pioneer 11* flyby of Saturn)

GLOSSARY

fission – flyby

focal length *(f)* The distance between the center of a LENS or mirror to the focus or focal point.

folded optics Any optical system containing reflecting components for the purpose of reducing the physical length of the system or for changing the path of the optical AXIS.

force *(F)* The cause of the ACCELERATION of material objects as measured by the rate of change of MOMENTUM produced on a free body. Force is a VECTOR quantity, mathematically expressed by the second of NEWTON'S LAWS OF MOTION: force = mass × acceleration.

fossa A long, narrow, shallow (ditchlike) depression found on the surface of a PLANET or a MOON.

free fall The unimpeded fall of an object in a gravitational field. For example, all the ASTRONAUTS and objects inside an EARTH-orbiting SPACECRAFT experience a continuous state of free fall and appear weightless as the FORCE of INERTIA counterbalances the force of Earth's GRAVITY. *See also* WEIGHTLESSNESS.

free-flying spacecraft *(free-flyer)* Any SPACECRAFT or PAYLOAD that can be detached from NASA's SPACE SHUTTLE or the INTERNATIONAL SPACE STATION and then operate independently in ORBIT.

frequency *(f* or *v)* The rate of repetition of a recurring or regular event; the number of cycles of a wave per second. For ELECTROMAGNETIC RADIATION, the frequency *(v)* equals the SPEED OF LIGHT *(c)* divided by the WAVELENGTH *(λ)*. *See also* HERTZ.

free rocket A ROCKET not subject to guidance or control in flight.

fuel cell A DIRECT-CONVERSION device that transforms chemical ENERGY directly into electrical energy by reacting continuously supplied chemicals. In a modern fuel cell, an electrochemical catalyst (like platinum) promotes a noncombustible reaction between a fuel (such as HYDROGEN) and an oxidant (such as oxygen).

fuselage The central part of an AEROSPACE VEHICLE or aircraft that accommodates crew, passengers, PAYLOAD, or cargo.

fusion *(nuclear)* The nuclear process by which lighter atomic NUCLEI join (or fuse) to create a heavier nucleus. The fusion of DEUTERIUM (D) with TRITIUM (T) results in the formation of a HELIUM (He) nucleus and a NEUTRON (n). This D-T reaction also involves the release of 17.6 million ELECTRON VOLTS (MeV) of ENERGY. Thermonuclear reactions are fusion reactions caused by very high temperatures (millions of degrees KELVIN). The energy of the SUN and other stars comes from thermonuclear fusion reactions.

g The symbol used for the ACCELERATION due to GRAVITY. At sea level on EARTH, *g* is approximately 9.8 meters per second squared (m/s^2)— that is, one *g*. This term is used as a unit of stress for bodies experiencing acceleration. When a ROCKET accelerates during LAUNCH, everything inside it experiences a FORCE that can be as high as several *g*s.

galactic cannibalism A postulated model of GALAXY interaction in which a more massive galaxy uses GRAVITATION and tidal FORCEs to pull away matter from a less massive, neighboring galaxy.

galactic cluster A diffuse collection of from 10 to perhaps several hundred STARS loosely held together by the FORCE of GRAVITATION. The term OPEN CLUSTER is now preferred by astronomers.

galactic cosmic rays (GCRs) Very energetic atomic PARTICLES that originate outside the SOLAR SYSTEM. *See also* COSMIC RAYS.

galactic nucleus The central region of a GALAXY.

galaxy A very large accumulation of STARS with from 1 million (10^6) to 1 million million (10^{12}) members. These ISLAND UNIVERSES come in a variety of sizes and shapes, from DWARF GALAXIES (like the MAGELLANIC CLOUDS) to majestic SPIRAL GALAXIES (like the ANDROMEDA GALAXY). Astronomers classify them as elliptical, spiral (or barred spiral), or irregular.

Galaxy When capitalized, humans' home GALAXY, the MILKY WAY GALAXY.

Galilean satellites The four largest and brightest MOONs of JUPITER, discovered by GALILEO GALILEI in 1610. They are IO, EUROPA, GANYMEDE, and CALLISTO.

Galilean telescope The early REFRACTING TELESCOPE assembled by GALILEO GALILEI in about 1610. It had a CONVERGING LENS as the OBJECTIVE and a DIVERGING LENS as the EYEPIECE.

Galileo Project NASA's highly successful scientific mission to JUPITER launched in October 1989. With electricity supplied by two RADIOISOTOPE THERMOELECTRIC GENERATOR (RTG) units, the *Galileo* SPACECRAFT has extensively studied the Jovian system since December 1995. Upon arrival, it also released a PROBE into the upper portions of Jupiter's ATMOSPHERE.

gamma ray *(γ)* Very short WAVELENGTH, high-frequency packets (or QUANTA) of ELECTROMAGNETIC RADIATION. Gamma ray PHOTONS are similar to X RAYS, except that they originate within the atomic NUCLEUS and have energies between 10,000 ELECTRON VOLTS (10 keV) and 10 million electron volts (10 MeV).

gamma ray astronomy Branch of ASTRONOMY based on the detection of the energetic gamma rays associated with SUPERNOVAS, EXPLODING GALAXIES, QUASARS, PULSARS, and phenomena near suspected BLACK HOLES.

Ganymede With a diameter of 5,262 km, the largest MOON of JUPITER and in the SOLAR SYSTEM. Discovered by GALILEO GALILEI in 1610.

Gemini Project The second U.S. crewed space project (1964–66) and the start of more sophisticated missions by pairs of American ASTRONAUTs in each Gemini SPACE CAPSULE. Through this project, NASA expanded the results of the MERCURY PROJECT and prepared the way for the ambitious lunar landing missions of the APOLLO PROJECT.

general relativity ALBERT EINSTEIN's theory, introduced in 1915, that GRAVITATION arises from the curvature of space and time—the more massive an object, the greater the curvature.

geocentric Relative to EARTH as the center; measured from the center of Earth.

geographic information system (GIS) A computer-assisted system that acquires, stores, manipulates, compares, and displays geographic data, often including MULTISPECTRAL SENSING data sets from EARTH-OBSERVING SPACECRAFT.

geomagnetic storm Sudden, often global fluctuations in EARTH's magnetic field, associated with the shock waves from SOLAR FLAREs that arrive at Earth within about 24 to 36 hours after violent activity on the SUN.

geosphere The solid (lithosphere) and liquid (HYDROSPHERE) portions of EARTH. Above the geosphere lies the ATMOSPHERE. At the interface between these two regions is found almost all of the BIOSPHERE, or zone of life.

geostationary orbit (GEO) A SATELLITE in a circular ORBIT around EARTH at an ALTITUDE of 35,900 km above the EQUATOR that goes around the PLANET at the same rate as Earth spins on its AXIS. COMMUNICATIONS SATELLITES, ENVIRONMENTAL SATELLITES, and SURVEILLANCE SATELLITES use this important orbit. If the SPACECRAFT's orbit is circular and lies in the equatorial plane, (to an observer on Earth) the spacecraft appears stationary over a given point on Earth's surface. If the satellite's orbit is inclined to the equatorial plane, (when observed from Earth) the spacecraft traces out a figure eight path every 24 hours. *See also* GEOSYNCHRONOUS ORBIT; SYNCHRONOUS SATELLITE.

geosynchronous orbit (GEO) An ORBIT in which a SATELLITE completes one REVOLUTION at the same rate as EARTH spins, namely 23 hours, 56 minutes, and 4.1 seconds. A satellite placed into such an orbit (at approximately 35,900 km ALTITUDE above the EQUATOR) revolves around Earth once per day. *See also* GEOSTATIONARY ORBIT; SYNCHRONOUS SATELLITE.

giant-impact model The hypothesis that the MOON originated when a MARS-sized object struck a young EARTH with a glancing blow. The giant (oblique) impact released material that formed an ACCRETION DISK around Earth out of which the Moon formed.

giant molecular cloud (GMC) Massive clouds of gas in INTERSTELLAR space composed primarily of MOLECULES of HYDROGEN (H_2) and dust. GMCs can contain enough MASS to make several million STARS like the SUN and are often the sites of star formation.

giant planets In this SOLAR SYSTEM, the large, gaseous, OUTER PLANETS: JUPITER, SATURN, URANUS, and NEPTUNE. Any detected or suspected EXTRASOLAR PLANETS as large or larger than Jupiter.

giant star A STAR near the end of its life that has swollen in size, such as a BLUE GIANT or a RED GIANT.

***Giotto* spacecraft** Scientific SPACECRAFT launched by the EUROPEAN SPACE AGENCY (ESA) in July 1985 that successfully encountered the NUCLEUS of COMET HALLEY in mid-March 1986 at a distance of about 600 km.

global change EARTH's environment is continuously changing. Many natural changes occur quite slowly, requiring thousands of years to achieve their full impact. Human-induced change can happen rapidly, in times as short as a few decades. Global change studies the interactive linkages among this PLANET's major natural and human-made systems that influence the planetary environment. *See also* EARTH SYSTEM.

Global Positioning System (GPS) The CONSTELLATION of over 20 U.S. Air Force SATELLITES in circular 20,350 km ALTITUDE ORBITS around EARTH that provide accurate navigation data to military and civilian users globally.

globular cluster Compact cluster of up to 1 million, generally older, STARS.

grain The integral piece of molded or extruded solid material that encompasses both fuel and OXIDIZER in a SOLID-PROPELLANT ROCKET motor. When ignited, the design and shape of the grain produce a specified THRUST over time.

gravitation The FORCE of attraction between two MASSES. From NEWTON'S LAW OF GRAVITATION, this attractive force operates along a line joining the CENTERS OF MASS, and its magnitude is inversely proportional to the square of the distance between the two masses. From ALBERT EINSTEIN'S GENERAL RELATIVITY theory, gravitation is viewed as a distortion of the space-time continuum.

gravitational collapse The unimpeded contraction of any MASS caused by its own GRAVITY.

graviton The hypothetical QUANTUM (or PARTICLE) of gravitational ENERGY predicted by ALBERT EINSTEIN in his GENERAL RELATIVITY theory.

gravity The attraction of a CELESTIAL BODY for any nearby MASS. Specifically, the downward FORCE imparted by EARTH on a mass near Earth or on its surface.

gravity anomaly A region on a CELESTIAL BODY where the local FORCE of GRAVITY is lower or higher than expected, assuming uniform DENSITY conditions. *See also* MASCON.

gravity assist The change in a SPACECRAFT's direction and speed achieved by a carefully calculated FLYBY through a PLANET's gravitational field. This change in spacecraft VELOCITY occurs without the use of supplementary propulsive ENERGY.

gravity well An analogy in which the gravitational field of a planetary body is considered as a deep well or pit out of which a SPACE VEHICLE has to climb to escape from this CELESTIAL BODY.

gray body A conceptual body in radiation heat transfer that absorbs some constant fraction, between zero and one, of all ELECTROMAGNETIC RADIATION incident upon it. *Compare with* BLACKBODY.

Great Dark Spot (GDS) A large, dark, oval-shaped feature in the clouds of NEPTUNE, discovered in 1989 by NASA's *VOYAGER 2* SPACECRAFT.

Great Red Spot (GRS) A distinctive, oval-shaped feature in the Southern Hemisphere clouds of JUPITER, first noted by SAMMUEL HEINRICH SCHWABE in 1831.

greenhouse effect The general warming of the lower layers of a PLANET's ATMOSPHERE caused by the presence of greenhouse gases, such as water vapor (H_2O), carbon dioxide (CO_2), and methane (CH_4), which prevent the escape of thermal radiation from the planet's surface to OUTER SPACE.

green satellite A SATELLITE in ORBIT around EARTH that collects a variety of environmental data. *See also* EARTH-OBSERVING SPACECRAFT.

Greenwich mean time (GMT) MEAN SOLAR TIME at the meridian of Greenwich, England, used as the basis for standard time throughout the world. It is normally expressed in four numerals, 0001 to 2400. Also called universal time *(UT)* or zulu time *(Z-time)*.

Gregorian calendar A more precise version of the JULIAN CALENDAR that was devised by CHRISTOPHER CLAVIUS and introduced by Pope Gregory XIII in 1582. The changes restored 21 March as the VERNAL EQUINOX. It is now the civil CALENDAR used in most of the world.

ground-elapsed time (GET) The time expired since LAUNCH.

ground support equipment (GSE) Any nonflight equipment used for LAUNCH, CHECKOUT, or in-flight support of an expendable ROCKET, REUSABLE LAUNCH VEHICLE, SPACECRAFT, or other type of PAYLOAD.

ground track The path followed by a SPACECRAFT over EARTH's surface.

guidance system A system that evaluates flight information; correlates it with target or destination data; determines the desired flight path of the MISSILE, SPACECRAFT, or AEROSPACE VEHICLE; and communicates the necessary commands to the vehicle's flight control system.

guided missile (GM) A self-propelled vehicle without a crew that moves above the surface of EARTH whose TRAJECTORY or course is capable of being controlled while in flight.

gun-launch to space (GLTS) An advanced LAUNCH concept involving the use of a long and powerful electromagnetic launcher to hurl small SATELLITES and PAYLOADS into ORBIT.

gyro A device that uses the ANGULAR MOMENTUM of a spinning MASS (rotor) to sense angular motion of its base about one or two AXES orthogonal (mutually perpendicular) to the spin axis. Also called a gyroscope.

H I region A diffuse region of neutral, predominantly atomic HYDROGEN in INTERSTELLAR space.

H II region A region in INTERSTELLAR space consisting mainly of ionized HYDROGEN and existing mostly in discrete clouds.

habitable payload A PAYLOAD with a pressurized compartment suitable for supporting an ASTRONAUT or COSMONAUT in a shirtsleeve environment.

Hadley Rille A long, ancient lava channel on the MOON that was the landing site for NASA's *APOLLO 15* mission. *See* APOLLO PROJECT.

half-life *(radioactive)* The time required for one-half of the ATOMS of a particular radioactive ISOTOPE population to disintegrate to another nuclear form. Values range from millionths of a second to billions of years.

halo orbit A circular or ELLIPTICAL ORBIT in which a SPACECRAFT remains in the vicinity of a LAGRANGIAN LIBRATION POINT.

hang fire A faulty condition in the ignition system of a ROCKET engine.

Halley's Comet *See* COMET HALLEY.

hard landing A relatively high-VELOCITY impact of a LANDER spacecraft or PROBE onto a solid planetary surface. The impact usually destroys all equipment, except perhaps a very rugged instrument package or PAYLOAD container.

Harvard classification system A method of classifying stars by their spectral characteristics (such as O, B, A, F, G, K, and M in order of decreasing surface temperature). It was developed by WILLAMINA PATON FLEMING while working for EDWARD CHARLES PICKERING at HARVARD COLLEGE OBSERVATORY in the 1880s.

Harvard College Observatory (HCO) The astronomical observatory founded in 1839 at Harvard University.

hatch A tightly sealed access door in the pressure hull of an AEROSPACE VEHICLE, SPACECRAFT, or SPACE STATION.

Hawking radiation A theory proposed in 1973 by STEPHEN WILLIAM HAWKING that suggests that due to a combination of properties of QUANTUM MECHANICS and GRAVITY, under certain conditions BLACK HOLES can seem to emit radiation.

heat death of the universe A possible ultimate fate of the UNIVERSE suggested by RUDOLF JULIUS EMMANUEL CLAUSIUS in the 19th century. As he evaluated the consequences of the second law of thermodynamics on a grand scale, he concluded that the universe would end (die) in a condition of maximum ENTROPY in which no ENERGY was available for useful work.

heat soak The increase in the temperature of ROCKET engine components after firing has ceased. This occurs because of HEAT TRANSFER through adjoining parts of the engine when no active cooling is present.

heat transfer The exchange of thermal ENERGY by CONDUCTION, CONVECTION, or RADIATION within an object and between an object and its surroundings. Heat is also transferred when a working fluid undergoes phase change, such as evaporation or condensation.

heavy-lift launch vehicle (HLLV) A conceptual, large-capacity, space-lift vehicle capable of carrying tons of cargo into LOW EARTH ORBIT (LEO) at substantially less cost than today's EXPENDABLE LAUNCH VEHICLES (ELVs).

heliocentric With the SUN as a center.

heliometer A former instrument used by astronomers to measure the diameter of the SUN or the angular separation of two STARS that appear close to each other.

heliopause The boundary of the HELIOSPHERE. It is thought to occur about 100 ASTRONOMICAL UNITS from the SUN and marks the edge of the Sun's influence and the beginning of INTERSTELLAR space.

heliosphere The region of OUTER SPACE within the boundary of the HELIOPAUSE in which the SOLAR WIND flows. Contains the SUN and the SOLAR SYSTEM.

heliostat A mirrorlike device designed to follow the SUN as it moves through the sky and to reflect the Sun's rays on a stationary collector.

helium (He) A noble gas, the second most abundant ELEMENT in the UNIVERSE. Natural helium is mostly the ISOTOPE helium 4, which contains two PROTONS and two NEUTRONS in the NUCLEUS. Helium 3 is a rare isotope of helium, containing two protons and one neutron in the nucleus. Helium was initially discovered in the SUN'S SPECTRUM in 1868 by SIR JOSEPH NORMAN LOCKYER before it was found on EARTH.

helium burning The release of ENERGY in STARS by the thermonuclear FUSION of HELIUM to form carbon.

hertz (Hz) The SI unit of frequency. One hertz equals one cycle per second. It is named in honor of HEINRICH RUDOLF HERTZ.

Hertzsprung-Russell (H-R) diagram A useful graphic depiction of the different types of STARS arranged according to their SPECTRAL CLASSIFICATION and LUMINOSITY. It is named in honor of EJNAR HERTZSPRUNG and HENRY NORRIS RUSSELL, who developed the diagram independently of one another.

high Earth orbit (HEO) An ORBIT around EARTH at an ALTITUDE greater than 5,600 km.

High Energy Astronomical Observatory (HEAO) A series of three NASA SPACECRAFT placed into EARTH's ORBIT (*HEAO-1* launched in August 1977; *HEAO-2* in November 1978; and *HEAO-3* in September 1979) to support X-RAY ASTRONOMY and GAMMA RAY ASTRONOMY. After LAUNCH, NASA renamed *HEAO-2* the *Einstein Observatory* to honor the famous physicist ALBERT EINSTEIN.

highlands The oldest exposed areas on the surface of the MOON; extensively cratered and chemically distinct from the MARIA.

Hohmann transfer orbit The most efficient orbit transfer path between two coplanar circular ORBITS. The maneuver consists of two impulsive high-THRUST burns (or firings) of a SPACECRAFT's PROPULSION SYSTEM. The technique was suggested by WALTER HOHMANN in 1925.

hold To stop the sequence of events during a COUNTDOWN until an impediment has been removed so that the countdown to LAUNCH can be resumed.

holddown test The test of a ROCKET while it is firing but restrained in a test stand.

Horsehead Nebula A DARK NEBULA in the CONSTELLATION of Orion that has the shape of a horse's head.

hot-fire test A liquid-fuel PROPULSION SYSTEM test conducted by actually firing the ROCKET engine(s) (usually for a short period of time) with the rocket vehicle secured to the LAUNCH PAD by holddown bolts. *Compare with* COLD-FLOW TEST.

housekeeping (*spacecraft*) The collection of routine tasks that must be performed to keep a SPACECRAFT functioning properly during an orbital flight or INTERPLANETARY mission.

Hubble classification EDWIN POWELL HUBBLE's widely used system for classifying GALAXIES based on their visual appearance, such as round ELLIPTICAL GALAXIES, SPIRAL GALAXIES, and IRREGULAR GALAXIES.

Hubble constant (H_0) The constant within HUBBLE's LAW proposed by EDWIN POWELL HUBBLE in 1929 that establishes an EMPIRICAL relationship between the distance to a GALAXY and its VELOCITY of recession due to the expansion of the UNIVERSE. A value of 70 kilometers per second per MEGAPARSEC (km/s/Mpc) ± 7 km/s/Mpc is currently favored by astronomers.

Hubble's law The hypothesis that the REDSHIFTS of distant GALAXIES are directly proportional to their distances from EARTH. EDWIN POWELL HUBBLE first proposed this relationship in 1929. It can be expressed as $V = H_0 \times D$, where V is the recessional VELOCITY of a distant galaxy, H_0 is the HUBBLE CONSTANT, and D is its distance from Earth.

***Hubble* Space Telescope** (HST) A cooperative EUROPEAN SPACE AGENCY (ESA) and NASA program to operate a long-lived, space-based optical observatory. Launched on 25 April 1990 by NASA's SPACE SHUTTLE *Discovery* (STS-31 mission), subsequent on-orbit repair and refurbishment missions have allowed this powerful EARTH-orbiting optical observatory to revolutionize humans' knowledge of the size, structure, and makeup of the UNIVERSE. It is named in honor of EDWIN POWELL HUBBLE.

Huygens Probe A scientific PROBE, sponsored by the EUROPEAN SPACE AGENCY (ESA) and named after CHRISTIAAN HUYGENS, that will be deployed into the ATMOSPHERE of SATURN'S MOON TITAN by NASA'S CASSINI MISSION SPACECRAFT sometime in 2004.

hydrazine (N_2H_4) A toxic, colorless liquid that is often used as a ROCKET PROPELLANT because it reacts violently with many OXIDIZERS. It is spontaneously ignitable with concentrated hydrogen peroxide (H_2O_2) and nitric acid.

hydrogen (H) A colorless, odorless gas that is the most abundant chemical ELEMENT in the universe. Hydrogen occurs as molecular hydrogen (H_2), atomic hydrogen (H), and ionized hydrogen (that is, broken down into a PROTON and its companion ELECTRON). Hydrogen has three isotopic forms: PROTIUM (ordinary hydrogen), DEUTERIUM (heavy hydrogen), and TRITIUM (radioactive hydrogen). LIQUID HYDROGEN (LH_2) is an excellent, high-performance cryogenic chemical PROPELLANT for ROCKET engines, especially those using liquid oxygen (LO_2 or LOX) as the OXIDIZER.

hydrosphere The water on EARTH's surface (including oceans, seas, rivers, lakes, ice caps, and glaciers) considered as an interactive system. *See also* EARTH SYSTEM.

hyperbolic orbit An ORBIT in the shape of a hyperbola. All INTERPLANETARY FLYBY SPACECRAFT follow hyperbolic orbits, both for EARTH departure and again upon arrival at the target PLANET. *See also* CONIC SECTION.

hypergolic fuel A ROCKET fuel that spontaneously ignites when brought into contact with an OXIDIZER. Also called hypergol.

hypergolic ignition An ignition that involves no external ENERGY source but results entirely from the spontaneous reaction of two materials (both liquid or a liquid-solid combination) when they are brought into contact.

hypersonic Pertaining to speeds much greater that the speed of sound, typically speeds of MACH NUMBER five ($M = 5$) and greater.

hypervelocity impact A collision between two objects that takes place at a very high relative VELOCITY—typically at a speed in excess of 5 km/s. A SPACECRAFT colliding with a piece of SPACE (ORBITAL) DEBRIS or an ASTEROID striking a PLANET are examples.

HZE particles Very damaging COSMIC RAYS, with high ATOMIC NUMBER *(Z)* and high KINETIC ENERGY *(KE)*.

ice catastrophe An extreme climate crisis in which all the liquid water on the surface of a life-bearing (or potentially life-bearing) PLANET has become frozen. *Compare with* RUNAWAY GREENHOUSE.

Ida A heavily cratered, irregularly shaped ASTEROID about $56 \times 24 \times 21$ km in size that has its own tiny natural SATELLITE, DACTYL.

igniter A device used to start the combustion of a ROCKET engine.

image The representation of a physical object or scene formed by a mirror, LENS, or electro-optical recording device.

imaging instrument *See* CHARGE-COUPLED DEVICE.

Imbrium basin Large (about 1,300 km across), ancient IMPACT CRATER on the MOON.

impact The event or moment when a high-speed object (such as an ASTEROID, COMET, METEOROID, ROCKET, SPACECRAFT, or PROBE) strikes the surface of a planetary body

impact crater The CRATER or basin formed on the surface of a planetary body as a result of the high-speed IMPACT of a METEROID, ASTEROID, or COMET.

inclination *(i)* One of the six KEPLERIAN ELEMENTS; inclination describes the angle of an object's orbital plane with respect to the central body's EQUATOR. For EARTH-orbiting objects, the orbital plane always goes

Impact crater (Courtesy of NASA)

through the center of Earth, but it can tilt at any angle relative to the equator. By general agreement, inclination is the angle between Earth's equatorial plane and the object's orbital plane measured counterclockwise at the ascending node.

incompressible fluid A fluid for which the DENSITY is assumed constant.

inertia The resistance of a body to a change in its state of motion. MASS is an inherent property of a body that helps to quantify inertia. *See also* NEWTON'S LAWS OF MOTION.

inertial upper stage (IUS) A versatile orbital transfer vehicle (OTV) developed by the U.S. Air Force that uses SOLID-PROPELLANT ROCKET motors to boost a PAYLOAD from LOW EARTH ORBIT (LEO) into higher-ALTITUDE destinations. *See also* UPPER STAGE.

inferior conjunction *See* CONJUNCTION.

inferior planets MERCURY and VENUS—the two PLANETS that have ORBITS that lie inside EARTH's orbit around the SUN. *Compare with* SUPERIOR PLANETS.

infinity (∞) A quantity beyond measurable limits.

in-flight phase The flight of a MISSILE or ROCKET from LAUNCH to detonation or impact; the flight of a SPACECRAFT from launch to the time of planetary FLYBY, ENCOUNTER and ORBIT, or impact.

infrared astronomy The branch of ASTRONOMY dealing with INFRARED RADIATION from relatively cool celestial objects, such as INTERSTELLAR clouds of dust and gas (typically 100 K) and STARS with surface temperatures below about 6,000 K.

infrared radiation (IR) That portion of the ELECTROMAGNETIC (EM) SPECTRUM between the optical (visible) and radio WAVELENGTHS. The infrared region extends from about one MICROMETER (μm) to 1,000 μm wavelength.

injector A device that propels (injects) fuel and/or OXIDIZER into the COMBUSTION CHAMBER of a LIQUID-PROPELLANT ROCKET ENGINE. It atomizes and mixes the PROPELLANTS so they can burn more completely.

inner planets The TERRESTRIAL PLANETS: MERCURY, VENUS, EARTH, and MARS—all of which have ORBITS around the SUN that lie inside the MAIN-BELT ASTEROID. *Compare with* OUTER PLANETS.

insertion The process of putting an ARTIFICIAL SATELLITE, AEROSPACE VEHICLE, or SPACECRAFT into ORBIT.

integration The collection of activities leading to the compatible assembly of PAYLOAD and LAUNCH VEHICLE into the desired final (flight) configuration.

intercontinental ballistic missile (ICBM) A BALLISTIC MISSILE with a range in excess of 5,500 km.

interferometer An instrument that achieves high angular resolution by combining signals from at least two widely separated TELESCOPES (optical interferometer) or a widely a separated ANTENNA ARRAY (radio interferometer). Radio interferometers are one of the basic instruments of RADIO ASTRONOMY. *See also* VERY LARGE ARRAY.

intergalactic Between or among the GALAXIES.

intermediate-range ballistic missile (IRBM) A BALLISTIC MISSILE with a range capability from about 1,000 to 5,500 km.

International Space Station (ISS) A major human spaceflight project, headed by NASA. Russia, Canada, Europe, Japan, and Brazil are also contributing key elements to this large, modular SPACE STATION in LOW EARTH ORBIT that represents a permanent human outpost in OUTER SPACE for MICROGRAVITY research and advanced space technology demonstrations. On-orbit assembly began in December 1998, with completion now anticipated by 2004. *See also* SECTION IV CHARTS & TABLES.

international system of units *See* SI UNITS.

interplanetary Between the PLANETs; within the SOLAR SYSTEM.

interplanetary dust (IPD) Tiny PARTICLES of matter (generally less than 100 MICROMETERS [μm] in diameter) that exist in OUTER SPACE within the confines of this SOLAR SYSTEM. *See also* ZODIACAL LIGHT.

interstage section A section of a MISSILE or ROCKET that lies between stages. *See also* STAGING.

interstellar Between or among the STARS.

interstellar medium (ISM) The gas and tiny dust PARTICLES found between the STARS in this GALAXY. Over 100 different MOLECULES have been discovered in INTERSTELLAR space, including many organic molecules.

interstellar probe A conceptual, highly automated, robotic INTERSTELLAR SPACECRAFT launched by the people of EARTH (or some other advanced alien civilization) in the mid-21st century to explore other STAR systems.

intravehicular activity (IVA) ASTRONAUT or COSMONAUT activities performed inside an orbiting SPACECRAFT or AEROSPACE VEHICLE. *Compare with* EXTRAVEHICULAR ACTIVITY.

inverse square law A relationship between physical quantities of the form x proportional to $1/y^2$, where y is usually a distance and x terms are of two kinds—FORCES and FLUXes.

inviscid fluid A hypothesized perfect fluid that has zero coefficient of viscosity (internal resistance to flow).

Io The pizza-colored, volcanic GALILEAN SATELLITE of JUPITER with a diameter of 3,630 km.

ion An ATOM or MOLECULE that has lost or (more rarely) gained one or more ELECTRONS. By this ionization process, it becomes electrically charged.

ion engine An electrostatic ROCKET engine in which a PROPELLANT (for example, cesium, mercury, argon, or xenon) is ionized and the propellant IONS are accelerated by an imposed electric field to very high exhaust VELOCITY. *See also* ELECTRIC PROPULSION.

ionizing radiation Any type of nuclear radiation that displaces ELECTRONS from ATOMs or MOLECULEs, thereby producing IONS within the irradiated material. Examples include alpha (α) radiation, beta (β) radiation, gamma (γ) radiation, PROTONS, NEUTRONS, and X RAYS. *See also* ALPHA PARTICLE; BETA PARTICLE; GAMMA RAY.

ionosphere That portion of EARTH's upper ATMOSPHERE, extending from about 50 to 1,000 km, in which IONS and free ELECTRONS exist in sufficient quantity to reflect RADIO WAVES.

irregular galaxy A GALAXY with a poorly defined structure or shape. *See also* HUBBLE CLASSIFICATION.

Ishtar Terra A very large highland plateau in the northern hemisphere of VENUS, about 5,000 km long and 600 km wide.

island universes Term introduced by IMMANUEL KANT in the 18th century to describe other GALAXIES.

isotope One of two or more ATOMs with the same ATOMIC NUMBER *(Z)* (that is, the same chemical ELEMENT) but with different ATOMIC WEIGHTS.

isotropic Having uniform properties in all directions.

jansky (Jy) A unit used to describe the strength of an incoming signal of ELECTROMAGNETIC RADIATION. It is named in honor of KARL GUTHE JANSKY and is commonly used in RADIO ASTRONOMY and INFRARED ASTRONOMY. One jansky (Jy) of signal strength equals 10^{-26} WATTS per meter squared per HERTZ ($W/[m^2\text{-}Hz]$).

jettison To discard or toss away.

joule (J) The SI UNIT of ENERGY or WORK. One joule is the work done by a FORCE of one NEWTON moving through a distance of one meter. Named after James Prescott Joule.

Jovian planet A large (Jupiter-like) planet characterized by a great total MASS, low average DENSITY, mostly liquid interior, and an abundance of the lighter ELEMENTS (especially HYDROGEN and HELIUM). In this SOLAR SYSTEM, the Jovian planets are JUPITER, SATURN, URANUS, and NEPTUNE.

Julian calendar The 12-month (approximately 365-day) CALENDAR introduced by the Roman Emperor Julius Caesar in 46 B.C.E. *See also* GREGORIAN CALENDAR.

Jupiter The fifth PLANET from the SUN and the largest in the SOLAR SYSTEM, with more than twice the MASS of all the other planets and their MOONS combined. Its thick, cold atmosphere consists primarily of HYDROGEN and HELIUM in approximately stellar composition—that is, about 89 percent HYDROGEN and 11 percent HELIUM. *See also* SECTION IV CHARTS & TABLES.

Kapustin Yar A minor, early Russian LAUNCH SITE that is located on the banks of the Volga River near Volgograd at approximately 48.4° north latitude and 45.8° east longitude.

Keck Observatory Located near the 4,200 m high summit of Mauna Kae, Hawaii, the W. M. Keck Observatory possesses two of the world's largest optical/infrared reflector TELESCOPES, each with a 10 m diameter primary mirror. Keck I started astronomical operations in May 1993 and its twin, Keck II, in October 1996. *See also* INFRARED ASTRONOMY.

kelvin (K) The SI UNIT of absolute thermodynamic temperature, honoring BARON WILLIAM THOMSON KELVIN. By international agreement, 1 K represents the fraction 1/273.16 of the thermodynamic temperature of the triple point of water.

Kennedy Space Center (KSC) Sprawling NASA SPACEPORT on the east-central coast of Florida adjacent to CAPE CANAVERAL Air Force

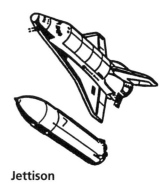

Jettison

Jupiter (Courtesy of NASA/JPL)

Station. It is the LAUNCH SITE (Complex 39) and primary landing/recovery site for the SPACE SHUTTLE.

Keplerian elements The six parameters that uniquely specify the position and path of a SATELLITE (natural or human made) in its ORBIT as a function of time. The elements and their characteristics are described in SECTION IV CHARTS & TABLES.

Kepler's laws The three EMPIRICAL laws describing the motion of the PLANETs in their ORBITs around the SUN, formulated by JOHANNES KEPLER in the early 17th century. The laws are (1) the orbits of the planets are ELLIPSEs, with the Sun at a common focus, (2) as a planet moves in its orbit, the line joining the planet and the Sun sweeps over equal areas in equal intervals of time, and (3) the square of the (orbital) PERIOD of any planet is proportional to the cube of its mean distance from the Sun (that is, the semimajor axis for the ELLIPTICAL ORBIT).

Kerr black hole As first proposed by ROY PATRICK KERR in 1963, a BLACK HOLE that is rotating, in contrast to a SCHWARZSCHILD BLACK HOLE, which does not rotate.

kinetic energy (*KE* or E_{KE}) The ENERGY an object possesses as a result of its motion. In Newtonian (nonrelativistic) mechanics, kinetic energy is one-half the product of MASS *(m)* and the square of its VELOCITY *(v)*, that is $E_{KE} = {}^1/_2 mv^2$.

Kirkwood gaps Regions in the main asteroid belt devoid of ASTEROIDs. This condition was explained by DANIEL KIRKWOOD in 1857 as being the result of complex orbital resonances with JUPITER (that is, periodic gravitational tugs in certain orbits). *See also* MAIN-BELT ASTEROID.

Kuiper belt A region in the outer SOLAR SYSTEM beyond NEPTUNE out to perhaps 1,000 ASTRONOMICAL UNITS that contains millions of icy PLANETESIMALs. These icy objects range in size from tiny particles to PLUTO-sized planetary bodies. GERARD PETER KUIPER first suggested the existence of a disk-shaped reservoir of icy objects in 1951. *See also* OORT CLOUD.

Lagrangian libration points The five points in OUTER SPACE (called L_1, L_2, L_3, L_4, and L_5) where a small object can experience a stable ORBIT in spite of the FORCE of GRAVITY exerted by two much more massive CELESTIAL BODIES when they orbit about a common CENTER OF MASS. JOSEPH LOUIS LAGRANGE calculated the existence and location of these points in 1772.

lander *(spacecraft)* A SPACECRAFT designed to reach the surface of a PLANET or MOON safely and to survive long enough on the planetary

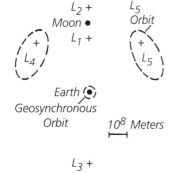

Lagrangian libration points (Courtesy of author)

body to collect useful scientific data that it sends back to EARTH by TELEMETRY.

LANDSAT The family of versatile, NASA-developed, EARTH-OBSERVING SPACECRAFT that have demonstrated numerous applications of space-based MULTISPECTRAL SENSING since 1972. *See also* REMOTE SENSING.

Large Magellanic Cloud (LMC) An IRREGULAR GALAXY about 20,000 LIGHT-YEARS in diameter and approximately 160,000 light-years from EARTH. *See also* MAGELLANIC CLOUDS.

launch (1) *(noun)* The action that occurs when a ROCKET or AEROSPACE VEHICLE propels itself from a planetary surface.
(2) *(verb)* To send off a rocket or MISSILE under its own propulsive power.

launch azimuth The initial compass heading of a ROCKET vehicle at LAUNCH.

launch pad The load-bearing base or platform from which a ROCKET, MISSILE, or AEROSPACE VEHICLE is launched.

launch site The extensive, well-defined area used to launch ROCKET vehicles for operational or for test purposes. Also called the launch complex.

launch vehicle (LV) An expendable (ELV) or reusable (RLV) ROCKET-propelled vehicle that provides sufficient THRUST to place a SPACECRAFT into ORBIT around EARTH or to send a PAYLOAD on an INTERPLANETARY TRAJECTORY to another CELESTIAL BODY. Sometimes called booster or space lift vehicle. *See also* SECTION IV CHARTS & TABLES.

launch window An interval of time during which a LAUNCH may be made to satisfy some mission objective. Sometimes it is just a short period each day for a certain number of days.

lens A curved piece of glass polished and carefully shaped to focus LIGHT from a distant object so as to form an IMAGE of that object (CONVERGING LENS) or to spread light out as if it came from a main focus (DIVERGING LENS).

life support system (LSS) The system that maintains life throughout the entire aerospace flight environment, including (as appropriate) travel in OUTER SPACE, activities on the surface of another world (for example, the LUNAR surface), and ascent and descent through EARTH's ATMOSPHERE. The LSS must reliably satisfy a human crew's daily needs for clean air, potable water, food, and effective waste removal.

Liftoff (NASA)

liftoff The action of a ROCKET or AEROSPACE VEHICLE as it separates from its LAUNCH PAD in a vertical ascent.

light The portion of the ELECTROMAGNETIC (EM) SPECTRUM that can be seen by the human eye. Visible light (radiation) ranges from approximately 750 NANOMETERS (nm) (long WAVELENGTH, red) to about 370 (nm) (short wavelength, violet).

light-gathering power (LGP) The ability of a TELESCOPE or other optical instrument to collect LIGHT.

light pollution Human-generated, artificial LIGHT sources (such as city lights) that interfere with EARTH-based optical ASTRONOMY.

light time The amount of time needed for LIGHT or RADIO WAVE signals to travel a certain distance at optical velocity (that is, $c = 299{,}792.5$ km/s). For example, one light-second corresponds to a distance of approximately 300,000 km.

light-year (ly) The distance LIGHT (or other forms of ELECTROMAGNETIC RADIATION) can travel in one year. One light-year equals a distance of approximately 9.46×10^{12} km or 63,240 ASTRONOMICAL UNITS (AU).

limb The visible outer edge or observable rim of the DISK of a CELESTIAL BODY.

line of apsides The line connecting the two points of an ORBIT that are nearest and farthest from the center of attraction, such as the PERIGEE and APOGEE of a SATELLITE in orbit around EARTH.

line of sight (LOS) The straight line between a SENSOR or the eye of an observer and the object or point being observed. Sometimes called the optical path.

liquid hydrogen (LH_2) A cryogenic liquid PROPELLANT used as the fuel, with LIQUID OXYGEN serving as the OXIDIZER, in high-performance ROCKET engines. HYDROGEN remains a liquid only at very low (cryogenic) temperatures, typically about 20 K (–253°C) or less, imposing special storage and handling requirements.

liquid oxygen (LOX or LO_2) A cryogenic liquid PROPELLANT often used as an OXIDIZER with RP-1 fuel in many EXPENDABLE LAUNCH VEHICLES or with LIQUID HYDROGEN fuel in high-performance LIQUID-PROPELLANT ROCKET ENGINES, like NASA's SPACE SHUTTLE main engines. LOX requires storage at temperatures below 90 K (–183°C).

liquid propellant Any combustible liquid fed into the COMBUSTION CHAMBER of a liquid-fueled ROCKET engine.

liquid-propellant rocket engine A ROCKET engine that uses chemical propellants in liquid form for both the fuel and OXIDIZER.

little green men (LGM) A science fiction expression for EXTRATERRESTRIAL beings, presumably intelligent.

Local Group A small cluster of about 30 galaxies, of which the MILKY WAY GALAXY (humans' home galaxy) and the ANDROMEDA GALAXY are dominant members.

Long March (LM) A family of EXPENDABLE LAUNCH VEHICLES developed by China.

long-period comet A COMET with an orbital PERIOD around the SUN greater than 200 years. *Compare with* SHORT-PERIOD COMET.

low Earth orbit (LEO) A circular ORBIT just above EARTH'S SENSIBLE ATMOSPHERE at an ALTITUDE of between 300 to 400 km.

luminosity *(L)* The rate at which a STAR or other luminous object emits ENERGY, usually in the form of ELECTROMAGNETIC RADIATION. The luminosity of the SUN is about 4×10^{26} WATTS. *See also* STEFAN-BOLTZMANN LAW.

Luna A series of Russian SPACECRAFT sent to explore the MOON in the 1960s and 1970s.

lunar Of or pertaining to EARTH's natural SATELLITE, the MOON.

lunar base A permanently inhabited complex on the surface of the MOON. It is the next logical step after brief human exploration expeditions, like NASA's APOLLO PROJECT.

lunar crater A depression, usually circular, on the surface of the MOON. *See also* IMPACT CRATER.

lunar day The period of time the MOON takes to make one complete ORBIT OF EARTH, about 27.3 Earth days. One lunar day is a SIDEREAL month.

lunar excursion module (LEM) The LANDER spacecraft used by NASA to deliver ASTRONAUTS to the surface of the MOON during the APOLLO PROJECT.

lunar highlands The light-colored, heavily cratered mountainous part of the MOON's surface.

lunar orbiter A SPACECRAFT placed into ORBIT around the MOON. Specifically, the series of five *Lunar Orbiter* spacecraft NASA used from 1966–67 to photograph the Moon's surface precisely in support of the APOLLO PROJECT.

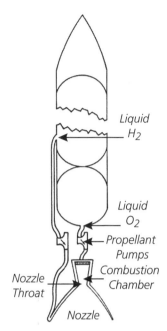

Liquid H_2

Liquid O_2

Propellant Pumps

Combustion Chamber

Nozzle Throat

Nozzle

Liquid-propellant rocket engine (Courtesy of NASA)

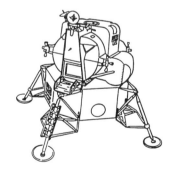

Lunar excursion module

lunar probe A PLANETARY PROBE for exploring and reporting conditions on or about the MOON. *See also* LUNA; RANGER PROJECT.

Lunar Prospector A NASA ORBITER SPACECRAFT that circled the MOON from 1998–99, searching for mineral resources. Data suggest the possible presence of lunar (water) ice deposits in permanently shadowed polar regions.

lunar rover Crewed or automated (robot) ROVER vehicles used to explore the MOON's surface. NASA's *Lunar Rover Vehicle (LRV)* served as a Moon car for APOLLO PROJECT ASTRONAUTS during the *Apollo 15,16,* and *17* expeditions. Russian *Lunokhod 1* and *2* robot rovers were operated on the Moon from EARTH between 1970–73.

Lunokhod A Russian eight-wheeled robot vehicle, controlled by RADIO WAVE signals from EARTH, and used to perform LUNAR surface exploration during the *LUNA 17* (1970) and *Luna 21* (1973) missions to the MOON.

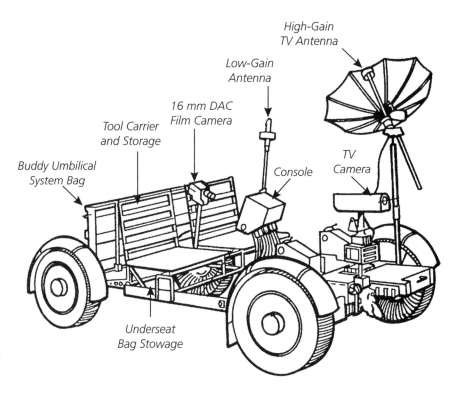

Lunar rover (Courtesy of NASA)

Mach number *(M)* The ratio of the speed of an object (with respect to the surrounding air) to the speed of sound in AIR. If M < 1, ERNST MACH called the flow subsonic. If $M > 1$, the flow is supersonic and disturbances cannot propagate ahead of the flow, so shock waves form. When $M = 1$, the flow is sonic.

MACHO The *ma*ssive *c*ompact *h*alo *o*bject hypothesized to populate the outer regions (or halo) of GALAXIES, accounting for most of the DARK MATTER. Current theories suggest that MACHOs might be low-luminosity STARS, JOVIAN PLANETS, or possibly BLACK HOLES. *See also* WIMP.

Magellanic Clouds The two dwarf, irregularly shaped neighboring GALAXIES that are closest to humans' MILKY WAY GALAXY. The LARGE MAGELLANIC CLOUD (LMC) is about 160,000 light-years away, and the SMALL MAGELLANIC CLOUD (SMC) is approximately 180,000 light-years away. Both can be seen with the NAKED EYE in the Southern Hemisphere. Their presence was first recorded in 1519 by the Portuguese explorer FERDINAND MAGELLAN, after whom they are named.

Magellan mission The NASA planetary ORBITER SPACECRAFT that used its radar-imaging system to make detailed surface maps of cloud-covered VENUS from 1990–94. It is named after the Portuguese explorer FERDINAND MAGELLAN. *See also* RADAR ASTRONOMY.

magma Molten rock beneath the surface of a PLANET or MOON. It may be ejected to the surface by volcanic activity.

magnetohydrodynamics (MHD) The branch of physics that studies the interactions between a magnetic field and a conducting fluid (such as a PLASMA).

magnetometer An instrument for measuring the strength and sometimes the direction of a magnetic field.

magnetosphere The region around a PLANET in which charged atomic PARTICLES are influenced (and often trapped) by the planet's own magnetic field rather than the magnetic field of the SUN as projected by the SOLAR WIND.

magnitude A number, measured on a logarithmic scale, that indicates the relative brightness of a celestial object. The small the magnitude number, the greater the brightness. Ancient astronomers called the brightest STARS of the night sky stars of the first magnitude, because they were the first visible after sunset. Other stars were called second-, third-, fourth-, fifth-, and sixth-magnitude stars according to

Magnitude (Courtesy of NASA)

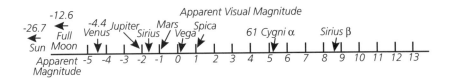

their relative brightness. Sixth-magnitude stars are the faintest stars visible to the NAKED EYE. In 1856, NORMAN ROBERT POGSON proposed a more precise logarithmic magnitude in which a difference of five magnitudes represents a relative brightness ratio of 100 to 1, while a difference of one magnitude is 2.512. This scale is now widely used in modern ASTRONOMY.

main-belt asteroid An asteroid located in the ASTEROID BELT between MARS and JUPITER.

main-sequence star A STAR in the prime of its life that shines with a constant LUMINOSITY achieved by steadily converting HYDROGEN into HELIUM through thermonuclear FUSION in its core.

main stage For a MULTISTAGE ROCKET vehicle, the stage that develops the greatest amount of THRUST.

manipulator The part of a robot capable of grasping or handling. It is a mechanical device designed to handle objects, such as the REMOTE MANIPULATOR SYSTEM (RMS) on NASA's SPACE SHUTTLE.

maria *(singular: mare)* Latin word for "seas." It was originally used by GALILEO GALILEI to describe the large, dark, ancient lava flows on the lunar surface, thought by early astronomers to be bodies of water on the MOON's surface. This term is still used by astronomers.

Mariner A series of NASA planetary exploration SPACECRAFT that performed FLYBY and orbital missions to MERCURY, MARS, and VENUS in the 1960s and 1970s.

Mars The red-colored fourth PLANET in the SOLAR SYSTEM that intrigued ancient astronomers and now serves as the focus of extensive investigation by a variety of FLYBY, ORBITER, and LANDER spacecraft seeking to answer the key question, Does (or did) Mars have life?

Mars base The surface base needed to support human explorers during a 21st-century MARS EXPEDITION.

Mars expedition The first crewed mission to visit MARS in the 21st century. Current concepts suggest a 600- to 1,000-day duration mission (starting from EARTH ORBIT), a total crew size of up to 15

Mars (Courtesy of NASA)

GLOSSARY

main-belt asteroid – Mars expedition

ASTRONAUTS, and about 30 days for surface excursion activities on Mars.

Mars Global Surveyor *(MGS)* A NASA ORBITER spacecraft launched in November 1996 that has been performing detailed studies of the Martian surface and atmosphere since March 1999.

Mars Odyssey Launched from CAPE CANAVERAL in April 2001, the *2001 Mars Odyssey* is NASA's latest ORBITER spacecraft to explore MARS, specifically searching for geologic features that could indicate the presence of (subsurface) water—past or present.

Mars Pathfinder An innovative NASA mission that successfully landed a MARS SURFACE ROVER—a small robot called *Sojourner*—in the Ares Vallis region of the RED PLANET in July 1997. For over 80 days, personnel on EARTH used TELEOPERATION and TELEPRESENCE to drive the six-wheeled minirover cautiously to interesting locations on the Martian surface.

Mars expedition (Courtesy of NASA)

Mars surface rovers Automated robot rovers and human-crewed mobility systems used to satisfy a number of surface exploration objectives on MARS in the 21st century.

Martian Of or relating to the planet MARS.

Martian meteorites The collection of a dozen or so unusual METEORITES considered to represent pieces of MARS that were blasted off the RED PLANET by ancient impact collisions, wandered through space for millions of years, and eventually landed on EARTH. In 1996, NASA scientists suggested that one particular specimen, called ALH84001, might contain fossilized evidence showing primitive life may have existed on Mars more than 3.6 billion years ago.

mascon An area of abnormal MASS concentration (mascon) or high DENSITY on the MOON beneath the LUNAR MARIA.

mass *(m)* Mass describes how much material makes up an object and gives rise to its INERTIA. The SI UNIT for mass is the kilogram (kg). An object that has 1 kg mass on Earth will also have 1 kg mass on the surface of MARS or anywhere else in the UNIVERSE.

mass fraction The fraction of a ROCKET's (or rocket stage's) MASS that is taken up by PROPELLANT. The remaining mass is structure and PAYLOAD.

mass number *(A)* The number of NUCLEONS (that is, PROTONS and NEUTRONS) in an atomic NUCLEUS. It is the nearest whole number to an atom's

ATOMIC WEIGHT. For example, the mass number of the ISOTOPE uranium 235 is 235.

mating The act of fitting together two major components of an AEROSPACE system, such as the mating of a LAUNCH VEHICLE and its PAYLOAD—a scientific SPACECRAFT. It is also the physical joining of two orbiting spacecraft either through a DOCKING or a BERTHING process.

Maxwell Montes A mountain range on VENUS located in ISHTAR TERRA, containing the highest peak (11 km ALTITUDE) on the PLANET. It is named after JAMES CLERK MAXWELL.

mean solar time The time shown on the clock that is based on EARTH'S ROTATION, which was originally assumed constant. *See also* GREENWICH MEAN TIME.

megaparsec (Mpc) 1 million PARSECs; a distance of approximately 3,260,000 LIGHT-YEARS.

Mercury The innermost PLANET in the SOLAR SYSTEM, orbiting the SUN at approximately 0.4 ASTRONOMICAL UNITS.

Mercury Project The initial United States ASTRONAUT program (1958–63) in which NASA selected seven military test pilots with the "right stuff" to become the first Americans to fly into OUTER SPACE. They flew in cramped, one-person SPACE CAPSULES, such as JOHN HERSCHEL GLENN, JR.'s *Friendship 7* Mercury capsule.

Messier catalogue A compilation of bright celestial objects, such as NEBULAS and GALAXIES, developed by CHARLES MESSIER in the late 18th century. Astronomers still use Messier object nomenclature—for example, the ORION NEBULA is M42.

meteor The luminous phenomenon that occurs when a METEOROID enters EARTH'S ATMOSPHERE. It is sometimes called a shooting star.

meteorite Metallic or stony EXTRATERRESTRIAL material that has passed through the ATMOSPHERE and reached EARTH's surface. *See also* METEOROID.

meteoroid An all-encompassing term that refers to solid objects found in OUTER SPACE, ranging in diameter from MICROMETERs to kilometers and in MASS from less than 10^{-12} g to more than 10^{+16} g. If the object has a mass of less than 1 g, it is called a micrometeoroid. When objects with a mass of more than 10^{-6} g reach EARTH'S ATMOSPHERE, they glow with heat and produce the visible effect popularly called a METEOR. If some of an original meteoroid survives its incandescent

Mercury (Courtesy of NASA/JPL)

plunge through Earth's atmosphere, the remaining unvaporized chunk of EXTRATERRESTRIAL matter is called a METEORITE.

meteorological rocket A SOUNDING ROCKET designed for routine observation of EARTH's upper ATMOSPHERE (as opposed to scientific research).

meteorological satellite An EARTH-OBSERVING SPACECRAFT that senses some or most of the atmospheric phenomena (such as wind and clouds) related to weather conditions on a local, regional, or hemispheric scale. These SATELLITES operate either close to Earth in POLAR ORBITS or else observe an entire hemisphere from GEOSTATIONARY ORBIT. Also called a weather satellite.

Metonic calendar Named for the ancient Greek astronomer METON (OF ATHENS), it is based on the Moon and counts each cycle of the PHASES OF THE MOON as one month (or one lunation). After a period of 19 years, the lunations will occur on the same days of the year.

metric system *See* SI UNITS.

microgravity *(μg)* Because its inertial TRAJECTORY compensates for the FORCE of GRAVITY, a SPACECRAFT in ORBIT around EARTH travels in a state of continual FREE FALL. All objects inside appear weightless—as if they were in a zero-gravity environment. However, the venting of gases, the minuscule DRAG exerted by Earth's residual ATMOSPHERE (at low orbital ALTITUDES), and crew motions tend to create nearly imperceptible forces on objects inside the orbiting vehicle. These tiny forces are collectively called microgravity. *See also* WEIGHTLESSNESS.

micrometer (μm) An SI UNIT of length equal to one-millionth (10^{-6}) of a meter. Also called a micron.

microwave background radiation (MBR) *See* COSMIC MICROWAVE BACKGROUND.

microwave radiation Comparatively short-wavelength ELECTROMAGNETIC RADIATION in the RADIO FREQUENCY (RF) portion of the SPECTRUM—from about 30 cm to 1 mm WAVELENGTH.

military satellite (MILSAT) A SATELLITE used primarily for military or defense purposes, such as intelligence gathering, MISSILE surveillance, or secure communications.

Milky Way Galaxy Humans' home GALAXY—a large SPIRAL GALAXY that contains between 200 and 600 billion SOLAR MASSES. The SUN lies some 30,000 LIGHT-YEARS from the galactic center.

**Milky Way Galaxy
(Courtesy of NASA)**

meteorological rocket – Milky Way Galaxy

GLOSSARY

minor planet *See* ASTEROID.

Mir *(Peace)* A third-generation Russian SPACE STATION of modular design that was assembled in ORBIT around a core module launched in February 1986. Although used extensively by many COSMONAUTS and guest researchers (including American ASTRONAUTS), the massive station was eventually abandoned because of economics and safely de-orbited into a remote area of the Pacific Ocean in March 2001.

missile Any object thrown, dropped, fired, launched, or otherwise projected with the purpose of striking a target. It is short for BALLISTIC MISSILE or GUIDED MISSILE. *Missile* should *not* be used loosely as an equivalent term for ROCKET or LAUNCH VEHICLE.

missing mass *See* DARK MATTER.

modulation The process of modifying a RADIO FREQUENCY (RF) signal by shifting its phase, FREQUENCY, or AMPLITUDE to carry information. The respective processes are called phase modulation (PM), frequency modulation (FM), and amplitude modulation (AM).

molecule A collection of ATOMS held together by chemical (bonding) FORCES. The atoms in a particular molecule may be identical, as in HYDROGEN (H_2), or different, as in carbon monoxide (CO) and carbon dioxide (CO_2). A molecule is the smallest unit of matter that can exist by itself and still retain all its chemical properties.

molniya orbit A highly elliptical, 12-hour ORBIT developed within the Russian space program for special COMMUNICATIONS SATELLITES. With an APOGEE of about 40,000 km and a PERIGEE of only 500 km, a SPACECRAFT in this orbit spends the bulk of its time above the horizon in view of high northern latitudes.

momentum *(linear)* The linear momentum (p) of a PARTICLE is the product of the particle's MASS (m) and its VELOCITY (v). NEWTON'S SECOND LAW OF MOTION states that the rate of change of momentum of a particle equals the resultant FORCE (F) on the particle. *See also* NEWTON'S LAWS OF MOTION.

monopropellant A LIQUID PROPELLANT for a ROCKET. It consists of a single chemical substance (such as HYDRAZINE) that decomposes in an EXOTHERMIC REACTION, producing a THRUST-generating heated exhaust jet without the use of a second chemical substance.

moon A small natural CELESTIAL BODY that orbits a larger one; a natural SATELLITE.

Moon   EARTH's only natural SATELLITE and closest celestial neighbor. It has an equatorial diameter of 3,476 km, keeps the same side (NEARSIDE) toward Earth, and orbits at an average distance (center to center) of 384,400 km.

mother spacecraft   Exploration SPACECRAFT that carries and deploys one or several ATMOSPHERIC PROBES and ROVER or LANDER spacecraft when arriving at a target PLANET. The mother spacecraft then relays its data back to EARTH and may orbit the planet to perform its own scientific mission. NASA's GALILEO PROJECT spacecraft to JUPITER and CASSINI MISSION spacecraft to SATURN are examples.

multispectral sensing   The REMOTE-SENSING method of simultaneously collecting several different bands (WAVELENGTH regions) of ELECTROMAGNETIC RADIATION (such as the visible, the near-INFRARED, and the thermal infrared bands) when observing a target.

multistage rocket   A vehicle that has two or more ROCKET units, each firing after the one behind it has exhausted its PROPELLANT. This type of rocket vehicle then discards (or jettisons) each exhausted stage in sequence. It is sometimes called a multiple-stage rocket or a step rocket.

nadir   (1) The direction from a SPACECRAFT directly down toward the center of a PLANET. It is the opposite of the ZENITH.
(2) That point on the CELESTIAL SPHERE directly beneath an observer and directly opposite the zenith.

naked eye   The normal human eye unaided by any optical instrument, such as a TELESCOPE. The use of corrective lenses (glasses) or contact lenses that restore an individual's normal vision are included in the concept of naked eye observing.

nanometer   (nm) A very small distance, namely 1×10^{-9} m.

NASA   The National Aeronautics and Space Administration, the civilian space agency of the United States. It was created in 1958 by an act of Congress. NASA's overall mission is to plan, direct, and conduct civilian (including scientific) aeronautical and space activities for peaceful purposes.

National Aeronautics and Space Administration (NASA) *See* NASA.

National Radio Astronomy Observatory (NRAO) A collection of government-owned RADIO ASTRONOMY facilities throughout the United States, including the RADIO TELESCOPE facility in Green Bank, West Virginia.

National Reconnaissance Office (NRO) The agency within the U.S. Department of Defense responsible for meeting the RECONNAISSANCE SATELLITE needs of various government organizations.

National Space Development Agency of Japan (NASDA) The government agency primarily responsible for implementing Japan's peaceful space development activities, including LAUNCH VEHICLES and EARTH-OBSERVING SPACECRAFT, and for participating in the INTERNATIONAL SPACE STATION.

natural satellite *See* MOON.

navigation satellite A SPACECRAFT placed into a well-known, stable ORBIT around EARTH that transmits precisely timed RADIO WAVE signals useful in determining locations on land, at sea, or in the air. Such satellites are deployed as part of an interactive CONSTELLATION. *See also* GLOBAL POSITIONING SYSTEM.

near-Earth asteroid (NEA) An inner SOLAR SYSTEM ASTEROID whose ORBIT around the SUN brings it close to EARTH, perhaps even posing a collision threat in the future. *See also* EARTH-CROSSING ASTEROID.

nearside The side of the MOON that always faces EARTH.

nebula *(plural: nebulas or nebulae)* A cloud of INTERSTELLAR gas or dust. It can be seen as either a dark hole against a brighter background (called a DARK NEBULA) or as a luminous patch of LIGHT (called a bright nebula).

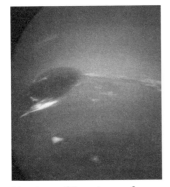

Neptune (Courtesy of NASA)

Neptune The eighth PLANET from the SUN and the outermost of the JOVIAN PLANETS. In 1846, it became the first to be discovered using theoretical predictions—a major triumph for 19th-century ASTRONOMY.

neutrino (ν) An uncharged fundamental PARTICLE with no (or very little) MASS that interacts only weakly with matter.

neutron (n) An uncharged elementary PARTICLE with a MASS slightly greater than that of the PROTON. Neutrons occur in the NUCLEUS of every ATOM heavier than simple HYDROGEN ($^1_1$H). Once outside the atomic nucleus, a free neutron becomes unstable and decays into an ELECTRON, a proton, and a NEUTRINO.

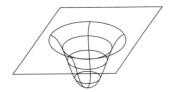

Neutron star (curving space-time)

neutron star A very small (typically 20–30 km in diameter), superdense stellar object—the gravitationally collapsed core of a massive STAR that has undergone a SUPERNOVA explosion. Astrophysicists hypothesize that PULSARs are rapidly spinning neutron stars that possess intense magnetic fields.

newton (N) The SI UNIT of FORCE, named after SIR ISAAC NEWTON. 1 N is the amount of force that gives a 1 kg MASS an ACCELERATION of 1 m/s^2.

Newtonian telescope A REFLECTING TELESCOPE in which a small plane mirror reflects the convergence beam from the OBJECTIVE (primary mirror) to an EYEPIECE at the side of the TELESCOPE. SIR ISAAC NEWTON designed the first reflecting telescope in about 1670.

Newton's law of gravitation The physical law proposed by SIR ISAAC NEWTON in 1687. It states that every PARTICLE of matter in the UNIVERSE attracts every other particle. The FORCE of gravitational attraction (F_G) acts along the line joining the two particles and is proportional to the product of the particle MASSes (m_1 and m_2) and inversely proportional to the square of the distance (r) between the particles. This law, expressed as an equation, is $F_G = Gm_1 m_2/r^2$, where G is the universal gravitational constant (approximately 6.6732×10^{-11} Nm^2/kg^2 in SI UNITs).

Newton's laws of motion The three postulates of motion formulated by SIR ISAAC NEWTON in about 1685. (1) His first law (the conservation of MOMENTUM) states that a body continues in a state of uniform motion (or rest) unless acted upon by an external FORCE. (2) The second law states that the rate of change of momentum of a body is proportional to the force acting upon the body and occurs in the direction of the applied force. (3) The third law (the action-reaction principle) states that for every force acting upon a body, there is a corresponding force of the same magnitude exerted by the body in the opposite direction. The third law is the basic principle by which every ROCKET operates.

nose cone The cone-shaped leading edge of a ROCKET vehicle, which contains and protects the PAYLOAD or WARHEAD.

nova *(plural: novas or novae)* From the Latin for "new," a highly evolved STAR that exhibits a sudden and exceptional brightness, usually temporary, and then returns to its former LUMINOSITY. A nova is now thought to be the outburst of a DEGENERATE STAR in a BINARY (DOUBLE) STAR SYSTEM.

nozzle A flow device that promotes the efficient expansion of the hot gases from the COMBUSTION CHAMBER of a ROCKET. As these gases leave at high VELOCITY, a propulsive (forward) THRUST also occurs in accordance with the third of NEWTON's LAWS OF MOTION (action-reaction principle).

nuclear-electric propulsion (NEP) A space-deployed PROPULSION SYSTEM that uses a space-qualified, compact NUCLEAR REACTOR to produce the electricity needed to operate a space vehicle's ELECTRIC PROPULSION engine(s).

nuclear fission *See* FISSION.

nuclear fusion *See* FUSION.

nuclear radiation IONIZING RADIATION consisting of PARTICLES (such as ALPHA PARTICLES, BETA PARTICLES, and NEUTRONS) and very energetic ELECTROMAGNETIC RADIATION (that is, GAMMA RAYS). Atomic NUCLEI emit this type of radiation during a variety of energetic nuclear reaction processes, including radioactive decay, FISSION, and FUSION. *See also* RADIOISOTOPE.

nuclear reactor An ENERGY-generating device in which a nuclear FISSION chain reaction is initiated, maintained, and controlled—thereby releasing large amounts of heat in a predictable way.

nuclear rocket A ROCKET vehicle that derives its propulsive THRUST from nuclear ENERGY. The nuclear thermal rocket uses a NUCLEAR REACTOR to heat HYDROGEN to extremely high temperatures before expelling it through a thrust-producing NOZZLE. *See also* NUCLEAR-ELECTRIC PROPULSION.

nucleon A NEUTRON or PROTON viewed as a fundamental constituent PARTICLE within the atomic NUCLEUS.

nucleosynthesis The production of heavier chemical ELEMENTs from the FUSION (joining together) of lighter chemical elements (such as HYDROGEN and HELIUM) in thermonuclear reactions in the interior of STARS.

nucleus (1) *(atomic)* The central portion of an ATOM, consisting of PROTONS and NEUTRONs bound together by the strong nuclear FORCE.
(2) *(cometary)* The small (few kilometer diameter), permanent, solid ice-rock central body of a COMET.
(3) *(galactic)* The central region of a GALAXY, a few LIGHT-YEARS in diameter, where matter is concentrated. It is a complex region characterized by a dense cluster of STARS or possibly even the hidden presence of a massive BLACK HOLE. *See also* ACTIVE GALACTIC NUCLEUS.

nuclide A general term used to describe the NUCLEUS of an ATOM as characterized by its ATOMIC NUMBER *(Z)* and number of NEUTRONS, which is the (atomic) MASS NUMBER *(A)* minus the atomic number *(Z)*.

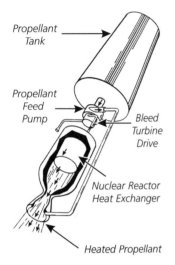

Propellant Tank

Propellant Feed Pump

Bleed Turbine Drive

Nuclear Reactor Heat Exchanger

Heated Propellant

Nuclear rocket (Courtesy of NASA and the U.S. Department of Energy)

objective The main LIGHT-gathering LENS or mirror of a TELESCOPE. It is sometimes called the primary lens or primary mirror of a telescope.

oblateness The degree of flattening of an oblate spheroid—a sphere flattened such that its polar diameter is smaller than its equatorial diameter.

observable universe The portions of the UNIVERSE that can be detected and studied by the LIGHT they emit.

observatory The place (or facility) from which astronomical observations are made. For example, the KECK OBSERVATORY is a ground-based observatory, while the *HUBBLE* SPACE TELESCOPE is a space-based (or EARTH-orbiting) observatory.

occult *(occultation)* The disappearance of one celestial object behind another. *See also* ECLIPSE.

Olympus Mons A huge mountain on MARS about 650 km wide and rising 26 km above the surrounding plains—the largest known single VOLCANO in the SOLAR SYSTEM.

one-*g* The downward ACCELERATION of GRAVITY at EARTH's surface (approximately 9.8 m/s^2). *See also* G.

Oort cloud The large number (about 10^{12}) or cloud of COMETs postulated by JAN HENDRIK OORT in 1950 to ORBIT the SUN at a distance of between 50,000 and 80,000 ASTRONOMICAL UNITS.

open cluster *See* GALACTIC CLUSTER.

open universe A UNIVERSE that will continue to expand forever. *See* COSMOLOGY.

opposition The alignment of a SUPERIOR PLANET with the SUN so that the planet appears to an observer on EARTH to be in opposite parts of the sky. *See also* CONJUNCTION.

optical radiation *See* ELECTROMAGNETIC RADIATION; LIGHT.

orbit The path followed by a body in space, generally under the influence of GRAVITY—as, for example, a SATELLITE around a PLANET.

orbital elements *See* KEPLERIAN ELEMENTS.

orbital injection The process of providing a SPACE VEHICLE or a SATELLITE with sufficient VELOCITY to establish an ORBIT.

orbital mechanics *See* CELESTIAL MECHANICS.

orbital period The interval between successive passages of a SATELLITE or SPACECRAFT through the same point in its ORBIT. Often called period.

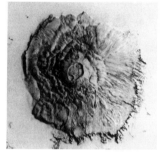

Olympus Mons (Courtesy of NASA)

orbital plane The imaginary plane that contains the ORBIT of a SATELLITE and passes through the center of its PRIMARY BODY. The angle of inclination (symbol θ) is defined as the angle between EARTH's equatorial plane and the orbital plane of the satellite

orbital transfer vehicle (OTV) A PROPULSION SYSTEM used to transfer a PAYLOAD from one orbital location to another. An expendable (one-shot) orbital transfer vehicle is often called an upper-stage unit, while a reusable OTV is called a space tug.

orbiter *(spacecraft)* A SPACECRAFT especially designed to travel through INTERPLANETARY space, achieve a stable ORBIT around the target PLANET (or other CELESTIAL BODY), and conduct a program of detailed scientific investigation.

Orbiter *(space shuttle)* The winged AEROSPACE VEHICLE portion of NASA's SPACE SHUTTLE. It carriers ASTRONAUTS and PAYLOAD into ORBIT and returns from OUTER SPACE by gliding and landing like an airplane. The operational orbiter vehicle (OV) fleet includes *Columbia (OV-102), Discovery (OV-103), Atlantis (OV-104),* and *Endeavour (OV-105).*

orbit inclination The angle between EARTH's equatorial plane and a SATELLITE's ORBITAL PLANE.

Orbiting Astronomical Observatory (OAO) A series of large, EARTH-orbiting, astronomical observatories developed by NASA in the 1960s to broaden humans' understanding of the UNIVERSE, especially as related to ULTRAVIOLET ASTRONOMY.

Orbiting Quarantine Facility (OQF) A proposed EARTH-orbiting laboratory in which soil and rock samples from MARS and other worlds could first be tested for potentially harmful alien microorganisms—before such EXTRATERRESTRIAL materials were allowed to enter Earth's BIOSPHERE.

order of magnitude A factor of 10; a value expressed to the nearest power of 10—for example, a cluster containing 9,450 STARs has approximately 10,000 stars in an order of magnitude estimate.

Orion Nebula A bright NEBULA about 1,500 LIGHT-YEARS away in the CONSTELLATION of Orion.

oscillating universe A CLOSED UNIVERSE in which gravitational collapse is followed by a new wave of expansion.

outer planets The PLANETs in the SOLAR SYSTEM with ORBITS greater than the orbit of MARS, including JUPITER, SATURN, URANUS, NEPTUNE, and PLUTO.

outer space Any region beyond EARTH's atmospheric envelope—usually considered to begin at between 100 and 200 km ALTITUDE.

oxidizer A substance whose main function is to supply oxygen for the burning of a ROCKET engine's fuel.

pad The platform from which a ROCKET vehicle is launched. *See also* LAUNCH PAD.

pair production The GAMMA RAY interaction process in which a high-ENERGY PHOTON disappears in the vicinity of an atomic NUCLEUS with the simultaneous appearance (formation) of a negative ELECTRON and a POSITRON (positive electron).

Pallas The large (about 540 km in diameter), MAIN-BELT ASTEROID discovered by HEINRICH WILHELM OLBERS in 1802—the second minor planet found.

Palomar Observatory An astronomical observatory on Mount Palomar (1.7 km altitude) in southern California. Founded by GEORGE ELLERY HALE and opened in 1948, the facility's main instrument is a 5 m (200 in.) REFLECTING TELESCOPE.

parallax The angular displacement in the apparent position of a CELESTIAL BODY when observed from two widely separated points. Astronomers define the trigonometric parallax (symbol: π) of a STAR at a distance *(d)* from the SUN as the angle it subtends when observed at two locations separated by a baseline of one ASTRONOMICAL UNIT. The use of the trigonometric parallax provided astronomers their first way of measuring stellar distances. *See also* PARSEC.

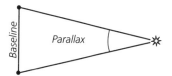

Parallax (Courtesy of NASA—modified by author)

Paris Observatory Founded in 1667 as the national observatory of France, this facility represented the first such national observatory. GIOVANNI DOMENICO CASSINI served as its first director.

parking orbit The temporary (but stable) ORBIT of a SPACECRAFT around a CELESTIAL BODY. It is used for the assembly and/or transfer of equipment or to wait for conditions favorable for departure from that orbit.

parsec (pc) An astronomical unit of distance corresponding to a trigonometric PARALLAX *(π)* of one second of arc. The term is a shortened form of *par*allax *sec*ond. 1 ps represents a distance of 3.26 LIGHT-YEARS (or 206,265 ASTRONOMICAL UNITS).

particle A minute constituent of matter, generally one with a measurable MASS; an elementary atomic particle such as a PROTON, NEUTRON, ELECTRON, or ALPHA PARTICLE.

pascal   (Pa) The SI UNIT of pressure, honoring Blaise Pascal, and defined as the pressure that results from a FORCE of one NEWTON (1 N) acting uniformly over an area of 1 m^2.

payload   That which a ROCKET, AEROSPACE VEHICLE, or SPACECRAFT carries over and above what is necessary for the operation of the vehicle during flight.

payload assist module   (PAM) A family of commercially developed UPPER-STAGE vehicles for use with NASA'S SPACE SHUTTLE or with EXPENDABLE LAUNCH VEHICLES, such as the DELTA.

payload bay   The large and long enclosed volume within NASA'S SPACE SHUTTLE ORBITER vehicle designed to carry a wide variety of PAYLOADS, including UPPER-STAGE vehicles, deployable SPACECRAFT, and attached equipment. Also called cargo bay.

Pegasus   *(space launch vehicle)* An aircraft-launched space BOOSTER ROCKET capable of placing small PAYLOADS and SPACECRAFT into LOW EARTH ORBIT.

perfect cosmological principle   The postulation that at all times, the UNIVERSE appears the same to all observers. *See also* COSMOLOGY.

peri-   A prefix meaning near.

periastron   The point of closest approach of two STARS in a BINARY (DOUBLE) STAR SYSTEM. *Compare with* APASTRON.

perigee   The point at which a SATELLITE'S ORBIT is the closest to its PRIMARY BODY; the minimum ALTITUDE attained by an EARTH-orbiting object. *Compare with* APOGEE.

perihelion   The point in an ELLIPTICAL ORBIT around the SUN that is nearest to the center of the Sun. *Compare with* APHELION.

perilune   The point in an ELLIPTICAL ORBIT around the MOON that is nearest to the LUNAR surface. *Compare with* APOLUNE.

period   *(orbital)* The time taken by a SATELLITE to travel once around its ORBIT.

periodic comet   A COMET with a period of less than 200 years. Also called a short-period comet.

permanently crewed capability   (PCC) A SPACE STATION or planetary surface base that can be continuously occupied and operated by a human crew.

perturbation   A disturbance in the ORBIT of a CELESTIAL BODY, most often caused by the FORCE of gravitational attraction of another body.

Atmospheric DRAG can cause a perturbation in the orbit of a SPACECRAFT if it has a sufficiently low-ALTITUDE orbit around a CELESTIAL BODY with an ATMOSPHERE.

phases of the Moon The changing illuminated appearance of the NEARSIDE surface of the MOON to an observer on EARTH. Major phases include the new Moon (not illuminated), first quarter, full Moon (totally illuminated), and last (third) quarter.

Phobos The larger, innermost of the two small MOONs of MARS—discovered in 1877 by ASAPH HALL. *See also* DEIMOS.

photoheliograph A TELESCOPE that produces a white-LIGHT photographic IMAGE of the SUN; developed by WARREN DE LA RUE in 1857.

photoionization The ionization of an ATOM or MOLECULE caused by collision with an energetic PHOTON.

photometer An instrument that measures LIGHT intensity and the brightness of celestial objects, such as STARS.

photon An elementary bundle (or packet) of ELECTROMAGNETIC RADIATION, such as a photon of LIGHT. Photons have no MASS and travel at the SPEED OF LIGHT. From QUANTUM THEORY, the ENERGY *(E)* of a photon equals the product of its FREQUENCY *(v)* and Planck's constant *(h)*, such that $E = hv$. Here h has the value of 6.626×10^{-34} joule-second, and the frequency is expressed in HERTZ. *See also* PLANCK'S RADIATION LAW.

photosphere The intensely bright (white-LIGHT), visible surface of the SUN or other STAR.

Pioneer 10, 11 spacecraft NASA's twin exploration SPACECRAFT that were the first to navigate the main ASTEROID BELT, the first to visit JUPITER (1973 and 1974), the first to visit SATURN *(Pioneer 11*—1979), and the first human-made objects to leave the SOLAR SYSTEM *(Pioneer 10*—1983). Each spacecraft is now on a different TRAJECTORY to the STARS, carrying a special message (the Pioneer plaque) for any intelligent alien civilization that might find it millions of years from now.

Pioneer Venus mission Two SPACECRAFT launched by NASA to VENUS in 1978. *Pioneer 12* was an ORBITER spacecraft that gathered data from 1978–92. The *Pioneer Venus Multiprobe* served as a MOTHER SPACECRAFT, launching one large and three identical small PLANETARY PROBEs into the Venusian atmosphere (December 1978).

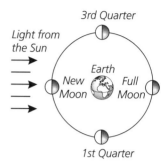

Phases of the Moon

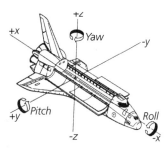

Pitch (Courtesy of NASA)

pitch The ROTATION of an AEROSPACE VEHICLE or SPACECRAFT about its lateral AXIS.

pitchover The programmed turn from the vertical that a LAUNCH VEHICLE (under power) takes as it describes an arc and points in a direction other than vertical.

pixel Contraction for *picture element;* the smallest unit of information on a screen or in an IMAGE.

plage A bright patch in the SUN's CHROMOSPHERE.

Planck's radiation law The physical principle, developed by MAX KARL PLANCK in 1900, that describes the distribution of ENERGY radiated by a BLACKBODY. With this law, Planck introduced his concept of the QUANTUM (or PHOTON) as a small unit of energy responsible for the transfer of ELECTROMAGNETIC RADIATION.

planet A nonluminous CELESTIAL BODY that ORBITS around the SUN or some other STAR; from the ancient Greek *planetes* ("wanderers")—since early astronomers identified the planets as the wandering points of light relative to the FIXED STARS. There are nine major planets in the SOLAR SYSTEM and numerous MINOR PLANETS (or ASTEROIDS). The distinction between a planet and a large SATELLITE is not always precise. The MOON is nearly the size of MERCURY and is very large in comparison to EARTH—suggesting the Earth-Moon system might be treated as a double-planet system.

planetary albedo The fraction of incident SOLAR RADIATION reflected by a PLANET (and its ATMOSPHERE) and returned to OUTER SPACE.

planetary nebula The shell of gas ejected from the outer layers of an extremely hot STAR (like a RED GIANT) at the end of its life cycle.

planetary probe An instrument-containing SPACECRAFT deployed in the ATMOSPHERE or on the surface of a planetary body in order to obtain environmental information.

planetesimals Small rock and rock/ice celestial objects found in the SOLAR SYSTEM, ranging from 0.1 km to about 100 km diameter. *See also* CENTAURS.

planet fall The act of a SPACECRAFT or SPACE VEHICLE landing on a PLANET or MOON.

planetoid *See* ASTEROID.

plasma An electrically neutral gaseous mixture of positive and negative IONS. It is sometimes called the fourth state of matter, because it behaves quite differently from solids, liquids, or gases.

Plesetsk The northern Russian LAUNCH SITE about 300 km south of Archangel that supports a wide variety of military space launches, BALLISTIC MISSILE testing, and scientific SPACECRAFT requiring a POLAR ORBIT.

plume The hot, bright exhaust gases from a ROCKET.

plutino *(little Pluto)* Any of the numerous, small (~100 km diameter), icy CELESTIAL BODIES that occupy the inner portions of the KUIPER BELT and whose orbital motion resonance with NEPTUNE resembles that of PLUTO—namely that each object completes two ORBITS around the SUN in the time Neptune takes to complete three orbits. *See also* TRANS-NEPTUNIAN OBJECT.

Pluto The smallest of the major PLANETS in the SOLAR SYSTEM and, despite its highly ECCENTRIC ORBIT, usually the most distant planet from the SUN. It was not discovered until 1930 by CLYDE WILLIAM TOMBAUGH, and its large companion MOON (CHARON) was detected in 1978. Frigid Pluto remains the only planet not yet visited by scientific SPACECRAFT.

Pluto-Kuiper Express Mission A proposed (but not funded) NASA FLYBY mission to PLUTO and the inner regions of the KUIPER BELT.

Pogson scale *See* MAGNITUDE.

polarization A distinct orientation of the wave motion and direction of travel of ELECTROMAGNETIC RADIATION, including LIGHT. Along with brightness and color, polarization is a special property of light. It represents a condition in which the planes of vibration of the various rays in a beam of light are at least partially (if not completely) aligned.

polar orbit An ORBIT around a PLANET (or PRIMARY BODY) that passes over or near its poles; an orbit with an INCLINATION of about 90°.

poles The poles for a rotating CELESTIAL BODY are located at the ends (usually called north and south) of the body's AXIS of ROTATION.

Population I stars Hot, luminous, young STARS, including those like the SUN, that reside in the disk of a SPIRAL GALAXY and are higher in heavy ELEMENT content (about 2 percent abundance) than POPULATION II STARS.

Population II stars Older STARS that are lower in heavy ELEMENT content than POPULATION I STARS and reside in GLOBULAR CLUSTERS as well

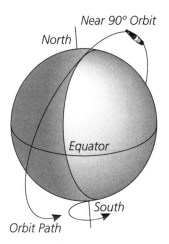

Polar orbit

as in the halo of a galaxy—that is, the distant spherical region that surrounds a GALAXY.

posigrade rocket An auxiliary ROCKET that fires in the direction in which the vehicle is pointed. It is often used in separating two stages of a LAUNCH VEHICLE or a PAYLOAD from the UPPER-STAGE propulsion stage. The firing of a posigrade rocket adds to a SPACECRAFT's speed, while the firing of a retrograde rocket (RETROROCKET) slows it down.

positron (*posi*tive elec*tron*) An elementary PARTICLE with the MASS of an ELECTRON but charged positively. Also called the antielectron.

potential energy *(PE)* The ENERGY or ability to do WORK possessed by an object by virtue of its position in a GRAVITY field above some reference position or datum.

power *(P)* The rate at which WORK is done or at which ENERGY is transformed per unit time. The WATT (W) is the fundamental SI UNIT of power.

precession The gradual, periodic change in the direction of the AXIS of ROTATION of a spinning body due to the application of an external FORCE; the wobbling of a spinning top is an example.

precession of equinoxes The slow westward motion of the EQUINOX points along the ECLIPTIC relative to the STARS of the ZODIAC caused by the slight wobbling of EARTH about its AXIS of ROTATION.

pressurized habitable environment Any module or enclosure in OUTER SPACE in which an ASTRONAUT may perform activities in a shirtsleeve environment.

primary body The CELESTIAL BODY around which a SATELLITE, MOON, or other object ORBITs or from which it is escaping or toward which it is falling.

prism A block (often triangular) of transparent material that disperses an incoming beam of white LIGHT into the visible SPECTRUM of rainbow colors—that is, red, orange, yellow, green, blue, indigo, and violet in order of decreasing WAVELENGTH.

probe *See* PLANETARY PROBE.

prograde orbit An ORBIT having an INCLINATION of between 0° and 90°.

Progress An uncrewed Russian supply SPACECRAFT configured to perform automated RENDEZVOUS and DOCKING operations with SPACE STATIONs and other orbiting spacecraft.

progressive burning A design condition for a SOLID-PROPELLANT ROCKET in which the surface area of burning PROPELLANT increases with time, thereby increasing THRUST for some specific period of operation.

Project Ozma The pioneering attempt to detect INTERSTELLAR RADIO WAVE signals from an intelligent EXTRATERRESTRIAL civilization. It was conducted in 1960 by FRANK DONALD DRAKE at the NATIONAL RADIO ASTRONOMY OBSERVATORY in Green Bank, West Virginia. No strong evidence was found after 150 hours of listening for intelligent signals from the vicinity of two SUNLIKE STARS about 11 LIGHT-YEARS away. *See also* SEARCH FOR EXTRATERRESTRIAL INTELLIGENCE.

propellant The material, such as a chemical fuel and OXIDIZER combination, carried in a ROCKET vehicle, energized, and then ejected at high VELOCITY as THRUST-producing reaction MASS. Chemical propellants are in either liquid or solid form. Modern LAUNCH VEHICLES use one of three general types of LIQUID PROPELLANTS: petroleum based, cryogenic (very cold), or hypergolic (self-igniting upon contact).

prominence A cloud of cooler PLASMA extending high above the SUN's visible surface, rising above the PHOTOSPHERE into the CORONA.

proper motion (μ) The apparent angular displacement of a STAR with respect to the CELESTIAL SPHERE. BARNARD'S STAR has the largest known proper motion.

propulsion system The LAUNCH VEHICLE (or SPACE VEHICLE) system that includes ROCKET engines, PROPELLANT tanks, fluid lines, and all associated equipment necessary to provide the propulsive FORCE (THRUST) as specified for the vehicle.

protium Ordinary HYDROGEN—one PROTON in the NUCLEUS surrounded by one ELECTRON.

protogalaxy A GALAXY in the early stages of evolution.

proton (p) A stable elementary PARTICLE with a single positive charge and a MASS of about 1.672×10^{-27} kg. A single proton makes up the NUCLEUS of an ordinary HYDROGEN atom (PROTIUM).

Proton A Russian LIQUID-PROPELLANT EXPENDABLE LAUNCH VEHICLE capable of placing 21,000 kg MASS PAYLOADS into LOW EARTH ORBIT (LEO) and sending SPACECRAFT on INTERPLANETARY TRAJECTORIES.

proton-proton reaction The series of thermonuclear FUSION reactions in stellar interiors by which four HYDROGEN NUCLEI are fused into a HELIUM nucleus. This is the main ENERGY liberation mechanism in STARS like the SUN.

proton storm The burst of PROTONs sent into interplanetary space by a SOLAR FLARE.

protoplanet Any of a star's PLANETS as such planets emerge during the process of ACCRETION in which PLANETESIMALS collide and coalesce into large objects.

protostar A STAR in the making. Specifically, the stage in a young star's evolution after it has separated from a gas cloud but prior to it collapsing sufficiently (due to GRAVITY) to support thermonuclear FUSION reactions.

Proxima Centauri The closest STAR to the SUN—the third member of the ALPHA CENTAURI triple-star system. It is some 4.2 LIGHT-YEARS away.

Ptolemaic system The ancient Greek model of an EARTH-centered (GEOCENTRIC) UNIVERSE as described by PTOLEMY in the ALMAGEST. *Compare with* COPERNICAN SYSTEM.

pulsar A rapidly spinning NEUTRON STAR that generates regular pulses of ELECTROMAGNETIC RADIATION. Although originally discovered by RADIO WAVE observations, they have since been observed at optical, X-RAY, and GAMMA RAY energies.

purge The process of removing residual fuel or OXIDIZER from the tanks or lines of a LIQUID-PROPELLANT ROCKET after a test firing.

quantum *(plural: quanta)* A discrete bundle of ENERGY possessed by a PHOTON. *See also* QUANTUM THEORY.

quantum mechanics The physical theory that emerged from MAX KARL PLANCK's original QUANTUM THEORY and developed into wave mechanics, matrix mechanics, and relativistic quantum mechanics in the 1920s and 1930s.

quantum theory A foundational theory of modern physics. MAX KARL PLANCK proposed in 1900 that all ELECTROMAGNETIC RADIATION was emitted and absorbed in QUANTA, or discrete ENERGY packets (called PHOTONs), instead of continuously.

quasar A mysterious, very distant object with a high REDSHIFT—that is, traveling away from EARTH at great speed. These objects appear almost like STARs but are far more distant than any individual star now observed. They might be the very luminous centers of active distant GALAXIES. When first identified in 1963, they were called *quasi*-stell*ar* radio sources—or quasars. Also called quasi-stellar object (QSO).

quiet Sun The collection of SOLAR phenomena and features, including the PHOTOSPHERE, the solar SPECTRUM, and the CHROMOSPHERE, that are always present. *Compare with* ACTIVE SUN.

radar astronomy The use of radar (RADIO WAVE reflections) to study objects in this SOLAR SYSTEM, such as the MOON, the PLANETs (especially VENUS), ASTEROIDS, and planetary RING systems.

Radarsat Canadian EARTH-OBSERVING SPACECRAFT launched in 1995 that uses an advanced SYNTHETIC APERTURE RADAR (SAR) to produce high-resolution IMAGES of EARTH's surface despite clouds and darkness.

radial burning A SOLID-PROPELLANT ROCKET GRAIN that burns in the radial direction, either outwardly (called an internally burning grain) or inwardly.

radian A unit of angle. One radian is the angle subtended at the center of a circle by an arc equal in length to the radius of the circle, approximately 57.3°.

radiant flux *See* FLUX.

radiation *(heat transfer)* The transfer of thermal ENERGY (heat) by ELECTROMAGNETIC RADIATION (EMR) due to the temperature of a body. The STEFAN-BOLTZMANN LAW determines the amount of radiant energy exchanged by a given surface area. This amount is also proportional to the fourth power of the ABSOLUTE TEMPERATURE of the radiating surface. It is called thermal radiation to distinguish it from other forms of EMR, such as RADIO WAVES, LIGHT, and X RAYS.

radiation belt The region(s) in a PLANET's MAGNETOSPHERE where there is a high DENSITY of trapped atomic PARTICLES from the SOLAR WIND. *See* EARTH'S TRAPPED RADIATION BELTS.

radiation sickness A potentially fatal illness resulting from excessive exposure to IONIZING RADIATION. Also called acute radiation syndrome.

radioactivity The spontaneous decay or disintegration of an unstable (atomic) NUCLEUS accompanied by the emission of IONIZING RADIATION, such as ALPHA PARTICLES, BETA PARTICLES, and GAMMA RAYS.

radio astronomy The branch of ASTRONOMY that collects and evaluates RADIO WAVE signals from a wide variety of celestial objects. It started in the 1930s when KARL GUTHE JANSKY detected the first EXTRATERRESTRIAL radio wave signals.

radio frequency (RF) The portion of the ELECTROMAGNETIC (EM) SPECTRUM useful for TELECOMMUNICATIONS with a FREQUENCY range between 10,000 and 3×10^{11} HERTZ.

radio galaxy A GALAXY (often dumbbell shaped) that produces very strong RADIO WAVE signals. Cygnus A is an example of an intense source about 650 million LIGHT-YEARS away.

radioisotope An unstable (radioactive) ISOTOPE of an ELEMENT that spontaneously decays at a predictable rate governed by its HALF-LIFE. *See also* RADIOACTIVITY.

radioisotope thermoelectric generator (RTG) Compact-space nuclear power system that uses DIRECT CONVERSION (based on the thermoelectric principle) to transform the thermal ENERGY from a RADIOISOTOPE source (generally plutonium 238) into electricity. All NASA SPACECRAFT that have explored the outer regions of the SOLAR SYSTEM have used RTGs for their electric power.

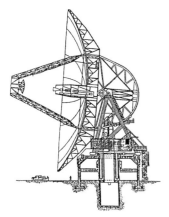

Radio telescope (Courtesy of NASA)

radio telescope A large, metallic device, generally parabolic (dish shaped), that collects RADIO WAVE signals from EXTRATERRESTRIAL objects (such as PULSARs and ACTIVE GALAXIES) or from distant SPACECRAFT and focuses these radio signals onto a sensitive RADIO-FREQUENCY (RF) receiver.

radio waves ELECTROMAGNETIC RADIATION (EMR) with a WAVELENGTH between about 1 mm (0.001 m) and several thousand kilometers.

Ranger Project The first NASA SPACECRAFT sent to the Moon in the 1960s. These IMPACT PLANETARY PROBES were designed to take a series of television IMAGES of the LUNAR surface before crash landing.

rays *(lunar)* Bright streaks extending across the surface from young IMPACT CRATERS on the MOON; also observed on MERCURY and on several large moons of the OUTER PLANETS.

reaction engine An engine that develops THRUST by its physical reaction to the ejection of a substance from it. Normally, a reaction engine ejects a stream of hot gases created by internally combusting a PROPELLANT—but more advanced reaction engine concepts involve the ejection of PHOTONS or nuclear PARTICLES. Both ROCKET and jet PROPULSION SYSTEMS involve reaction engines that obey the third of NEWTON'S LAWS OF MOTION (the action-reaction principle).

reconnaissance satellite A military SATELLITE that ORBITS EARTH and performs intelligence gathering against enemy nations and potential adversaries. Also called a SPY SATELLITE.

red dwarf *(star)* Reddish MAIN-SEQUENCE STARS (spectral type K and M) that are relatively cool (~ 4000 K surface temperature) and have low MASS (about 0.5 SOLAR MASS or less). These faint, low-LUMINOSITY stars are inconspicuous, yet they represent the most common type of star in the UNIVERSE and the longest lived. BARNARD'S STAR is an example.

red giant *(star)* A large, cool STAR with a surface temperature of about 2,500 K and a diameter 10 to 100 times that of the SUN. This type of highly luminous star is at the end of its evolutionary life, departing the main sequence after exhausting the HYDROGEN in its core. It is often a VARIABLE STAR. Some 5 billion years from now, the Sun will evolve into a massive red giant. *See also* MAIN-SEQUENCE STAR.

Red Planet The planet MARS—so named because of its distinctive reddish soil.

redshift The apparent increase in the WAVELENGTH of a source of LIGHT (toward longer-wavelength red light) caused by the DOPPLER SHIFT effect in a receding object. *Compare with* BLUESHIFT.

reentry The return of objects, originally launched from EARTH, back into the SENSIBLE ATMOSPHERE; the action involved in this event. The major types of reentry are ballistic, gliding, and skip. When a piece of SPACE (ORBITAL) DEBRIS undergoes an uncontrolled ballistic reentry, it usually burns up in the ATMOSPHERE due to excessive AERODYNAMIC HEATING. An AEROSPACE VEHICLE, like NASA's SPACE SHUTTLE, makes a safe, controlled atmospheric reentry by using a gliding TRAJECTORY designed to dissipate its KINETIC ENERGY and POTENTIAL ENERGY carefully prior to landing.

reentry vehicle (RV) The part of a ROCKET or SPACE VEHICLE designed to reenter EARTH'S ATMOSPHERE in the terminal portion of its TRAJECTORY. For example, the NOSE CONE portion of an INTERCONTINENTAL BALLISTIC MISSILE is designed to survive the AERODYNAMIC HEATING of REENTRY and to protect its PAYLOAD (a nuclear WARHEAD) while it descends on a BALLISTIC TRAJECTORY to its target.

reflecting telescope An optical TELESCOPE that collects and focuses LIGHT from distant objects by means of a mirror (called the OBJECTIVE or primary mirror). SIR ISAAC NEWTON developed the first reflecting telescope in about 1670—using a design now called the NEWTONIAN TELESCOPE or the Newtonian reflector.

reflection The return of all or part of a beam of LIGHT when it encounters the interface (boundary) between two different media, like air and water. A mirrorlike surface reflects most of the light falling onto it.

refracting telescope An optical TELESCOPE that collects and focuses LIGHT from distant objects by means of a LENS (called the OBJECTIVE or primary lens) or system of lens. In 1609, GALILEO GALILEI began improving the first refracting telescope invented by HANS LIPPERSHEY and then applied his own instrument to astronomical observations. It was this refracting Galilean telescope that helped launch the scientific revolution of the 17th century.

refraction The change in direction (bending) of a beam of LIGHT at the interface (boundary) between two different transparent media, such as air and water or a pair of LENSes that possess different optical properties.

regolith *(lunar)* The unconsolidated MASS of surface debris that overlies the MOON's bedrock. This blanket of pulverized lunar dust and soil was created by millions of years of meteoric and cometary IMPACTS.

regressive burning For a SOLID-PROPELLANT ROCKET, the condition in which the burning surface of the PROPELLANT decreases with time—thereby decreasing the pressure and THRUST.

relativity The theory of space and time developed by ALBERT EINSTEIN early in the 20th century and one of the foundations of modern physics. Special relativity, introduced in 1905, involves Einstein's fundamental postulate that the SPEED OF LIGHT *(c)* has the same value for all observers. One major consequence of special relativity is the famous MASS-ENERGY equivalence formula: $E = \Delta mc^2$ in which E is the energy equivalent of an amount of matter (Δm) that is annihilated or converted into pure energy and c is the speed of light. In 1915, Einstein introduced his general theory of relativity, postulating that GRAVITATION is not really a FORCE between two masses (as SIR ISAAC NEWTON proposed) but, rather, arises as a consequence of the curvature of space and time. For Einstein, in a four-dimensional UNIVERSE (that is x-, y-, z-space and time) space-time becomes curved in the presence of matter—the more massive the object, the more curvature and therefore the more gravitation.

remote manipulator system (RMS) The dexterous, Canadian-built, 15.2 m long articulated arm that is remotely controlled by ASTRONAUTs from the aft flight deck of NASA's SPACE SHUTTLE.

remote sensing The sensing of an object or phenomenon, using different portions of the ELECTROMAGNETIC (EM) SPECTRUM, without having the SENSOR in direct contact with the object being studied. In ASTRONOMY, characteristic ELECTROMAGNETIC RADIATION signatures often carry distinctive information about an interesting celestial object that is LIGHT-YEARS away from the sensor.

rendezvous The close approach of two or more SPACECRAFT in the same ORBIT so that DOCKING can take place. Orbiting objects meet at a preplanned location and time. They slowly come together with essentially zero relative velocity.

retrograde motion Motion in a reverse or backward direction.

retrorocket *(retrograde rocket)* An auxiliary ROCKET that fires in the direction opposite to which a SPACE VEHICLE is traveling (pointed). Low-THRUST retrograde rockets produce a retarding FORCE that opposes the vehicle's forward motion and reduces its VELOCITY. *Compare with* POSIGRADE ROCKET.

resolution The smallest detail (measurement) that can be distinguished by a sensor system under specific conditions, such as its spatial resolution or spectral resolution.

reusable launch vehicle (RLV) A conceptual space LAUNCH VEHICLE that includes simple, fully reusable designs that support flexible airline-type operations and greatly reduced costs per kilogram of PAYLOAD delivered into LOW EARTH ORBIT (LEO). These design goals would be achieved primarily through the use of advanced space technology and innovative operational techniques. In 2001, NASA canceled the X-33 prototype RLV program, citing technical difficulties and cost overruns for its decision.

revolution One complete cycle of movement of a SATELLITE or a CELESTIAL BODY around its PRIMARY BODY; the orbital motion of a celestial body or SPACECRAFT about a center of gravitational attraction such as the SUN or a PLANET as distinct from ROTATION about an internal AXIS.

rift valley A depression in a PLANET's surface due to crustal MASS separation.

rill A deep, narrow depression on the LUNAR surface that cuts across all other types of topographical features on the MOON.

ring *(planetary)* A DISK of matter that encircles a PLANET. Such rings usually contain ice and dust PARTICLES, ranging in size from microscopic fragments up to chunks that are tens of meters in diameter.

ringed world A PLANET with a RING or set of rings encircling it. In the SOLAR SYSTEM, JUPITER, SATURN, URANUS, and NEPTUNE all have

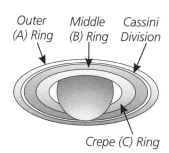

Outer (A) Ring   Middle (B) Ring   Cassini Division

Crepe (C) Ring

Ring (Saturn, South Pole view)

ring systems of varying degrees of composition and complexity. Ring systems may be a common feature of EXTRASOLAR JOVIAN PLANETS.

robot spacecraft A semiautomated or fully automated SPACECRAFT capable of executing its primary exploration mission with minimal or no human supervision.

Roche limit As postulated by EDOUARD ALBERT ROCHE in the 19th century, the smallest distance from a PLANET at which gravitational FORCES can hold together a SATELLITE or MOON that has the same average DENSITY as the PRIMARY BODY. If the moon's ORBIT falls within the Roche limit, it will be torn apart by tidal forces.

rocket A completely self-contained projectile or flying vehicle propelled by a REACTION ENGINE. Since a rocket carries all of its required PROPELLANT, it can function in the vacuum of OUTER SPACE and represents the key to space travel. This fact was independently recognized early in the 20th century by the founders of ASTRONAUTICS: KONSTANTIN EDUARDOVICH TSIOLKOVSKY, ROBERT HUTCHINGS GODDARD, and HERMANN J. OBERTH. Rockets obey the third of NEWTON'S LAWS OF MOTION (the action-reaction principle). There are CHEMICAL ROCKETS, NUCLEAR ROCKETS, and ELECTRIC-PROPULSION rockets. Chemical rockets are further divided into SOLID-PROPELLANT ROCKETS and LIQUID-PROPELLANT ROCKETS.

rogue star A wandering STAR that passes close to a SOLAR SYSTEM, disrupting the CELESTIAL BODIES in the system and triggering cosmic catastrophes on life-bearing PLANETS.

roll The rotational or oscillatory movement of an AEROSPACE VEHICLE or ROCKET about its longitudinal (lengthwise) AXIS. *See also* PITCH; YAW.

rotation The turning of an object (especially a CELESTIAL BODY) about an internal AXIS.

rover A crewed or robot SPACE VEHICLE used to explore a planetary surface.

Rover Program The overall U.S. NUCLEAR ROCKET development program from 1959–73.

Royal Greenwich Observatory (RGO) The English observatory founded by King Charles II in 1675 at Greenwich in London. The original structure is now a museum, while the RGO is presently located at Cambridge. Until 1971, the ASTRONOMER ROYAL served as the RGO director, but these positions are now separate appointments.

15° per hour

Rotation (of Earth)

RP-1   *Rocket propellant number one*—a commonly used hydrocarbon-based, LIQUID-PROPELLANT ROCKET ENGINE fuel that is refined kerosene.

rumble   A form of combustion instability in a LIQUID-PROPELLANT ROCKET ENGINE, characterized by a low-pitched, low-FREQUENCY rumbling noise.

runaway greenhouse   An environmental catastrophe during which the GREENHOUSE EFFECT produces excessively high global temperatures that cause all the liquid (surface) water on a life-bearing PLANET to evaporate permanently. *Compare with* ICE CATASTROPHE.

Salyut   *(salute)* An evolutionary series of Russian SPACE STATIONS placed into ORBIT around EARTH in the 1970s and early 1980s to support a variety of military and civilian missions.

satellite   A secondary (smaller) CELESTIAL BODY in ORBIT around a larger PRIMARY BODY. For example, EARTH is a natural satellite of the SUN, while the MOON is a natural satellite of Earth. A human-made SPACECRAFT placed into orbit around Earth is called an artificial satellite—or more commonly just a satellite.

satellite power system (SPS)   A conceptual CONSTELLATION of very large (kilometers on a side) GEOSTATIONARY ORBIT space structures, constructed in space, that continuously harvest SOLAR ENERGY and transmit it as MICROWAVE RADIATION to special receiving/converter stations on EARTH's surface.

Saturn   (1) *(launch vehicle)* Family of powerful, EXPENDABLE LAUNCH VEHICLES developed for NASA by WERNHER VON BRAUN to carry ASTRONAUTS to the MOON in the APOLLO PROJECT.
(2) *(planet)* The sixth PLANET from the SUN and the JOVIAN PLANET with the most extensive and beautiful RINGS. Its major SATELLITE, TITAN, is the only known moon with a dense ATMOSPHERE.

Saturnian   Of or pertaining to the PLANET SATURN.

scarp   A cliff produced by erosion or faulting.

scattering   The collision (or deflection) process that changes the TRAJECTORY of a PARTICLE or a PHOTON.

Schwarzschild black hole   An uncharged BLACK HOLE that does not rotate; KARL SCHWARZSCHILD predicted this basic model of a black hole in 1916.

Schwarzschild radius   The radius of the EVENT HORIZON of a BLACK HOLE.

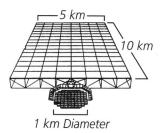

Satellite power system (Courtesy of NASA)

science payload The collection of scientific instruments on a SPACECRAFT.

scientific airlock A special opening in a crewed SPACECRAFT or SPACE STATION through which experiments and research equipment can be extended outside (into OUTER SPACE) without violating the atmospheric integrity of the pressurized interior of the space vehicle.

Scout A four-stage, SOLID-PROPELLANT ROCKET developed by NASA and used as an EXPENDABLE LAUNCH VEHICLE to place small-MASS (~ 200 kg or less) PAYLOADS into LOW EARTH ORBIT or on suborbital TRAJECTORIES.

screaming For a LIQUID-PROPELLANT ROCKET ENGINE, a relatively high-FREQUENCY form of combustion instability, characterized by a high-pitched noise.

scrub To cancel or postpone a ROCKET firing, either before or during the COUNTDOWN.

search for extraterrestrial intelligence (SETI) An attempt to answer the important philosophical question, Are we alone in the universe? The goal of contemporary SETI programs (now being conducted by private foundations) is to detect coherent RADIO FREQUENCY (microwave) signals generated by intelligent EXTRATERRESTRIAL civilizations—should they exist.

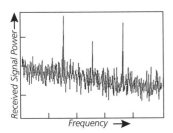

Search for extraterrestrial intelligence (simulated signal) (Courtesy of NASA)

secondary crater The CRATER formed when a large chunk of material from a primary-IMPACT CRATER strikes the surrounding planetary surface.

self-replicating system (SRS) An advanced space robot system, first postulated by JOHN VON NEUMANN, that would be capable of gathering materials, maintaining itself, manufacturing desired products, and even making copies of itself (self-replication).

semimajor axis One-half the major AXIS of an ELLIPSE. For a PLANET, this corresponds to its average orbital distance from the SUN.

sensible atmosphere That portion of a PLANET's ATMOSPHERE that offers resistance to a body passing through it.

sensor The portion of a scientific instrument that detects and/or measures some physical phenomenon.

Seyfert galaxy A type of SPIRAL GALAXY with a very bright GALACTIC NUCLEUS—first observed by Carl Seyfert in 1943.

shepherd moon A small inner MOON (or pair of moons) that shapes and forms a particular RING around a (ringed) PLANET. For example, the shepherds, Ophelia and Cordelia, tend the Epsilon Ring of URANUS.

shield volcano A wide, gently sloping VOLCANO formed by the gradual outflow of molten rock; many occur on VENUS.

short-period comet A COMET with an orbital PERIOD of less than 200 years.

sidereal Of or pertaining to the STARS.

sidereal month The average amount of time the MOON takes to complete one orbital REVOLUTION around EARTH when using the FIXED STARS as a reference; approximately 27.32 days.

sidereal period The period of time required by a CELESTIAL BODY to complete one REVOLUTION around another celestial body with respect to the FIXED STARS.

sidereal year The period of one REVOLUTION of EARTH around the SUN with respect to the fixed STARS; some 365.25636 days.

singularity The hypothetical central point in a BLACK HOLE at which the curvature of space and time becomes infinite; a theoretical point that has infinite DENSITY and zero volume.

SI units The international system of units (the metric system) that uses the meter (m), kilogram (kg), and second (s) as its basic units of length, MASS, and time, respectively. *See also* SECTION IV CHARTS & TABLES.

Skylab The first U.S. SPACE STATION that NASA placed into ORBIT in 1973 and was visited by three ASTRONAUT crews between 1973–74. It reentered the ATMOSPHERE on 11 July 1979 as a large, abandoned derelict—with surviving SPACE (ORBITAL) DEBRIS pieces impacting in the Indian Ocean and remote portions of Australia.

sloshing The back-and-forth movement of a LIQUID PROPELLANT in its tank(s), creating stability and control problems for the ROCKET vehicle. Engineers often use ANTISLOSH BAFFLES in the PROPELLANT tanks to avoid this problem.

Small Magellanic Cloud (SMC) An IRREGULAR GALAXY about 9,000 LIGHT-YEARS in diameter and 180,000 light-years from EARTH. *See also* MAGELLANIC CLOUDS.

soft landing The act of landing onto the surface of a PLANET without damaging any portion of a SPACECRAFT or its PAYLOAD, except possibly an expendable landing gear structure. *Compare with* HARD LANDING.

Sun as Seen

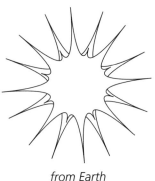

from Earth

from Jupiter

from Saturn

Solar constant

sol A MARTIAN day (about 24 hours, 37 minutes, and 23 seconds in duration). Seven sols equal about 7.2 EARTH days.

Sol The SUN.

solar Of or pertaining to the SUN; caused by the Sun.

solar activity Any variation in the appearance or ENERGY output of the SUN.

solar cell A DIRECT-CONVERSION device that transforms incoming sunlight (SOLAR ENERGY) directly into electricity. It is used extensively (in combination with rechargeable storage batteries) as the prime source of electric POWER for SPACECRAFT orbiting EARTH or on missions within the inner SOLAR SYSTEM. Also called photovoltaic cell.

solar constant The total average amount of the SOLAR ENERGY (in all WAVELENGTHS) crossing perpendicular to a unit area at the top of EARTH'S ATMOSPHERE. It is measured by SPACECRAFT at about 1,370 WATTS per square meter at one ASTRONOMICAL UNIT from the SUN.

solar cycle The approximately 11-YEAR period in the variation of the number of SUNSPOTS and the levels of SOLAR activity. *See* ACTIVE SUN; QUIET SUN; SUNSPOT CYCLE.

solar eclipse See ECLIPSE.

solar-electric propulsion (SEP) A low-THRUST PROPULSION SYSTEM that uses SOLAR CELLS to provide the electricity for a SPACECRAFT'S ELECTRIC PROPULSION ROCKET engines.

solar energy ENERGY from the SUN; radiant energy in the form of sunlight. *See also* SOLAR CONSTANT; SOLAR RADIATION.

solar flare A highly concentrated, explosive release of ELECTROMAGNETIC RADIATION and nuclear PARTICLES within the SUN'S ATMOSPHERE near an active SUNSPOT.

solar mass The MASS of the SUN, about 1.99×10^{30} kg. It is commonly used as a reference mass in stellar ASTRONOMY.

solar nebula The cloud of dust and gas from which the SUN, the PLANETs, and other minor bodies of this SOLAR SYSTEM are thought to have formed about 5 billion YEARS ago.

solar panel The winglike assembly of SOLAR CELLS used by a SPACECRAFT to convert sunlight (SOLAR ENERGY) directly into electrical ENERGY. Also called a solar array.

solar photovoltaic conversion The DIRECT CONVERSION of sunlight (SOLAR ENERGY) into electrical ENERGY by means of the photovoltaic effect in SOLAR CELLS. Engineers combine a number of solar cells to form a SOLAR PANEL that increases a SPACECRAFT's electric power supply.

solar radiation The ELECTROMAGNETIC RADIATION emitted by the SUN, often approximated as a BLACKBODY RADIATION source at approximately 5,770 K. Therefore, 99.9 percent of the radiated SOLAR ENERGY lies within the WAVELENGTH interval 0.15 to 4.0 MICROMETERS (μM). About 50 percent of that is in the visible portion of the ELECTROMAGNETIC (EM) SPECTRUM and the remainder in the near-infrared portion. *See also* INFRARED RADIATION; SOLAR CONSTANT.

solar storm A major disturbance in the space environment triggered by an intense SOLAR FLARE (or flares) that produces bursts of ELECTROMAGNETIC RADIATION and charged PARTICLES, threatening unprotected SPACECRAFT and ASTRONAUTS alike.

solar system (1) Any STAR and its gravitationally bound collection of nonluminous objects, such as PLANETs, ASTEROIDs, and COMETs. (2) Humans' home solar system, consisting of the SUN and all the objects bound to it by GRAVITATION. This includes nine major PLANETs with more than 60 known MOONs, over 2,000 MINOR PLANETs, and a very large number of COMETs. Except for the comets, all the CELESTIAL OBJECTs travel around the Sun in the same direction.

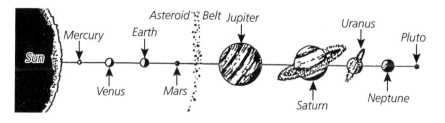

Solar system (Courtesy of NASA)

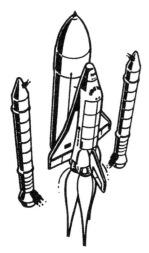

Solid-rocket booster

solar wind The variable stream of PLASMA (that is, ELECTRONS, PROTONS, ALPHA PARTICLES, and other atomic NUCLEI) that flows continuously outward from the SUN into INTERPLANETARY space.

solid Earth The lithosphere portion of the EARTH SYSTEM, including this PLANET's core, mantle, crust, and all the surface rocks and unconsolidated rock fragments.

solid-propellant rocket A ROCKET propelled by a chemical mixture of fuel and OXIDIZER in solid form and intimately mixed into a monolithic (but not powdered) GRAIN. Sometimes called a solid rocket.

solid-rocket booster (SRB) The two very large SOLID-PROPELLANT ROCKETS that operate in parallel to augment the THRUST of the SPACE SHUTTLE's three main engines for the first two minutes after LAUNCH. Each SRB develops about 11,800-kiloNEWTONS (kN) of thrust at LIFTOFF. After burning for about 120 seconds, the depleted SRBs are JETTISONed from the Space Shuttle and recovered in the Atlantic Ocean DOWNRANGE of CAPE CANAVERAL for refurbishment and PROPELLANT reloading.

solstice The two times of the YEAR when the SUN's position in the sky is the most distant from the celestial EQUATOR. For the Northern Hemisphere, the summer solstice (longest day) occurs about 21 June and the winter solstice (shortest day) about 21 December.

sounding rocket A SOLID-PROPELLANT ROCKET used to carry scientific instruments on parabolic TRAJECTORIES into the upper regions of EARTH's SENSIBLE ATMOSPHERE and into near-Earth space.

Soyuz (union) spacecraft The evolutionary family of crewed Russian SPACECRAFT used by COSMONAUTS on a wide variety of EARTH-orbiting missions since 1967. Unfortunately, malfunctions with this spacecraft caused the two major Russian space tragedies: *Soyuz 1* and the death of cosmonaut Vladimir Komarov (1967), and *Soyuz 11* and the deaths of cosmonauts Georgi Dobrovolsky, Victor Patseyev, and Vladislav Volkov (1971). The *Soyuz 19* was successfully used by cosmonauts Leonov and Kubasov during the international APOLLO-SOYUZ TEST PROJECT (1975).

space OUTER SPACE; all the regions of the UNIVERSE that lie beyond the limits of EARTH's SENSIBLE ATMOSPHERE.

space base A large, permanently inhabited space facility located in ORBIT around a CELESTIAL BODY or on its surface that would serve as the

Sounding rocket

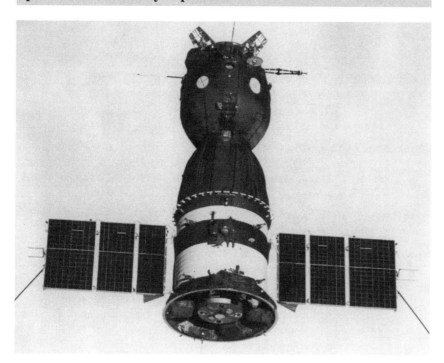

**Soyuz spacecraft
(Courtesy of NASA)**

center of future human operations in some particular region of the SOLAR SYSTEM. *See also* LUNAR BASE; MARS BASE.

space-based astronomy The use of astronomical instruments on SPACECRAFT in ORBIT around EARTH and in other locations throughout the SOLAR SYSTEM to view the UNIVERSE from above Earth's ATMOSPHERE. Major breakthroughs in astronomy, ASTROPHYSICS, and COSMOLOGY have occurred because of the unhampered viewing advantages provided by space platforms.

space capsule The family of small, container-like, tear-shaped SPACECRAFT used to carry American ASTRONAUTS into OUTER SPACE and return them to EARTH as part of NASA's MERCURY PROJECT, GEMINI PROJECT, and APOLLO PROJECT.

spacecraft A platform that can function, move, and operate in OUTER SPACE or on a planetary surface. Spacecraft can be human occupied or uncrewed (robot) platforms. They can operate in ORBIT around EARTH or while on an INTERPLANETARY TRAJECTORY to another CELESTIAL BODY. Some spacecraft travel through space and orbit another PLANET. Others descend to a planet's surface, making a HARD

LANDING (collision IMPACT) or a (survivable) SOFT LANDING. Exploration spacecraft are often categorized as either FLYBY, ORBITER, ATMOSPHERIC PROBE, LANDER, or ROVER spacecraft.

spacecraft clock The time-keeping component within a SPACECRAFT's command and data-handling system. It meters the passing time during a mission and regulates nearly all activity within the spacecraft.

space launch vehicle (SLV) The expendable or reusable ROCKET-propelled vehicle(s) used to lift a PAYLOAD or SPACECRAFT from the surface of EARTH and place it into ORBIT around the PLANET or on an INTERPLANETARY TRAJECTORY.

spaceman A person, male or female, who travels in OUTER SPACE. The term *ASTRONAUT* is preferred.

Space (orbital) debris (Courtesy of USAF and NASA)

space (orbital) debris Space junk; abandoned or discarded human-made objects in ORBIT around EARTH. It includes operational debris (items discarded during SPACECRAFT deployment), used or failed ROCKETs, inactive or broken SATELLITES, and fragments from collisions and space object breakup. When a spacecraft collides with an object or a discarded rocket spontaneously explodes, thousands of debris fragments become part of the orbital debris population.

space physics The branch of physics that studies the magnetic and electric phenomena that occur in OUTER SPACE, in the upper ATMOSPHERE of various PLANETS, and on the SUN.

space platform An uncrewed, free-flying platform in orbit around EARTH that is dedicated to a specific mission, such as the long-duration exposure of test materials to the space environment.

spaceport A facility that serves as both a doorway to OUTER SPACE from the surface of a PLANET and as a port of entry for AEROSPACE VEHICLES returning from space to the planet's surface. NASA's KENNEDY SPACE CENTER with its SPACE SHUTTLE LAUNCH SITE and landing complex is an example.

space probe See PLANETARY PROBE.

space resources The resources available in OUTER SPACE that could be used to support an extended human presence and eventually become the physical basis for a thriving SOLAR SYSTEM-level civilization. These EXTRATERRESTRIAL resources include unlimited SOLAR ENERGY, mineral resources (from the MOON, ASTEROIDS, MARS, and numerous OUTER PLANET MOONS), LUNAR (water) ice, and special environmental conditions like access to a high vacuum and physical isolation from the TERRESTRIAL BIOSPHERE.

space settlement A proposed very large, human-made habitat in OUTER SPACE within which from 1,000 to 10,000 people would live, work, and play while supporting various research and commercial activities, such as the construction of SATELLITE POWER SYSTEMS.

spaceship An INTERPLANETARY SPACECRAFT that carries a human crew.

space shuttle The major spaceflight component of NASA's Space Transportation System (STS). It consists of a winged orbiter vehicle, three space shuttle main engines (SSMEs), the giant EXTERNAL TANK (ET)—which feeds LIQUID HYDROGEN and LIQUID OXYGEN to the shuttle's three main LIQUID-PROPELLANT ROCKET ENGINES—and the two SOLID-ROCKET BOOSTERS (SRBs). *See also* SECTION IV CHARTS & TABLES.

space sickness The space-age form of motion sickness whose symptoms include nausea, vomiting, and general malaise. This temporary condition lasts no more than a day or so. However, it affects 50 percent of the ASTRONAUTs or COSMONAUTs when they encounter the MICROGRAVITY environment (WEIGHTLESSNESS) of an orbiting SPACECRAFT after a launch. Also called space adaptation syndrome.

space station An EARTH-orbiting facility designed to support long-term human habitation in OUTER SPACE. *See also* INTERNATIONAL SPACE STATION.

spacesuit The flexible, outer garmentlike structure (including visored helmet) that protects an ASTRONAUT in the hostile environment of OUTER SPACE. It provides portable life support functions, supports communications, and accommodates some level of movement and flexibility so the astronaut can perform useful tasks during an EXTRA-VEHICULAR ACTIVITY or while exploring the surface of another world.

space transportation system (STS) The official name for NASA's SPACE SHUTTLE.

space vehicle The general term describing a crewed or robot vehicle capable of traveling through OUTER SPACE. An AEROSPACE VEHICLE can operate both in outer space and in EARTH'S ATMOSPHERE.

space walk The popular term for an EXTRAVEHICULAR ACTIVITY (EVA).

special relativity Theory introduced by ALBERT EINSTEIN in 1905. *See* RELATIVITY.

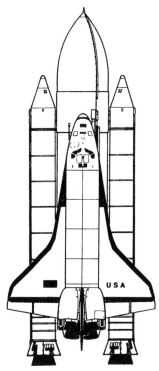

Space shuttle

specific impulse *(I_{sp})* An index of performance for ROCKET engines and their various PROPELLANT combinations. It is defined as the THRUST produced by propellant combustion divided by the propellant mass flow rate.

spectral classification The system in which STARS are given a designation. It consists of a letter and a number according to their SPECTRAL LINES, which correspond roughly to surface temperature. Astronomers classify stars as O (hottest), B, A, F, G, K, and M (coolest). The numbers represent subdivisions within each major class. The SUN is a G2 star. M stars are numerous but very dim, while O and B stars are very bright but rare. *See also* HARVARD CLASSIFICATION SYSTEM.

spectral line A narrow range of spectral color, emitted (or absorbed) by a specific ATOM or MOLECULE. Characteristic emission line PHOTONS are produced when ELECTRONS of the atom or molecule transition (fall) from a higher (excited) ENERGY state to a lower energy state. *See also* ABSORPTION LINE.

spectrogram The photographic IMAGE of a SPECTRUM.

spectrometer An optical instrument that splits incoming visible LIGHT (or other ELECTROMAGNETIC RADIATION) from a celestial object into a SPECTRUM by DIFFRACTION and then measures the relative amplitudes of the different WAVELENGTHS. INFRARED RADIATION and ULTRAVIOLET RADIATION spectrometers are often carried on scientific spacecraft.

spectroscopy The study of SPECTRAL LINES from different ATOMS and MOLECULES. Astronomers use emission spectroscopy to infer the material composition of the objects that emitted the light and absorption spectroscopy to infer the composition of the intervening medium.

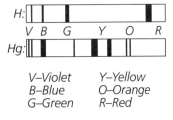

V–Violet Y–Yellow
B–Blue   O–Orange
G–Green R–Red

Spectrum

spectrum *(plural: spectra)* A plot of the intensity of incident ELECTROMAGNETIC RADIATION as a function of WAVELENGTH (or FREQUENCY). Originally, it was the dispersion of white LIGHT into the spread of pure colors visible to the eye as seen in a rainbow. *See also* ELECTROMAGNETIC (EM) SPECTRUM.

speed of light *(c)* The speed at which ELECTROMAGNETIC RADIATION (including LIGHT) moves through a vacuum; a universal constant equal to approximately 300,000 km/s.

spin stabilization Directional stability of a MISSILE or SPACECRAFT obtained as a result of spinning the moving body about its AXIS of symmetry.

spiral galaxy A GALAXY with spiral arms, similar to the MILKY WAY GALAXY or the ANDROMEDA GALAXY.

SPOT A family of EARTH-OBSERVING SPACECRAFT built by the French Space Agency (CNES). The ACRONYM SPOT stands for *s*atellite *p*our l'*o*bservation de la *t*erre.

Sputnik 1 Launched by the former Soviet Union on 4 October 1957, it was the first SATELLITE to ORBIT EARTH. *Sputnik* means "fellow traveler." This simple, spherically shaped, 84 kg Russian SPACECRAFT inaugurated the space age.

spy satellite Popular term for a military reconnaissance satellite. *See also* RECONNAISSANCE SATELLITE.

staging The practice of placing smaller ROCKETs on top of larger ones, thereby increasing the ability of the combination to lift larger PAYLOADS or to give a particular payload a higher final VELOCITY. In MULTISTAGE ROCKETs, the stages are numbered chronologically in the order of burning (that is, first stage, second stage, third stage, and so on). When the first stage stops burning, it separates from the rest of the vehicle and falls away. Then the second-stage rocket ignites, fires until burnout, and also separates. The staging process continues up to the last stage, which contains the PAYLOAD. In this way, the MASS of empty PROPELLANT tanks is discarded during the ascent.

star A self-luminous ball of very hot gas that liberates ENERGY through thermonuclear FUSION reactions within its core. Stars are classified as either normal or abnormal. Normal stars, like the SUN, shine steadily—exhibiting one of a variety of distinctive colors such as red, orange, yellow, blue, and white (in order of increasing surface temperature). There are also several types of abnormal stars, including GIANT STARS, DWARF STARS, and VARIABLE STARS. Stars experience an evolutionary life cycle from birth in an INTERSTELLAR cloud of gas to death as a compact WHITE DWARF, NEUTRON STAR, or BLACK HOLE.

star cluster A group of STARs (numbering from a few to perhaps thousands) that were formed from a common gas cloud and are now bound together by their mutual gravitational attraction.

star probe A conceptual NASA robot scientific SPACECRAFT, capable of approaching within 1 million km of the SUN's surface (PHOTOSPHERE) and providing the first in situ measurements of its CORONA (outer ATMOSPHERE).

Star cluster (Courtesy of NASA and USNO)

starship A conceptual, very advanced SPACE VEHICLE capable of traveling the great distances between STAR systems within decades or less. The term *starship* is generally reserved for vehicles that could carry intelligent beings, while INTERSTELLAR PROBE applies to an advanced ROBOT SPACECRAFT capable of interstellar travel at 10 percent or more of the SPEED OF LIGHT.

station keeping The sequence of maneuvers that maintains a SPACE VEHICLE or SPACECRAFT in a predetermined ORBIT or on a desired TRAJECTORY.

steady state The condition of a physical system in which parameters of importance do not vary significantly with time.

steady-state universe A COSMOLOGY model (based on the PERFECT COSMOLOGICAL PRINCIPLE) suggesting that the UNIVERSE looks the same to all observers at all times.

Stefan-Boltzmann law The RADIATION HEAT TRANSFER relationship developed by LUDWIG BOLTZMANN and JOSEF STEFAN. It states that the ELECTROMAGNETIC RADIATION emitted by an ideal BLACKBODY radiator per unit time per unit area is proportional to the fourth power of its ABSOLUTE TEMPERATURE.

stellar evolution The different phases in the lifetime of a STAR from its formation out of INTERSTELLAR gas and dust to the time after its nuclear FUSION fuel is exhausted.

steradian (sr) The SI UNIT of solid angle, defined as the solid angle subtended at the center of a sphere by an area of surface equal to the square of the radius. The total surface of a sphere subtends a solid angle of 4π Sr about its center.

Stonehenge A circular ring of large vertical stones topped by capstones located in southern England. It is believed to have been built between 3000 B.C.E. and 1000 B.C.E. as an ancient astronomical CALENDAR. *See also* ARCHAEOLOGICAL ASTRONOMY.

storable propellant ROCKET PROPELLANT (usually liquid) that can be stored for prolonged periods without special temperature or pressure environments.

Sun Humans' parent STAR (SOL), the central CELESTIAL BODY in the SOLAR SYSTEM. It is a MAIN-SEQUENCE STAR of SPECTRAL CLASSIFICATION G2 that derives most of its ENERGY from the FUSION of HYDROGEN into HELIUM. *See also* SOLAR CONSTANT; SOLAR ENERGY; SOLAR RADIATION.

sunlike star A yellow, G SPECTRAL CLASSIFICATION, MAIN-SEQUENCE STAR with a surface temperature between 5,000 K and 6,000 K.

sunspot A relatively dark, sharply defined region on the SUN's visible surface that represents a magnetic area. The sunspot's umbra (darkest region) is about 2,000 K cooler than the effective temperature of the PHOTOSPHERE (some 5,800 K). It is surrounded by a less dark region called the penumbra. Sunspots generally occur in groups of two or more and have diameters ranging from 4,000 km to over 200,000 km.

sunspot cycle The approximately 11-year cycle in the variation of the number of SUNSPOTs. A reversal in the Sun's magnetic polarity also occurs with each successive sunspot cycle, creating a 22-year solar magnetic cycle.

sun-synchronous orbit A very useful POLAR ORBIT that allows a SATELLITE's SENSOR to maintain a fixed relation to the SUN during each local data collection—an important feature for EARTH-OBSERVING SPACECRAFT.

supergiant The largest and brightest type of STAR, with a LUMINOSITY of 10,000 to 100,000 times that of the SUN.

superior conjunction *See* CONJUNCTION.

superior planets Planets that have an ORBIT around the SUN outside EARTH's orbit—MARS, JUPITER, SATURN, URANUS, NEPTUNE, and PLUTO.

superluminal With a (hypothetical) speed greater than the SPEED OF LIGHT.

supernova The catastrophic explosion of a massive STAR at the end of its life cycle. As the star collapses and explodes, it creates (by NUCLEOSYNTHESIS) heavier ELEMENTs that are scattered into space. Its brightness increases several million times in a matter of days and outshines all other objects in its GALAXY.

surveillance satellite An EARTH-orbiting military SATELLITE that watches regions of the PLANET for hostile military activities, such as BALLISTIC MISSILE LAUNCHes and nuclear weapons detonations. *See also* DEFENSE SUPPORT PROGRAM.

Surveyor Project The NASA MOON exploration effort in which five LANDER SPACECRAFT softly touched down onto the LUNAR surface between 1966–68—the robot precursor to the APOLLO PROJECT human expeditions.

synchronous orbit An ORBIT around a PLANET (or PRIMARY BODY) in which a SATELLITE (secondary body) moves around the planet in the same amount of time the planet takes to rotate on its AXIS. *See also* GEOSYNCHRONOUS ORBIT.

synchronous rotation The ROTATION of a MOON around its PRIMARY BODY (PLANET) in which the orbital PERIOD equals the period of rotation of the NATURAL SATELLITE about its own AXIS. As a result, the moon always presents the same side (face) to the parent planet. The EARTH-Moon orbital relationship is an example.

synchronous satellite An equatorial west-to-east satellite orbiting EARTH at an ALTITUDE of approximately 35,900 km. At this altitude, the satellite makes one REVOLUTION in 24 hours and remains synchronous with Earth's ROTATION. *See also* GEOSYNCHRONOUS ORBIT.

synchrotron radiation ELECTROMAGNETIC RADIATION given off when very high-energy ELECTRONS traveling at nearly the SPEED OF LIGHT encounter magnetic fields.

synthetic aperture radar A space-based radar system that computer correlates the echoes of signals emitted at different points along a SATELLITE'S ORBIT—thereby mimicking the performance of a radar antenna system many times larger than the one actually being used. *See also* RADARSAT.

tail *(cometary)* The long, wispy portion of some COMETS, containing the gas (plasma tail) and dust (dust tail) streaming out of the comet's head (COMA) as it approaches the SUN. The plasma tail interacts with the SOLAR WIND and points straight back from the Sun, while the dust tail can be curved and fan shaped.

telecommunications The transmission of information over great distances using RADIO WAVES or other portions of the ELECTROMAGNETIC (EM) SPECTRUM. *See also* COMMUNICATIONS SATELLITE.

telemetry The process of taking measurements at one point and transmitting the information via RADIO WAVES over some distance to another location for evaluation and use. Telemetered data on a SPACECRAFT'S communications DOWNLINK often includes scientific data as well as spacecraft state-of-health data.

teleoperation The technique by which a human controller operates a versatile robot system that is at a distant, often hazardous, location. High-resolution visual and tactile sensors on the robot, reliable TELECOMMUNICATIONS links, and computer-generated virtual reality displays enable the human worker to experience TELEPRESENCE.

telepresence The process, supported by an information-rich control station environment, that enables a human controller to manipulate a distant robot through TELEOPERATION and almost feel physically present in the robot's remote location.

telescope An instrument that collects ELECTROMAGNETIC RADIATION from a distant object so as to form an IMAGE of the object or to permit the radiation signal to be analyzed. Optical (astronomical) TELESCOPES are divided into two general classes: REFRACTING TELESCOPES and REFLECTING TELESCOPES. EARTH-based astronomers also use large RADIO TELESCOPES. Orbiting observatories use optical, INFRARED RADIATION, ULTRAVIOLET RADIATION, X-RAY, and GAMMA RAY telecopes to study the UNIVERSE.

terminator The distinctive boundary line separating the illuminated (that is, sunlit) and dark portions of a nonluminous CELESTIAL BODY like the MOON.

Terra The first in a new family of sophisticated NASA EARTH-OBSERVING SPACECRAFT, successfully placed into POLAR ORBIT on 18 December 1999 from VANDENBERG AIR FORCE BASE, California.

terrestrial Of or relating to EARTH.

terrestrial planets In addition to EARTH, the PLANETS MERCURY, VENUS, and MARS—all of which are relatively small, high-DENSITY CELESTIAL BODIES, composed of metals and silicates, and with shallow or no ATMOSPHERES in comparison with the JOVIAN PLANETS.

thermonuclear *See* FUSION.

Thor An INTERMEDIATE-RANGE BALLISTIC MISSILE (IRBM) developed by the U.S. Air Force in the late 1950s; also used by NASA and the military as a SPACE LAUNCH VEHICLE.

throttling The variation of the THRUST of a ROCKET engine during powered flight.

thrust *(T)* The forward FORCE provided by a REACTION ENGINE, such as a ROCKET.

Titan (1) *(launch vehicle)* The family of powerful U.S. Air Force BALLISTIC MISSILES and EXPENDABLE LAUNCH VEHICLES that began in 1955 with the *Titan I*—the first American two-stage INTERCONTINENTAL BALLISTIC MISSILE (ICBM). The *Titan IV* is the newest and most powerful member.

Titan (launch vehicle) (Courtesy of USAF/Lockheed—Martin Company)

(2) *(moon)* The largest MOON of SATURN, discovered in 1655 by CHRISTIAAN HUYGENS. It is the only SATELLITE in the SOLAR SYSTEM with a significant ATMOSPHERE.

tracking Following the movement of a SATELLITE, ROCKET, or AEROSPACE VEHICLE. It is usually performed with optical, INFRARED RADIATION, RADAR ASTRONOMY, or RADIO WAVE systems.

trajectory The three-dimensional path traced by any object moving because of an externally applied FORCE; the flight path of a SPACE VEHICLE.

transfer orbit An elliptical INTERPLANETARY TRAJECTORY tangent to the ORBITs of both the departure PLANET and target planet (or MOON). *See also* HOHMANN TRANSFER ORBIT.

transit *(planetary)* The passage of one CELESTIAL BODY in front of another (larger-diameter) celestial body, such as VENUS across the face of the SUN.

Trans-Neptunian object (TNO) Any of the numerous small, icy CELESTIAL BODIES that lie in the outer fringes of the SOLAR SYSTEM beyond NEPTUNE. TNOs include PLUTINOS and KUIPER BELT objects.

tritium (T or $^3_1$H) The RADIOISOTOPE of HYDROGEN with two NEUTRONS and one PROTON in the NUCLEUS. It has a HALE-LIFE of 12.3 years.

Triton The largest MOON of NEPTUNE.

Trojan Group The collection of ASTEROIDS found near the two LAGRANGIAN LIBRATION POINTS in JUPITER's ORBIT around the SUN. Many of these MINOR PLANETS were named after the mythical heroes of the Trojan War.

Tunguska event A violent explosion that occurred in a remote area of Siberia in late June 1908 that some scientists now attribute to the IMPACT of an extinct cometary NUCLEUS or a large, stony METEORITE.

turbopump system The high-speed pumping equipment in a LIQUID-PROPELLANT ROCKET ENGINE, designed to raise the pressure of the PROPELLANTs (fuel and OXIDIZER) so they can go from the tanks into the COMBUSTION CHAMBER at specified flow rates.

ullage The amount that a container, such as a PROPELLANT tank, lacks of being full.

ultraviolet astronomy The branch of ASTRONOMY, conducted primarily from space-based observatories, that uses the ultraviolet portion of the ELECTROMAGNETIC (EM) SPECTRUM to study unusual INTERSTELLAR and intergalactic phenomena.

ultraviolet radiation (UV) The region of the ELECTROMAGNETIC (EM) SPECTRUM between visible (violet) LIGHT and X RAYS, with wavelengths from 400 NANOMETERS (nm) (just past violet light) down to about 10 nm (the extreme ultraviolet cutoff).

umbilical An electrical or fluid-servicing line between the ground or tower and an upright ROCKET vehicle before LAUNCH.

universal time (UT) The worldwide civil time standard, equivalent to GREENWICH MEAN TIME.

universe Everything that came into being at the moment of the BIG BANG and everything that has evolved since then. It includes all ENERGY, all matter, and the space-time continuum that contains them.

uplink The TELEMETRY signal sent from a ground station to a SPACECRAFT or PLANETARY PROBE.

upper stage The second, third, or later ROCKET stage of a MULTISTAGE ROCKET vehicle. Once lifted into LOW EARTH ORBIT, a spacecraft often uses an attached upper stage to reach its final destination—a higher ALTITUDE ORBIT around EARTH or an INTERPLANETARY TRAJECTORY. *See also* INERTIAL UPPER STAGE; PAYLOAD ASSIST MODULE; STAGING.

Uranus The seventh PLANET from the SUN, unknown until discovered with a TELESCOPE in 1781 by SIR (FREDERICK) WILLIAM HERSCHEL. Its AXIS of ROTATION lies in the plane of its orbit around the Sun rather than vertical to the orbital plane, as occurs with the other planets.

U.S. Naval Observatory (USNO) The astronomical observatory founded by the U.S. government in Washington, D.C., in 1844.

Valles Marineris An extensive canyon system on MARS near the PLANET's EQUATOR, discovered in 1971 by NASA's *MARINER 9* SPACECRAFT.

Van Allen radiation belts *See* EARTH'S TRAPPED RADIATION BELTS; *EXPLORER I.*

Vandenberg Air Force Base (VAFB) Located on the central California coast north of Santa Barbara, this U.S. Air Force facility is the LAUNCH SITE of all military, NASA, and commercial space launches that require high INCLINATION, especially POLAR ORBITS.

variable star A STAR that does not shine steadily but whose brightness (LUMINOSITY) changes over a short period of time.

vector A physical quantity, such as FORCE, VELOCITY, or ACCELERATION, that has both magnitude and direction at each point in space. It is contrasted with a scalar quantity, which has just magnitude.

velocity A VECTOR quantity that describes the rate of change of position. Velocity has both magnitude (speed) and direction. It is expressed in units of length per unit of time (for example, kilometers per second).

Vandenberg Air Force Base (Courtesy of USAF)

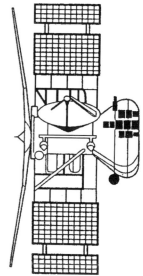

Venera 15

Venera The family of Russian SPACECRAFT (FLYBYS, ORBITERS, PROBES, and LANDERS) that successfully explored VENUS, including its inferno-like surface, between 1961–84.

vengence weapon 2 (V-2) The V-2 or *Vergeltungwaffe 2* was the first modern military BALLISTIC MISSILE. This LIQUID-PROPELLANT ROCKET was designed and flown by the German Army during World War II and then became the technical ancestor for many large American and Russian rockets constructed during the COLD WAR.

Venus (Courtesy of NASA)

Venus The cloud-enshrouded second PLANET from the SUN whose ATMOSPHERE portrays an inferno-like RUNAWAY GREENHOUSE effect.

Venusian Of or pertaining to the PLANET VENUS.

vernal equinox The spring EQUINOX, which occurs on or about 21 March.

vernier engine A ROCKET engine of small THRUST used primarily to obtain a fine adjustment in the VELOCITY and TRAJECTORY or in the ATTITUDE of a rocket or AEROSPACE VEHICLE.

Very Large Array (VLA) A spatially extended RADIO TELESCOPE complex near Socorro, New Mexico, operated by the NATIONAL RADIO ASTRONOMY OBSERVATORY.

Viking Project NASA's highly successful MARS exploration effort in the 1970s in which two ORBITER and two LANDER SPACECRAFT conducted the first detailed study of the Martian environment and the first (albeit inconclusive) scientific search for life on the RED PLANET.

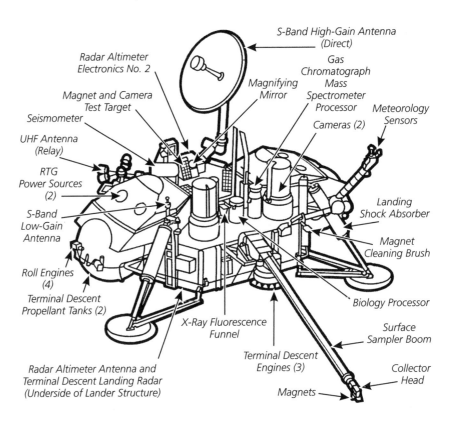

**Viking Project (lander)
(Courtesy of NASA)**

visible radiation *See* LIGHT.

volcano A vent in the crust of a PLANET or MOON from which molten lava, gases, and other pyroclastic materials flow.

Vostok The first Russian crewed SPACECRAFT. With room for just a single COSMONAUT, the spacecraft's small spherical cabin housed YURI A. GAGARIN in 1961 when he became the first human to fly in OUTER SPACE.

Voyager NASA's twin SPACECRAFT that explored the outer regions of the SOLAR SYSTEM, visiting all the JOVIAN PLANETS. *Voyager 1* encountered JUPITER (1979) and SATURN (1980) before departing on an INTERSTELLAR TRAJECTORY. *Voyager 2* performed the historic grand tour by visiting Jupiter (1979), Saturn (1981), URANUS (1986), and NEPTUNE (1989). Both RTG-powered spacecraft are now involved in the Voyager Interstellar Mission (VIM). Each carries a special recording ("Sounds of Earth")—a digital message for any intelligent species that finds them drifting between the STARS millennia from now.

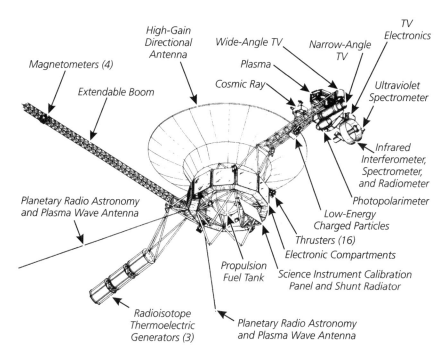

Voyager (Courtesy of NASA)

Vulcan The hypothetical (but nonexistent) planet that some 19th-century astronomers believed existed in an extremely hot orbit between MERCURY and the SUN.

warhead The PAYLOAD of a BALLISTIC MISSILE or military ROCKET, usually a nuclear weapon or high explosive.

watt (W) The SI UNIT of POWER (that is, WORK per unit time); 1 W represents one JOULE of ENERGY per second.

wavelength (λ) The mean distance between maxima (or minima) of a periodic pattern; the distance between two crests of a propagating wave of a single FREQUENCY (v). The wavelength is related to frequency and phase speed (c) by the formula $\lambda = c/v$. (Note, here c is the speed of propagation of the wave disturbance.)

wave number The reciprocal of the WAVELENGTH, $1/\lambda$.

weather satellite An EARTH-OBSERVING SPACECRAFT that carries a variety of special environmental sensors to observe and measure atmospheric properties and processes. Operational weather satellites are in GEOSTATIONARY ORBIT and in POLAR ORBIT—each with a different capability and purpose. Also called METEOROLOGICAL SATELLITE or ENVIRONMENTAL SATELLITE.

Wavelength

Wavelength

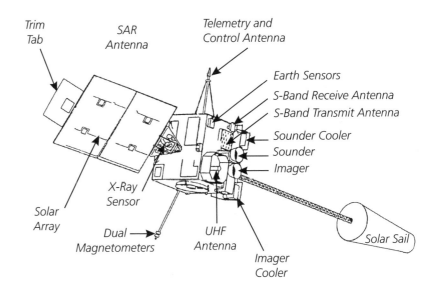

Trim Tab

SAR Antenna

Telemetry and Control Antenna

Earth Sensors

S-Band Receive Antenna

S-Band Transmit Antenna

Sounder Cooler

Sounder

Imager

Solar Array

X-Ray Sensor

Dual Magnetometers

UHF Antenna

Imager Cooler

Solar Sail

Weather satellite (Courtesy of NASA and NOAA)

weight *(w)* The FORCE of attraction with which an object is pulled toward EARTH (or another planetary body) by its GRAVITY. It is the product of the object's MASS *(m)* and the local ACCELERATION OF GRAVITY *(g)*.

weightlessness The condition of FREE FALL (or ZERO *G* in which objects inside an EARTH-orbiting, unaccelerated SPACECRAFT appear weightless even though the objects and the spacecraft are still under the influence of Earth's GRAVITY. It is the condition in which no ACCELERATION, whether of GRAVITY or other FORCE, can be detected by an observer within the system in question. *See also* MICROGRAVITY.

wet emplacement A LAUNCH PAD that provides a deluge of water for cooling the FLAME BUCKET and other equipment during a ROCKET firing.

white dwarf *(star)* A compact STAR at the end of its life cycle. Once a star of one SOLAR MASS or less exhausts its nuclear fuel, it collapses under GRAVITY into a very dense object about the size of EARTH.

WIMP A hypothetical PARTICLE, called the *w*eakly *i*nteracting *m*assive *p*article, thought by some scientists to pervade the UNIVERSE as it is hard to observe DARK MATTER. *See also* MACHO.

W. M. Keck Observatory *See* KECK OBSERVATORY.

work *(w)* The ENERGY associated with a FORCE acting through a specific distance. The SI UNIT of work is the JOULE.

X ray A penetrating form of ELECTROMAGNETIC RADIATION of very short WAVELENGTH (approximately 0.01 to 10 NANOMETERS) and high PHOTON ENERGY (approximately 100 ELECTRON VOLTS to some 100 kiloelectron volts.)

X-ray astronomy The branch of ASTRONOMY, primarily space based, that uses characteristic X-RAY emissions to study very energetic and violent processes throughout the UNIVERSE. X-ray emissions carry information about the temperature, density, age, and other physical conditions of celestial objects that produced them—including SUPERNOVA remnants, PULSARS, ACTIVE GALAXIES, and energetic SOLAR FLARES. *See also* CHANDRA X-RAY OBSERVATORY.

yaw The rotation or oscillation of a MISSILE or AEROSPACE VEHICLE about its vertical axis so as to cause the longitudinal axis of the vehicle to deviate from the flight line or heading in its horizontal plane. *See also* PITCH; ROLL.

year The period of one REVOLUTION of EARTH around the SUN; but the choice of the celestial reference point determines the precise length of the year. The civil calendar year (based on the GREGORIAN CALENDAR) is 365.2425 days.

Yohkoh   A Japanese SOLAR X-RAY observation SATELLITE launched in 1991.

young stellar object (YSO) Any celestial object in an early stage of STAR formation, from a PROTOSTAR to a MAIN-SEQUENCE STAR.

zenith   The point on the CELESTIAL SPHERE vertically overhead. *Compare with* NADIR.

Zenith   A three-stage Russian EXPENDABLE LAUNCH VEHICLE capable of placing about 14,000 kg MASS PAYLOADS into LOW EARTH ORBIT.

zero-g   A common (but imprecise) term for the condition of continuous FREE FALL and apparent WEIGHTLESSNESS experienced by ASTRONAUTS and objects in an EARTH-ORBITING SPACECRAFT. *See also* MICROGRAVITY.

zodiac   From the Greek word meaning "circle of figures." Early astronomers described the band in the sky about 9° on each side of the ECLIPTIC, which they divided into 30° intervals—each representing a sign of the zodiac. Within their GEOCENTRIC COSMOLOGY, the SUN appeared to enter a different CONSTELLATION of the zodiac each month. So the signs of the zodiac helped them mark the annual REVOLUTION of EARTH around the SUN. *See also* SECTION IV CHARTS & TABLES.

zodiacal light A faint cone of LIGHT extending upward from the horizon in the direction of the ECLIPTIC caused by the REFLECTION of sunlight from tiny pieces of INTERPLANETARY DUST in ORBIT around the SUN.

Zond   A family of Russian SPACECRAFT that explored the MOON, MARS, VENUS, and INTERPLANETARY SPACE in the 1960s.

zulu time The U.S. military expression for 24-hour clock time based on GREENWICH MEAN TIME (GMT).

SECTION TWO
BIOGRAPHIES

Adams, John Couch (1819–92) English astronomer who is cocredited with the mathematical discovery of NEPTUNE. From 1843 to 1845, he investigated irregularities in the orbit of URANUS and predicted the existence of a PLANET beyond. However, his work was ignored until the French astronomer URBAIN JEAN JOSEPH LEVERRIER made similar calculations that enabled the German astronomer JOHANN GOTTFRIED GALLE to discover Neptune on 23 September 1846. He was a professor of astronomy and also the director of the Cambridge Observatory.

Adams, Walter Sydney (1876–1956) American astronomer who specialized in stellar spectroscopic studies and who developed a technique for determining a STAR's real MAGNITUDE from its SPECTRUM. In 1932, his spectroscopic studies of the Venusian ATMOSPHERE showed that it was rich in carbon dioxide (CO_2). From 1923 to 1946, he was the director of the Mount Wilson Observatory in California.

Airy, Sir George Biddell (1801–92) English astronomer who modernized the ROYAL GREENWICH OBSERVATORY and served as the seventh ASTRONOMER ROYAL (1835–81). Despite many professional accomplishments, he is often remembered for his failure to recognize the value of the theoretical work of JOHN COUCH ADAMS, thereby delaying the discovery of NEPTUNE until 1846.

Aitken, Robert Grant (1864–1951) American astronomer who worked at the Lick Observatory in California and specialized in BINARY (DOUBLE) STAR SYSTEMS. In 1932, he published *New General Catalog of Double Stars,* a seminal work that contains measurements of over 17,000 double stars and is often called the *Aitken Double Star Catalog (ADSC).*

Al-Battani (aka Albategnius) (858–929) Arab astronomer and mathematician who improved solar, lunar, and planetary (five) motion data found in PTOLEMY's *ALMAGEST* with more accurate measurements. He also introduced the use of trigonometry in observational astronomy. In 880, he produced a major STAR catalog and refined the length of the YEAR to approximately 365.24 days. His precise observations influenced the medieval astronomers of western Europe. His full Arab name is Abu-'Abdullah Muhammad Ibn Jabir Ibn Sinan Al-Battani.

Aldrin, Edwin E. "Buzz," Jr. (b. 1930) American ASTRONAUT and United States Air Force officer (colonel) who served as the lunar module pilot for NASA's *APOLLO 11* mission, which took place 16–24 July 1969. This mission was the first human-landing mission on the MOON. He followed Astronaut Neil Armstrong onto the lunar surface on 20 July 1969, becoming the second person in history to walk on another world.

Alfonso X of Castile (1221–84) Scholarly Spanish monarch who ordered the compilation of revised planetary tables based on Arab astronomy (the *ALMAGEST* of PTOLEMY) but updated with observations made at Toledo, Spain. The *Alfonsine Tables* were completed in 1272 and served medieval astronomers for over three centuries.

Edwin E. "Buzz" Aldrin, Jr.
(Courtesy of NASA)

Alfvén, Hannes Olof Gösta (1908–95) Swedish physicist who developed the theory of magnetohydrodynamics that helped explain sunspot formation and magnetic field-plasma interactions (Alfvén waves) taking place in the outer regions of the SUN and other STARS. He shared the 1970 Nobel Prize in physics for this work.

Alpher, Ralph Asher (b. 1921) American theoretical physicist who collaborated with GEORGE GAMOW (and other scientists) in 1948 to develop the BIG BANG theory of the origin of the UNIVERSE. He also extended the nuclear transmutation theory (NUCLEOSYNTHESIS) within a very hot, early fireball to predict the existence of a now-cooled, residual, COSMIC MICROWAVE BACKGROUND RADIATION.

Al-Sufi, Abd al-Rahman (903–86) Arab astronomer who produced a STAR catalog in about 964 based on PTOLEMY's *ALMAGEST* but that included some of his own observations—including the first reference to the ANDROMEDA GALAXY (reported as a fuzzy NEBULA).

Alvarez, Luis Walter (1911–88) American physicist and Nobel laureate who collaborated with his son Walter (b. 1940), a geologist, to develop the EXTRATERRESTRIAL CATASTROPHE THEORY (or Alvarez hypothesis) in which a large ASTEROID or COMET struck EARTH some 65 million years ago, causing a mass extinction of life, including the dinosaurs. In 1980, he discovered a worldwide enrichment of iridium in the thin sediment layer between the Cretaceous-Tertiary periods—an unusual elemental abundance attributed to the impact of an EXTRATERRESTRIAL mineral source.

Al-Wefa, Abu'l (940–98) Arab mathematician and astronomer who developed spherical trigonometry and worked in the Baghdad Observatory, constructed by Muslim Prince Sharaf al-Dawla. He introduced the use of tangent and cotangent functions in his astronomical activities, which included careful observations of SOLSTICEs and EQUINOXes.

Ambartsumian, Viktor Amazaspovich (1908–96) Armenian astrophysicist who founded BYURAKAN ASTROPHYSICAL OBSERVATORY in 1946 on Mount Aragatz near Yerevan, Armenia. This facility served as one of the major astronomical observatories in the former Soviet Union. His major contributions involved theories concerning the origin and evolution of STARS.

Amici, Giovanni Battista (1786–1863) Italian astronomer and optician who developed better mirrors for REFLECTING TELESCOPEs. While serving as astronomer to the grand duke of Tuscany, he used his improved telescopes to make more refined observations of the Jovian moons and of selected BINARY (DOUBLE) STAR SYSTEMS.

Anaxagoras (c. 500 B.C.E.–c. 428 B.C.E.) Greek philosopher and early cosmologist who speculated that the SUN was really a huge, incandescent rock; the planets and stars flaming rocks; and the MOON had materials similar to EARTH, possibly including inhabitants. For these bold hypotheses, he was charged with religious impiety by Athenian authorities and banished from the city. His COSMOLOGY included the concept that "mind" (Greek word *nous,* νους) formed the material objects of the UNIVERSE.

Anaximander (c. 610 B.C.E.–c. 546 B.C.E.) Greek philosopher and astronomer who was the earliest Hellenistic thinker to propose a systematic worldview. Within his COSMOLOGY, a stationary, cylindrically shaped EARTH floated freely in space at the center of the UNIVERSE. By recognizing that the heavens appeared to revolve around the North Star (Polaris), he used a complete sphere to locate CELESTIAL BODIES (that is, STARs and PLANETs) in the night sky. He also proposed *apeiron* (the Greek word for unlimited or infinite) as the source element for all material objects—a formless, imperceptible substance from which all things originate and back into which all things return.

Anders, William A. (b. 1933) American ASTRONAUT and United States Air Force officer who was a member of NASA's *APOLLO 8* mission (21–27 December 1968), history's first crewed flight to the vicinity of the MOON. Along with FRANK BORMAN and JAMES ARTHUR LOVELL, JR., he flew to the Moon in the Apollo SPACECRAFT, completed 10 orbital REVOLUTIONs of the Moon, and then returned safely to EARTH.

Ångström, Anders Jonas (1814–74) Swedish physicist and astronomer who performed pioneering spectral studies of the SUN. In 1862, he discovered that hydrogen was present in the Sun's ATMOSPHERE and went on to publish a detailed map of the solar SPECTRUM, covering other elements as well. A special unit of WAVELENGTH, the ANGSTROM (Å), now honors his accomplishments in SPECTROSCOPY.

Antoniadi, Eugene Michael (1870–1944) French astronomer who made detailed studies of the PLANETs, especially MARS, in the post–GIOVANNI VIRGINIO SCHIAPARELLI period. He developed a popular viewing index (the Antoniadi scale) that qualitatively describes the suitability of terrestrial atmospheric motion conditions for observing the planets.

Apollonius of Perga (c. 262 B.C.E.–c. 190 B.C.E.) Greek mathematician who developed the theory of CONIC SECTIONS (including the circle, ELLIPSE, parabola, and hyperbola) that allowed JOHANNES KEPLER and SIR ISAAC NEWTON to describe the motion of CELESTIAL BODIES in the SOLAR SYSTEM accurately. He also created the key mathematical treatment that enabled HIPPARCHUS OF NICAEA and (later) PTOLEMY to promote a geocentric EPICYCLE theory of planetary motion.

Arago, Dominique François (1786–1853) French scientist and statesman who developed instruments to study the

POLARIZATION of LIGHT. He had a special interest in polarized light from COMETs and determined through careful observation that COMET HALLEY (1835 passage) and other comets were not self-luminous. His compendium, *Popular Astronomy,* extended scientific education to a large portion of the European middle class.

Argelander, Friedrich Wilhelm August (1799–1875) German astronomer who investigated VARIABLE STARS and compiled a major telescopic (but prephotography) survey of all the STARS in the Northern Hemisphere brighter than the ninth MAGNITUDE. From 1859 to 1862, he published the four-volume *Bonn Survey (Bonner Durchmusterung),* containing over 324,000 stars.

Aristarchus of Samos (c. 320 B.C.E.–c. 250 B.C.E.) Greek mathematician and astronomer who was the first to suggest that EARTH not only revolved on its AXIS but also traveled around the SUN along with the other known PLANETs. Unfortunately, he was severely criticized for his bold hypothesis of a moving Earth. At the time, Greek society favored the teachings of ARISTOTLE and others who advocated a geocentric COSMOLOGY with an immovable Earth at the center of the UNIVERSE. Almost 18 centuries would pass before NICHOLAS COPERNICUS revived HELIOCENTRIC (Sun-centered) cosmology.

Aristotle (384 B.C.E.–322 B.C.E.) Greek philosopher who was one of the most influential thinkers in the development of Western civilization. He endorsed and embellished the geocentric (Earth-centered) COSMOLOGY of EUDOXUS OF CNIDUS. In Aristotelian cosmology, a nonmoving EARTH was surrounded by a system of 49 concentric, transparent (crystal) spheres, each helping to account for the motion of all visible CELESTIAL BODIES. The outermost sphere contained the FIXED STARS. PTOLEMY later replaced Aristotle's spheres with a system of EPICYCLES. Modified, but unchallenged, Aristotelian cosmology dominated Western thinking for almost two millennia—until finally surrendering to the HELIOCENTRIC (Sun-centered) cosmology of NICHOLAS COPERNICUS, JOHANNES KEPLER, and GALILEO GALILEI.

Armstrong, Neil A. (b. 1930) American ASTRONAUT who served as the commander for NASA's *APOLLO 11* lunar-landing mission in July 1969. As he became the first human being to set foot on

Neil A. Armstrong
(Courtesy of NASA)

the MOON (20 July 1969), he uttered these historic words, "That's one small step for a man, one giant leap for mankind."

Arp, Halton Christian (b. 1927) American astronomer who published the *Atlas of Peculiar Galaxies* in 1966 and also proposed a controversial theory concerning REDSHIFT phenomena associated with distant QUASARs.

Arrhenius, Svante August (1859–1927) Swedish physical chemist and Nobel laureate (Chemistry 1903) with multidisciplinary interests who suggested that life could be abundant in the UNIVERSE. In his book *Worlds in the Making* (1908), he introduced the panspermia hypothesis. This was a bold speculation that life could be spread through outer space from PLANET to planet or even from STAR system to star system by the diffusion of spores, bacteria, or other microorganisms. He was also one of the first scientists (circa 1895) to associate the presence of heat-trapping gases, such as carbon dioxide, in a planet's ATMOSPHERE with the GREENHOUSE EFFECT.

Baade, (Wilhelm Heinrich) Walter (1893–1960) German-American astronomer who carefully studied CEPHEID VARIABLES, enabling him to double the estimated distance, age, and scale of the UNIVERSE. In 1942, he showed that the ANDROMEDA GALAXY was over 2 million LIGHT-YEARS away—more than twice the previously accepted distance. He also discovered that STARS occupy two basic populations or groups. POPULATION I STARS are younger, bluish stars found in the outer regions of a GALAXY. POPULATION II STARS are older, reddish stars found in the central regions of a galaxy. This categorization significantly advanced stellar evolution theory.

Babcock, Harold Delos (1882–1968) American astronomer and physicist who, along with his son Horace (b. 1912), invented the SOLAR magnetograph in 1951. This instrument allowed them and other solar physicists to make detailed (pre–space age) investigations of the SUN's magnetic field.

Baily, Francis (1774–1844) English astronomer and stockbroker who discovered the beads of light phenomenon (now called BAILY'S BEADS) that occurs around the LUNAR disk during a total

ECLIPSE of the SUN. He made this discovery on 15 May 1836, and his work stimulated a great deal of interest in eclipses.

Barnard, Edward Emerson (1857–1923) American astronomer who discovered JUPITER's fifth MOON, AMALTHEA, in 1892 and pioneered the use of photography in astronomy. He was the first to find a COMET by photographic means. He also discovered BARNARD'S STAR, a faint RED DWARF with a large PROPER MOTION, situated about six LIGHT-YEARS from the SUN.

Bayer, Johann (1572–1625) German astronomer who published *Uranometria* in 1603—the first major STAR catalog for the entire CELESTIAL SPHERE. He charted over 2,000 stars visible to the NAKED EYE and introduced the practice of assigning Greek letters (such as alpha α, beta β, and gamma γ) to the main stars in each CONSTELLATION, usually in an approximate (descending) order of their brightness.

Bean, Alan L. (b. 1932) American ASTRONAUT and U.S. Navy officer who commanded the LUNAR EXCURSION MODULE (LEM), during NASA's *APOLLO 12* mission. Along with Pete Conrad, he walked on the LUNAR surface and deployed instruments in November 1969 as part of the second MOON landing mission. He also served as SPACECRAFT commander for the *SKYLAB II* SPACE STATION mission in 1973. These spaceflight experiences inspired his work as a space artist.

Bell Burnell, (Susan) Jocelyn (b. 1943) English astronomer who, in August 1967, while still a doctoral student at Cambridge University under the supervision of ANTHONY HEWISH, discovered an unusual, repetitive radio signal that proved to be the first PULSAR. She received her Ph.D. for her pioneering work on pulsars, while Hewish and a Cambridge colleague (SIR MARTIN RYLE) eventually shared the 1974 Nobel Prize in physics for their work on pulsars. Possibly because she was a student at the time of her discovery, the Nobel committee rather unjustly ignored her contribution.

Bernal, John Desmond (1910–71) Irish physicist and writer who speculated about the colonization of space and the construction of very large, spherical SPACE SETTLEMENTs (now called

BERNAL SPHERES) in his futuristic 1929 work *The World, the Flesh and the Devil.*

Bessel, Friedrich Wilhelm (1784–1846) German astronomer and mathematician who pioneered precision astronomy and was the first to measure accurately the distance to a STAR (other than the SUN). In 1818, he published *Fundamenta Astronomiae,* a catalog of over 3,000 stars. By 1833, he completed a detailed study from the Königsberg Observatory of 50,000 stars. His greatest accomplishment occurred in 1838, when he carefully observed the BINARY (DOUBLE) STAR SYSTEM 61 Cygni and used its PARALLAX (annual angular displacement) to estimate its distance at about 10.3 LIGHT-YEARS from the Sun (the current value is 11.3 light-years). His mathematical innovations and rigorous observations greatly expanded the scale of the UNIVERSE and helped shift astronomical interest beyond the SOLAR SYSTEM.

Bethe, Hans Albrecht (b. 1906) German-American physicist and Nobel laureate who proposed the mechanisms by which STARS generate their vast quantities of energy through the NUCLEAR FUSION of hydrogen into helium. In 1938, he worked out the sequence of nuclear fusion reactions, called the CARBON CYCLE, that dominate energy liberation in stars more massive than the SUN. Then, while working with a Cornell University colleague (Charles Critchfield), he proposed the PROTON-PROTON REACTION as the nuclear fusion process for stars up to the size of the Sun. For this astrophysical work, he received the 1967 Nobel Prize in physics.

Biela, Wilhelm von (1782–1856) Austrian military officer and amateur astronomer who rediscovered a short-period COMET in 1826 and then calculated its orbital period (about 6.6 years). During the predicted 1846 PERIHELION passage of Biela's comet, its NUCLEUS split into two. Following the 1852 return of Biela's double comet, the object disappeared, apparently disintegrating into an intense METEOR shower. The dynamics and disappearance of Biela's comet demonstrated to astronomers that comets were transitory, finite objects.

Biermann, Ludwig (1907–86) German astronomer who theoretically predicted in 1951 the existence of the SOLAR WIND and its influence on the dynamics of ionized COMET tails.

Biot, Jean-Baptiste (1774–1862) French physicist who examined specimens of the METEORITE that fell onto L'Aigle, France, in 1803 and concluded that meteorites are of extraterrestrial origin. During a balloon flight in 1804, he investigated how EARTH's magnetic field varied with ALTITUDE.

Birkeland, Kristian Olaf Bernhard (1867–1917) Norwegian physicist who, early in the 20th century, deployed instruments during polar region expeditions and then suggested that AURORAS were electromagnetic in nature, created by an interaction of EARTH's MAGNETOSPHERE with a flow of charged particles from the SUN. In the 1960s, instruments onboard early Earth-orbiting SATELLITES confirmed his hypothesis.

Bliss, Nathanial (1700–64) English astronomer who worked with JAMES BRADLEY at the ROYAL GREENWICH OBSERVATORY and briefly served as the fourth ASTRONOMER ROYAL from 1762 to 1764.

Bluford, Guion S., Jr. (b. 1942) American ASTRONAUT and U.S. Air Force officer who served as a NASA mission specialist on three SPACE SHUTTLE missions. During the space shuttle STS-8 mission launched from the KENNEDY SPACE CENTER (KSC) on 30 August 1983, he became the first African American to fly in space.

Bode, Johann Elert (1747–1826) German astronomer who publicized an empirical formula that approximated the average distance to the SUN of each of the six PLANETs known in 1772. This formula, often called Bode's law, is only a convenient mathematical relationship and does not describe a physical principle or natural phenomenon. Furthermore, Bode's empirical formula was actually discovered in 1766 by JOHANN DANIEL TITIUS. In 1801, Bode published *Uranographia,* a comprehensive listing of over 17,000 STARS and NEBULAS.

Bok, Bartholomeus "Bart" Jan (1906–83) Dutch-American astronomer who investigated the STAR-forming regions of the

MILKY WAY GALAXY in the 1930s and discovered the small, dense, cool (~10 K) DARK NEBULAS, now called Bok globules.

Boltzmann, Ludwig (1844–1906) Austrian physicist who developed statistical mechanics and key thermophysical principles that enabled astronomers to interpret a STAR's SPECTRUM and its LUMINOSITY better. In the 1870s and 1880s, he collaborated with JOSEF STEFAN in discovering an important physical law (now called the STEFAN-BOLTZMANN LAW) that relates the total radiant energy output (luminosity) of a star to the fourth power of its ABSOLUTE TEMPERATURE.

Bond, George Phillips (1825–65) American astronomer who succeeded his father (WILLIAM CRANCH BOND) as the director of the HARVARD COLLEGE OBSERVATORY. He specialized in SOLAR SYSTEM observations, including the 1848 codiscovery (with his father) of Hyperion, a MOON of SATURN. In the 1850s, he pursued developments in ASTROPHOTOGRAPHY, demonstrating that STAR photographs could be used to estimate stellar MAGNITUDES.

Bond, William Cranch (1789–1859) American astronomer who was the founder and first director of the HARVARD COLLEGE OBSERVATORY. While collaborating with his son (GEORGE PHILLIPS BOND), he codiscovered Hyperion, a MOON of SATURN, in 1848 and pioneered ASTROPHOTOGRAPHY by taking images of JUPITER and the STAR Vega in 1850 and the BINARY (DOUBLE) STAR SYSTEM Mizar in 1857.

Bondi, Sir Hermann (b. 1919) Austrian-born British astronomer and mathematician who collaborated with THOMAS GOLD and SIR FRED HOYLE in 1948 to propose a steady-state model of the UNIVERSE. Although scientists now have generally abandoned this STEADY-STATE UNIVERSE hypothesis in favor of the BIG BANG theory, his work stimulated a great deal of beneficial technical discussion within the astrophysical community.

Borman, Frank (b. 1928) American ASTRONAUT and U.S. Air Force officer who commanded NASA's *APOLLO 8* mission (December 1968)—the first mission to take human beings to the vicinity of the MOON. He also served as the commander of NASA's *GEMINI 7* mission in 1965.

Boscovich, Ruggero Giuseppe (1711–87) Croatian Jesuit mathematician and astronomer who developed new methods for determining the ORBITS of PLANETS and their AXES of ROTATION. As a creative, multitalented scientist, he was also an influential science adviser to Pope Benedict XIV, a strong advocate for NEWTON'S LAW OF GRAVITATION, and a pioneer of geodesy and modern atomic theory.

Boss, Benjamin (1880–1970) American astronomer who, in 1937, published the popular, five-volume *Boss General Catalogue,* a modern listing of over 33,000 STARS. His publication culminated the precise star position work initiated in 1912 by his father, LEWIS BOSS, when the elder Boss published the *Preliminary (Boss) General Catalogue,* containing about 6,200 stars.

Boss, Lewis (1846–1912) American astronomer who published the popular *Preliminary (Boss) General Catalogue* in 1912. This work was the initial version of an accurate, modern STAR catalog and contained about 6,200 stars. His son (BENJAMIN BOSS) completed the task by preparing a more extensive edition of the catalog in 1937.

Bouguer, Pierre (1698–1758) French physicist who established the field of experimental photometry, the scientific measurement of light. In 1748, he developed the heliometer, an instrument used by astronomers to measure the diameter of the SUN, the PARALLAX of STARS, and the ANGULAR DIAMETER of PLANETS.

Bowen, Ira Sprague (1898–1973) American physicist who performed detailed investigations of the light spectra from NEBULAS, including certain strong green spectral lines that were originally attributed by other astronomers to a hypothesized new element they called nebulium. In 1927, his detailed studies revealed that the mysterious green lines were actually special (forbidden) transitions of the ionized gases oxygen and nitrogen. His work supported important advances in the spectroscopic study of the SUN, other STARS, and nebulas.

Bradley, James (1693–1762) English astronomer who discovered the ABERRATION OF STARLIGHT in 1728 while attempting to detect stellar PARALLAX. Upon the death of EDMOND HALLEY (1742), he was appointed as the third ASTRONOMER ROYAL. As a skilled observer, in 1748 he announced his discovery of the nutation (small variation in tilt) of EARTH's AXIS.

Brahe, Tycho (1546–1601) Quarrelsome Danish astronomer who is considered the greatest pretelescope astronomical observer. His precise, NAKED EYE records of planetary motions enabled JOHANNES KEPLER to prove that all the PLANETs move in elliptical ORBITs around the SUN. Brahe discovered a SUPERNOVA in 1572 and reported his detailed observations in *De Nova Stella* (1573). In 1576, Danish King Frederick II provided support for him to build a world-class observatory on an island in the Baltic Sea. For two decades, Brahe's Uraniborg (Castle of the Sky) was the great center for observational astronomy. In 1599, German Emperor Rudolf II invited him to move to Prague, where he was joined by Kepler. Brahe, a non-Copernican, advocated his own Tychonic system—a geocentric model in which all the planets (except EARTH) revolved around the Sun, and the Sun and its entire assemblage revolved around a stationary Earth.

Brahmagupta (c. 598–c. 660) Indian mathematician and astronomer who wrote the *Brahmaisiddhanta Siddhanta* around 628, introducing algebraic and trigonometric methods for solving astronomical problems. His work, and that of other Indian scientists of the period, greatly influenced later Arab astronomers and mathematicians.

Braun, Wernher von *See* VON BRAUN, WERNHER.

Bredichin, Fyodor (aka Fedor) Aleksandrovich (1831–1904) Russian astronomer who performed pioneering studies concerning the length, composition, and shape (curvature) of the tails of COMETs. He also investigated the structure and behavior of METEORs.

Brezhnev, Leonid I. (1906–82) Russian political leader who was the First Secretary of the Communist Party of the (former) Soviet Union between 1964 and 1982 and Soviet leader during the country's COLD WAR era lunar exploration program. He was also responsible for the development of the initial Soviet SPACE STATIONs that were constructed and launched in the 1970s.

Brown, Ernest Williams (1866–1938) English mathematician and CELESTIAL MECHANICS expert who specialized in lunar orbital theory. In 1919, he published improved tables of the MOON's motion. These superior lunar tables built upon the work of GEORGE WILLIAM HILL and PHILIP HERBERT COWELL and included over 150 years of precision observations made at the ROYAL GREENWICH OBSERVATORY.

Bruno, Giordano (1548–1600) Italian philosopher, writer, and former Dominican monk who managed to antagonize authorities throughout western Europe by adamantly supporting such politically unpopular concepts as the HELIOCENTRIC theory of NICHOLAS COPERNICUS, the infinite size of the UNIVERSE, and the existence of intelligent life on other worlds. His self-destructive, belligerent manner eventually brought him before the Roman Inquisition. After an eight-year-long trial, an uncompromising Bruno was convicted of heresy and burned to death at the stake on 17 February 1600.

Bunsen, Robert Wilhelm (1811–99) German chemist who collaborated with GUSTAV ROBERT KIRCHHOFF in 1859 to develop SPECTROSCOPY, the process of identifying individual chemical ELEMENTs from the light each emits or absorbs when heated to incandescence. SPECTRUM analysis, based on Bunsen's work with Kirchhoff, revolutionized ASTRONOMY by allowing scientists to determine the chemical composition of distant CELESTIAL BODIES. He also made other contributions in chemistry, including the development of the popular gas burner that bears his name.

Burbidge, (Eleanor) Margaret (née Peachey) (b. 1922) English astrophysicist who collaborated in 1957 with her husband (GEORGE RONALD BURBIDGE), SIR FRED HOYLE, and WILLIAM ALFRED FOWLER to publish an important scientific paper describing how NUCLEOSYNTHESIS creates the ELEMENTs of higher MASS in the interior of evolved STARs. In 1967, she coauthored a fundamental book on QUASARs with her astrophysicist husband. From 1972 to 1973, she was the first woman to serve as the director of the ROYAL GREENWICH OBSERVATORY.

Burbidge, Geoffrey Ronald (b. 1925) English astrophysicist who collaborated in 1957 with his astrophysicist wife ([ELEANOR] MARGARET BURBIDGE), SIR FRED HOYLE, and WILLIAM ALFRED FOWLER as they developed the detailed theory explaining how NUCLEOSYNTHESIS creates heavier ELEMENTs in the interior of STARS. In 1967, he coauthored an important book on QUASARs with his wife. From 1978 to 1984, he served as the director of the Kitt Peak National Observatory in Arizona.

Callippus of Cyzicus (c. 370 B.C.E.–c. 300 B.C.E.) Greek astronomer and mathematician who modified the GEOCENTRIC system of COSMIC spheres developed by his teacher, EUDOXUS OF CNIDUS. Callippus was a skilled observer and proposed that a stationary EARTH was surrounded by 34 rotating spheres upon which all CELESTIAL BODIES moved. ARISTOTLE liked this geocentric COSMOLOGY but proposed a system of 55 solid crystalline (transparent) spheres, whereas the heavenly spheres of Callippus and Eudoxus were assumed geometric but not material in nature.

Campbell, William Wallace (1862–1938) American astronomer who made pioneering measurements of the radial (LINE-OF-SIGHT) VELOCITIES of STARS by measuring the DOPPLER SHIFT of their spectral lines. In other words, a star's SPECTRUM is BLUESHIFTed when it is approaching Earth and REDSHIFTed when it is receding. In 1913, he published *Stellar Motions* and also prepared a major catalog that listed the radial velocities of over 900 stars (later expanded to 3,000). He measured the subtle deflection of a light beam from a star as it just grazed the SUN's surface during a 1922 SOLAR ECLIPSE. His work confirmed ALBERT EINSTEIN's theory of GENERAL RELATIVITY and the previous measurements made by SIR ARTHUR STANLEY EDDINGTON.

Cannon, Annie Jump (1863–1941) American astronomer who worked at the HARVARD COLLEGE OBSERVATORY under EDWARD CHARLES PICKERING and was instrumental in developing a widely used system for classifying stellar spectra. She designed the HARVARD CLASSIFICATION SYSTEM by arranging STARS into categories according to their temperatures (for example, type O stars are the hottest). Her efforts culminated in the publication of the *Henry Draper Catalogue,* a nine-volume work (completed in 1924) that contained the spectral classification of 225,300 stars.

Carpenter, Scott (b. 1925) American ASTRONAUT and U.S. Navy officer who flew the second American crewed orbital spaceflight on 24 May 1962. His *Aurora 7* SPACE CAPSULE made three REVOLUTIONS of EARTH as part of NASA's MERCURY PROJECT.

Carrington, Richard Christopher (1826–75) English astronomer who made important studies of the ROTATION of the SUN by carefully observing the number and positions of SUNSPOTS. Between 1853 and 1861, he discovered that sunspots at the solar EQUATOR rotated in about 25 days, while those at 45° solar latitude rotated in about 27.5 days. In 1859 (without special viewing equipment), he reported the first visual observation of a SOLAR FLARE—although he thought the phenomenon he had just witnessed was the result of a large METEOR falling into the Sun.

Cassegrain, Guillaume (c. 1629–93) The Frenchmen who, in 1672, invented a special REFLECTING TELESCOPE configuration, now called the CASSEGRAIN TELESCOPE, that has become the most widely used reflecting telescope in ASTRONOMY. Unfortunately, except for this important fact, very little else is known about this man who is rumored to have been possibly a priest, a teacher, an instrument maker, or a physician.

Cassini, César François (aka Cesare Francesco Cassini) (1714–84) French astronomer who was grandson of GIOVANNI DOMENICO CASSINI and who succeeded his father, JACQUES CASSINI, as director of the PARIS OBSERVATORY. He started the development of the first topographical map of France—a task completed by his son, JACQUES DOMINIQUE CASSINI. In 1761, he traveled to Vienna, Austria, to observe a TRANSIT of VENUS across the SUN's disk.

Cassini, Giovanni Domenico (aka Jean Dominique Cassini) (1625–1712) Italian-French astronomer who was invited in 1669 by King Louis XIV of France to establish and direct the PARIS OBSERVATORY. As an accomplished astronomical observer, he studied MARS, JUPITER, and SATURN. In 1672, he determined the PARALLAX of Mars by using observations he made in Paris and those made by JEAN RICHER in Cayenne, French Guiana. The simultaneous measurements of Mars allowed Cassini to make the first credible determination of distances in the SOLAR SYSTEM, including the EARTH-SUN distance, which he estimated as 140 million km (87 million mi). In 1675, he discovered a distinctive division or gap in the rings of Saturn, a feature now called the Cassini Division. By using improved telescopes, between 1671 and 1684 he discovered four new MOONS of Saturn: Iapetus (1671), Rhea (1672), Dione (1684), and Tethys (1684).

Cassini, Jacques (aka Giacomo Cassini) (1677–1756) French astronomer who was the son of GIOVANNI DOMENICO CASSINI and continued his father's work as director of the PARIS OBSERVATORY. By carefully observing Arcturus in 1738, he became one of the first astronomers to determine the PROPER MOTION of a STAR accurately.

Cassini, Jacques Dominique (1748–1845) French astronomer who was the great-grandson of GIOVANNI DOMENICO CASSINI and who succeeded his father, CÉSAR FRANÇOIS CASSINI, as director of the PARIS OBSERVATORY. He published *Voyage to California,* a discussion of an expedition to observe the 1769 TRANSIT of VENUS across the SUN's disk. He also completed the topographical map of France started by his father.

Celsius, Anders (1701–44) Swedish astronomer who published a detailed account of the aurora borealis (northern lights) in 1733 and was the first scientist to associate the AURORA with the EARTH's magnetic field. In 1742, he introduced a temperature scale (the CELSIUS TEMPERATURE SCALE) that is still widely used today. He initially selected 100° on this scale as the freezing point of water and 0° as the boiling point of water. However, after discussions with his colleagues, he soon reversed the order of these reference point temperatures.

Cernan, Eugene A. (b. 1934) American ASTRONAUT and U.S. Navy officer who was the lunar module pilot on NASA's *APOLLO 10* mission (16–26 May 1969) that circumnavigated the MOON and then the SPACECRAFT commander for the *Apollo 17* lunar landing mission (6–19 December 1972). He was the last human being to walk on the lunar surface in the 20th century.

Chaffee, Roger B. (1935–67) American ASTRONAUT and U.S. Navy officer who was assigned to the crew of the first NASA APOLLO PROJECT flight (called *Apollo 1*). Unfortunately, he and his crewmates (VIRGIL "GUS" I. GRISSOM and EDWARD H. WHITE II), died on 27 January 1967 in a tragic LAUNCH PAD accident in CAPE CANAVERAL when an oxygen-fed flash fire swept through the interior of their Apollo SPACE CAPSULE as they were performing a full-scale launch simulation test.

Chandler, Seth Carlo (1846–1913) American astronomer who discovered irregular movements of EARTH's geographical poles (called the Chandler wobble) and the 14-month oscillation of Earth's polar AXIS (called the Chandler period).

Chandrasekhar, Subrahmanyan (aka Chandra) (1910–95) Indian-American astrophysicist who made important contributions to the theory of stellar evolution—especially the role of WHITE DWARF stars as the last stage of evolution of many stars that are about the MASS of the SUN. He shared the 1983 Nobel Prize in physics with WILLIAM ALFRED FOWLER for his theoretical studies of the physical processes important to the structure and evolution of stars. In July 1999, NASA successfully launched the *Advanced X-Ray Astrophysics Facility* that was renamed the *CHANDRA X-RAY OBSERVATORY* (CXO) in his honor.

Chladni, Ernst Florens Friedrich (1756–1827) German physicist who suggested in 1794 that METEORITES were of EXTRA-TERRESTRIAL origin, possibly the debris from a PLANET that exploded in ancient times. In 1819, he also postulated that a physical relationship might exist between METEORS and COMETS.

Christie, William Henry Mahoney (1845–1922) English astronomer who served as the eighth ASTRONOMER ROYAL from 1881 to 1910. Under his leadership as Astronomer Royal, the role of the ROYAL GREENWICH OBSERVATORY grew in both precision astronomical measurements and as a leading institution for research. He acquired improved, new TELESCOPES for the facility and expanded its activities to include SPECTROSCOPY and ASTROPHOTOGRAPHY.

Clairaut, Alexis Claude (1713–65) French mathematician and astronomer who accurately calculated the PERIHELION date for the 1759 return of COMET HALLEY. In a brilliant application of SIR ISAAC NEWTON's physical principles, Clairaut compensated for the gravitational influence of both SATURN and JUPITER on the COMET's TRAJECTORY. His detailed calculations predicted a perihelion date of 13 April 1759, while the observed perihelion date for this famous comet was 14 March 1759. Prior to this activity, he participated in an expedition to Lapland (northern Scandinavia) to collect geophysical evidence of the OBLATE-NESS (flattened poles) of a rotating EARTH.

Clark, Alvan (1804–87) American optician and TELESCOPE maker who founded the famous 19th-century telescope-manufacturing company Alvan Clark & Sons in Cambridge, Massachusetts. While testing an 18 in (0.46 m) diameter OBJECTIVE (lens) REFRACTING TELESCOPE in 1862, his son (ALVAN GRAHAM CLARK) pointed the new instrument at the bright STAR Sirius and detected its faint companion Sirius B, a WHITE DWARF. In 1877, ASAPH HALL used the 26 in (0.66 m) Clark-built refracting telescope at the U.S. NAVAL OBSERVATORY to discover the two tiny MOONS of MARS, PHOBOS and DEIMOS.

Clark, Alvan Graham (1832–97) American optician and TELESCOPE maker who, while testing a new 18 in (0.46 m) OBJECTIVE (lens) REFRACTING TELESCOPE in 1862 for his father (ALVAN CLARK), pointed the instrument at the bright STAR Sirius and discovered its WHITE DWARF companion, Sirius B. In 1844, FRIEDRICH WILHELM BESSEL had examined the irregular motion of Sirius and hypothesized that the star must have a dark (unseen) companion. The younger Clark's discovery earned him an award from the French Academy of Science and represented the first example of the important small dense object that is the end product of stellar evolution for all but the most massive stars.

Clarke, Sir Arthur C. (b. 1917) English science writer and science fiction author who is widely known for his enthusiastic support of space exploration. In 1945, he published the technical article

"Extra Terrestrial Relays" in which he predicted the development of COMMUNICATIONS SATELLITES that operated in GEOSYNCHRONOUS ORBIT. In 1968, he worked with film director Stanley Kubrick in developing the movie version of his book *2001: A Space Odyssey*. This motion picture is still one of the most popular and realistic depictions of human spaceflight across vast interplanetary distances. He was knighted by Queen Elizabeth II of England in 1998.

Clausius, Rudolf Julius Emmanuel (1822–88) German theoretical physicist who developed the first comprehensive understanding of the second law of thermodynamics in 1865 by introducing the concept of ENTROPY. His work had a major impact on 19th-century COSMOLOGY. He assumed that the total ENERGY of the UNIVERSE (considered a closed system) was constant, so the entropy of the universe must then strive to achieve a maximum value in accordance with the laws of thermodynamics. The end state of the universe in this model is one of complete temperature equilibrium, with no energy available to perform any useful work. This condition is called the HEAT DEATH OF THE UNIVERSE.

Clavius, Christopher (1537–1612) The German Jesuit mathematician and astronomer who reformed the JULIAN CALENDAR and enabled Pope Gregory XIII to introduce the GREGORIAN CALENDAR in 1582. This calendar is widely used throughout the world as the international civil calendar.

Collins, Eileen Marie (b. 1956) American ASTRONAUT and U.S. Air Force officer who was the first woman to serve as commander of NASA's SPACE SHUTTLE. In July 1999, she commanded the STS-93 mission, which successfully deployed the *CHANDRA X-RAY OBSERVATORY*. Prior to that flight, she served as the first woman space shuttle pilot during the STS-63 mission (February 1995)—the initial RENDEZVOUS mission with the Russian *MIR* SPACE STATION.

Collins, Michael (b. 1930) American ASTRONAUT and U.S. Air Force officer who served as the command module pilot on NASA's historic *Apollo 11* mission to the MOON in July 1969. He remained in lunar ORBIT while APOLLO PROJECT astronauts NEIL A. ARMSTRONG and EDWIN E. "BUZZ" ALDRIN, JR.,

Eileen Marie Collins
(Courtesy of NASA)

became the first human beings to walk on the Moon's surface. In April 1971, he joined the Smithsonian Institution as director of the National Air and Space Museum, where he remained for seven years.

Common, Andrew Ainslie (1841–1903) English TELESCOPE maker who pursued improvements in ASTROPHOTOGRAPHY in the 1880s and 1890s and is credited with recording the first detailed photographic image of the ORION NEBULA.

Compton, Arthur Holly (1892–1962) American physicist who shared the 1927 Nobel Prize in physics for his pioneering work on the scattering of high-energy PHOTONs by ELECTRONs. This phenomenon, called COMPTON SCATTERING or the Compton effect, is the foundation of many GAMMA RAY detection techniques used in high-energy ASTROPHYSICS. In his honor, NASA named the large astrophysics observatory launched in April 1991 the *COMPTON GAMMA RAY OBSERVATORY* (CGRO).

Condon, Edward Uhler (1902–74) American theoretical physicist who served as the director of an investigation sponsored by the U.S. Air Force (USAF) concerning unidentified flying object (UFO) sighting reports. These reports were accumulated between 1948 and 1966 under USAF Project Blue Book (and its predecessors). Condon's team at the University of Colorado investigated various cases and then wrote the report *Scientific Study of Unidentified Flying Objects.* This document, sometimes called the *Condon Report,* helped the secretary of the air force decide to terminate Project Blue Book in 1969, citing that there was no evidence to indicate that any of the sightings categorized as unidentified were extraterrestrial in origin or posed a threat to national security.

Congreve, Sir William (1772–1828) English colonel of artillery who examined black-powder (gunpowder) ROCKETs captured during battles in India and then supervised the development of a series of improved British military rockets. In 1804, he wrote *A Concise Account of the Origin and Progress of the Rocket System.* British forces used Congreve's rockets quite effectively in large-scale bombardments during the Napoleonic Wars and the War of 1812. His pioneering work on these early

SOLID-PROPELLANT ROCKETS represents an important technical step in the overall evolution of the modern military rocket.

Conrad, Charles (Pete), Jr. (1930–99) American ASTRONAUT and U.S. Navy officer who served as the SPACECRAFT commander during NASA's *APOLLO 12* lunar-landing mission (14–24 November 1969). He was the third human being to walk on the surface of the MOON. As a prelude to his Moon walk, he flew into space during the GEMINI PROJECT in 1965 (*Gemini-Titan V* mission) and in 1966 *(Gemini-Titan XI)*. He served as the commander of NASA's *SKYLAB* SL-2 mission (25 May to 22 June 1973), making mission-saving repairs on the first American SPACE STATION.

Cooper, Leroy Gordon, Jr. (b. 1927) American ASTRONAUT and U.S. Air Force officer who was selected as one of the seven original Mercury astronauts and who flew NASA's last MERCURY PROJECT mission, *Faith-7 (Mercury-Atlas 9)* on 15–16 May 1963. During that mission, he became the first human to perform a pilot-controlled REENTRY of a SPACE CAPSULE. In 1965, he also flew into space as part of NASA's GEMINI PROJECT in 1965 (*Gemini-Titan V* mission).

Copernicus, Nicholas (aka Nicolaus) (1473–1543) Polish astronomer and church official who triggered the scientific revolution of the 17th century with his book *On the Revolution of Celestial Spheres.* When published in 1543 while he lay on his deathbed, this book overthrew the PTOLEMAIC SYSTEM by boldly suggesting a HELIOCENTRIC model for the SOLAR SYSTEM in which EARTH and all the other PLANETs moved around the SUN. His heliocentric model (possibly derived from the long-forgotten ideas of ARISTARCHUS OF SAMOS) caused much technical, political, and social upheaval before finally displacing two millennia of Greek GEOCENTRIC COSMOLOGY.

Cowell, Philip Herbert (1870–1949) English scientist who specialized in CELESTIAL MECHANICS and who cooperated with ANDREW CLAUDE CROMMELIN in calculating the precise time of the 1910 appearance of COMET HALLEY.

Crippen, Robert L. (b. 1937) American ASTRONAUT and U.S. Navy officer who served as pilot and accompanied astronaut JOHN W. YOUNG (the SPACECRAFT commander) on the inaugural flight of NASA's SPACE SHUTTLE. This first (called STS-1) took place from 12–14 April 1981 and used orbiter vehicle (OV) 102, the *Columbia*. He also served as spacecraft commander on three other space shuttle missions: STS-7 (18–24 June 1983), STS 41-C (6–13 April 1984), and STS 41-G (5–13 October 1984).

Crommelin, Andrew Claude (1865–1939) French-Irish astronomer who specialized in calculating the ORBITS of COMETS. He collaborated with PHILIP HERBERT COWELL in computing and predicting the precise date of the 1910 return of COMET HALLEY. Based on this success, he then calculated the dates of previous appearances of this famous comet down through history back to the third century B.C.E.

Cunningham, R. Walter (b. 1932) American ASTRONAUT and U.S. Marine Corps officer who served as the LUNAR EXCURSION MODULE pilot during NASA's *Apollo 7* mission (11–22 October 1968). Although confined to ORBIT around EARTH, this important mission was the first human-crewed flight of APOLLO PROJECT SPACECRAFT hardware. Its success paved the way for the first lunar landing on 20 July 1969 by the *Apollo 11* astronauts.

Curtis, Heber Doust (1872–1942) American astronomer who proposed in 1920 that spiral NEBULAS were actually ISLAND UNIVERSES—that is, other SPIRAL GALAXIES that existed beyond humans' home galaxy, the MILKY WAY GALAXY. In 1924, EDWIN POWELL HUBBLE confirmed Curtis's hypothesis by showing that the ANDROMEDA GALAXY (a spiral nebula) was another large galaxy well beyond the Milky Way.

Danjon, André (1890–1967) French astronomer who was noted for the development of precise astronomical instruments (such as the Danjon prismatic ASTROLABE); the calculation of accurate ALBEDOS of the MOON, VENUS, and MERCURY; and studies of EARTH'S ROTATION.

Darwin, Sir George Howard (1845–1912) English mathematical astronomer (second son of the famous naturalist) who investigated the influence of tidal phenomena on the dynamics of the EARTH-MOON system. In 1879, he proposed that the Moon was formed from material thrown from Earth's CRUST when Earth was a newly formed, fast rotating, young PLANET. His tidal theory of lunar formation remained plausible until the 1960s, when general scientific support shifted to the GIANT-IMPACT MODEL of lunar formation.

Dawes, William Rutter (1799–1868) English amateur astronomer and physician who made precision measurements of BINARY (DOUBLE) STAR SYSTEMS and who independently discovered the major inner C or crêpe ring of SATURN in 1850. This ring is often called the crêpe ring because of its light and delicate appearance.

De La Rue, Warren (1815–89) English astronomer and physicist who developed the PHOTOHELIOGRAPH in 1858, enabling routine photography of the SUN's outer surface.

De Laval, Carl Gustaf Patrik (1845–1913) Swedish engineer and inventor who developed the CONVERGING-DIVERGING (CD) NOZZLE that he applied to steam turbines. The DE LAVAL NOZZLE has also become an integral part of most modern ROCKET engines.

Delporte, Eugène Joseph (1882–1955) Belgian astronomer who, in 1932, discovered the 1 km diameter Amor ASTEROID that now gives its name to an entire group of NEAR-EARTH ASTEROIDS that cross the ORBIT of MARS but not EARTH's orbit around the SUN.

Denning, William Frederick (1848–1931) English amateur astronomer who focused his efforts on the study of METEORS.

DeSitter, Willem (1872–1934) Dutch astronomer, cosmologist, and mathematician who explored the consequences of ALBERT EINSTEIN's GENERAL RELATIVITY theory and then proposed in 1917 an expanding-UNIVERSE model. While the specific physical details of his model proved unrealistic, his basic hypothesis served as an important stimulus to other scientists. For example, EDWIN POWELL HUBBLE's observations in the early 1920s proved the universe was indeed expanding.

DeVaucouleurs, Gérard Henri (1918–95) French-American astronomer who performed detailed studies of distant GALAXIES and the MAGELLANIC CLOUDS.

Dewar, Sir James (1842–1923) Scottish physicist and chemist who specialized in the study of low-temperature phenomena and invented a special double-walled vacuum flask that now carries his name. In 1892, he started using the DEWAR FLASK to store liquified gases at very low (cryogenic) temperatures. In 1898, he successfully produced liquified HYDROGEN.

Dicke, Robert Henry (1916–97) American physicist who, in 1964, revived the hypothesis that the BIG BANG event that began the UNIVERSE should have a detectable MICROWAVE RADIATION remnant. However, before he could make his own experimental observations, two other scientists (ARNO ALLEN PENZIAS and ROBERT WOODROW WILSON) detected this COSMIC MICROWAVE BACKGROUND and provided direct experimental evidence of Dicke's hypothesis.

Dirac, Paul Adrien Maurice (1902–84) English theoretical physicist who made major contributions to QUANTUM MECHANICS and shared the 1933 Nobel Prize in physics. With his theory of PAIR PRODUCTION, he postulated the existence of the POSITRON. In 1938, he proposed a link between the HUBBLE CONSTANT (a physical measure that describes the size and age of the UNIVERSE) and the fundamental physical constants of subatomic PARTICLES. This hypothesis formed the basis of DIRAC COSMOLOGY.

Disney, Walter Elias (Walt) (1901–66) American entertainment visionary who popularized the concept of spaceflight in the early 1950s, especially through a widely acclaimed, three-part television series. Because of Disney's commitment to excellence, space visionaries like WERNHER VON BRAUN inspired millions of Americans with credible images of space exploration in *Man in Space* (1955), *Man and the MOON* (1955), and finally *MARS and Beyond* (1957). The last episode premiered on 8 December 1957—barely two months after the former Soviet Union launched *SPUTNIK 1*.

Dolland, John (1706–61) English optician who invented a practical HELIOMETER in the early 1750s and then developed a composite lens from two types of glass for an achromatic TELESCOPE in 1758. An achromatic telescope (or lens) is one that corrects for CHROMATIC ABERRATION.

Donati, Giovanni Battista (1826–73) Italian astronomer who specialized in discovering COMETs (for example, Comet Donati in 1858) and then became the first scientist to observe the SPECTRUM of a comet (Comet Tempel in 1864).

Doppler, Christian Johann (1803–53) Austrian physicist who published a paper in 1842 that mathematically described the interesting phenomenon of how sound from a moving source changes pitch as the source approaches (FREQUENCY increases) and then goes away from (frequency decreases) an observer. This phenomenon, now called the DOPPLER SHIFT, or the Doppler effect, is widely used by astronomers to tell whether a distant celestial object is coming toward Earth (BLUESHIFT) or going away (receding) from Earth (REDSHIFT).

Douglass, Andrew Ellicott (1867–1962) American astronomer and environmental scientist who proposed a relationship between SUNSPOT activity and Earth's climate. He supported this hypothesis by examining tree ring patterns and then developed the field of dendrochronology (tree dating), which relates the rate of tree ring development with weather effects and, ultimately, solar activity. He was director of the Steward Observatory at the University of Arizona from 1918 to 1938.

Drake, Frank Donald (b. 1930) American astronomer who, while working at the NATIONAL RADIO ASTRONOMY OBSERVATORY (NRAO) in Green Bank, West Virginia, conducted the first organized attempt to detect RADIO WAVE signals from an alien intelligent civilization across INTERSTELLAR distances. This initial SEARCH FOR EXTRATERRESTRIAL INTELLIGENCE (SETI) was performed under PROJECT OZMA. It led to the formulation of a speculative, semiEMPIRICAL mathematical expression (the DRAKE EQUATION) that tries to estimate the number of intelligent alien civilizations that might now be capable of communicating with each other in this GALAXY. *See also* SECTION IV CHARTS & TABLES.

Draper, Charles Stark (1901–87) American physicist and instrumentation expert who used the principle of the GYRO to develop intertial GUIDANCE SYSTEMS for BALLISTIC MISSILES, SATELLITES, and the SPACECRAFT used in NASA's APOLLO PROJECT.

Draper, John William (1811–82) American scientist who, like his son HENRY DRAPER, made pioneering contributions to the field of ASTROPHOTOGRAPHY. He was the first person to photograph the MOON (1840) and then became the first to make a spectral photograph of the SUN (1844).

Draper, Henry (1837–82) American physician and amateur astronomer who pioneered key areas of ASTROPHOTOGRAPHY. He was the first astronomer to photograph the SPECTRUM of a STAR (Vega) in 1872 and then the first to photograph successfully a NEBULA (the ORION NEBULA) in 1880. His widow financed publication of the famous *Henry Draper Catalogue of Stellar Spectra.*

Dreyer, Johan Ludvig Emil (aka John Lewis Emil Dreyer) (1852–1926) Danish astronomer who compiled an extensive catalog of NEBULAS and STAR CLUSTERs, which was first published in 1888.

Duke, Charles Moss, Jr. (b. 1935) American ASTRONAUT and U.S. Air Force officer who served as the LUNAR EXCURSION MODULE pilot on NASA's *APOLLO 16* mission in April 1972. Along with fellow astronaut JOHN W. YOUNG, he explored the rugged lunar highlands in the Descartes region during the fifth MOON-landing mission.

Dyson, Sir Frank Watson (1868–1939) English astronomer who participated with SIR ARTHUR STANLEY EDDINGTON on the 1919 eclipse expedition that observed the bending of a STAR's light by the SUN's GRAVITATION—providing the first experimental evidence to support ALBERT EINSTEIN's theory of GENERAL RELATIVITY. He also served as England's ASTRONOMER ROYAL from 1910–33.

Dyson, Freeman John (b. 1923) English-American theoretical physicist who participated in Project Orion, a nuclear FISSION, pulsed-ROCKET concept studied by the U.S. government in the early 1960s as a means of achieving rapid INTERPLANETARY space travel.

Eddington, Sir Arthur Stanley (1882–1944) English astronomer, mathematician, and physicist who helped create modern ASTROPHYSICS. In May 1919, he led a solar ECLIPSE expedition to Principe Island (West Africa) to measure the gravitational deflection of a beam of starlight as it passed close to the SUN— thereby providing support for ALBERT EINSTEIN'S GENERAL RELATIVITY theory. In his 1933 publication *The Expanding Universe,* he popularized the notion that the outer GALAXIES (spiral nebulas) were receding from one another as the UNIVERSE expanded.

Ehricke, Krafft A. (1917–84) The German-American ROCKET engineer who designed advanced propulsion systems for the American space program, including the CENTAUR UPPER-STAGE vehicle. As a technical visionary, he also expounded upon the positive consequences of space technology for the human race.

Einstein, Albert (1879–1955) German-Swiss-American physicist whose revolutionary theory of RELATIVITY (special relativity in 1905 and then GENERAL RELATIVITY in 1915) shaped modern physics. Like GALILEO GALILEI and SIR ISAAC NEWTON before him, Einstein changed forever people's view of the UNIVERSE and how it functions. He was awarded the 1921 Nobel Prize in physics and escaped to the United States when Hitler rose to power in Germany in 1933. While fearing nuclear weapon developments by Nazis in Germany, in 1939 he encouraged President Franklin D. Roosevelt to start the American nuclear weapons program later known as the Manhattan Project.

Eisele, Donn F. (1930–87) American ASTRONAUT and U.S. Air Force officer who flew onboard the *Apollo 7* mission in October 1968—the pioneering voyage in ORBIT around EARTH of NASA'S APOLLO PROJECT that ultimately landed human beings on the MOON between 1969 and 1972.

Eisenhower, Dwight D. (1890–1969) American army general and 34th president of the United States who was deeply interested in the use of space technology for national security and directed that INTERCONTINENTIAL BALLISTIC MISSILES (ICBMs) and

RECONNAISSANCE SATELLITES be developed on the highest national priority basis.

Encke, Johann Franz (1791–1865) German astronomer and mathematician who, in 1819, established the common identity of the COMET (now called Comet Encke) with the shortest known PERIOD (3.3 years). Its discovery by JEAN PONS in 1818 prompted Encke to perform calculations that proved that the comet was previously observed by other astronomers, including CAROLINE HERSCHEL.

Eratosthenes of Cyrene (c. 276 B.C.E.–c. 194 B.C.E.) Greek astronomer, mathematician, and geographer who made a remarkable attempt at measuring the circumference of EARTH in about 250 B.C.E. After recognizing that Earth curved, he used the difference in latitude between Alexandria and Aswan, Egypt, (about 800 km apart) and the corresponding angle of the SUN at ZENITH at both locations during the summer SOLSTICE. Some historic interpretations of the *stadia* (an ancient unit) suggest his results were about 47,000 km or less versus the true value of 40,000 km.

Eudoxus of Cnidus (c. 400 B.C.E.–c. 347 B.C.E.) Early Greek astronomer and mathematician who first suggested a GEOCENTRIC COSMOLOGY in which the SUN, MOON, PLANETS, and FIXED STARS moved around EARTH on a series of 27 giant, geocentric spheres. CALLIPPUS OF CYZICUS, ARISTOTLE, and then PTOLEMY embraced geocentric cosmology though each offered modifications to Eudoxus's model.

Euler, Leonhard (1707–83) Swiss mathematician and physicist who developed advanced mathematical methods for observational ASTRONOMY that supported precise predictions of the motions of the MOON and the PLANETS.

Evans, Ronald E. (1933–90) American ASTRONAUT and U.S. Navy officer who served as the command module pilot during NASA's *APOLLO 17* mission (December 1972). He maintained a solo vigil in ORBIT around the MOON, while fellow astronauts EUGENE A. CERNAN and HARRISON H. SCHMITT completed their explorations of the Taurus-Littrow landing area on the lunar surface.

Fabricius, David (1564–1617) German astronomer and clergyman who was a skilled NAKED EYE observer, corresponded with JOHANNES KEPLER, and discovered the first VARIABLE STAR (Mira) in 1596. His son JOHANNES FABRICIUS was a student in the Netherlands and, in 1611, introduced his father to the recently invented TELESCOPE. Soon father and son followed GALILEO GALILEI by making their own telescopic observations of the heavens. Unfortunately, the son's death at age 29 and the father's murder brought their early contributions to the era of telescope-based ASTRONOMY to a sudden halt. *See also* HANS LIPPERSHEY.

Fabricius, Johannes (1587–1616) German astronomer and son of DAVID FABRICIUS whose early telescopic observations of the heavens with his father discovered SUNSPOTs in 1611. He investigated the number and dynamic behavior of sunspots, writing an early paper on the subject. However, his work is generally overshadowed by GALILEO GALILEI and CHRISTOPH SCHEINER, who independently performed sunspot studies at about the same time.

Fermi, Enrico (1901–54) Italian-American physicist who helped create the nuclear age and won the 1938 Nobel Prize in physics. In addition to making numerous contributions to nuclear physics, he is also credited with the famous Fermi paradox: where are they?—a popular speculative inquiry concerning the diffusion of an advanced alien civilization through the GALAXY on a wave of exploration.

Flamsteed, John (1646–1719) English astronomer who used both the TELESCOPE and the clock to assemble a precise STAR catalog. In 1675, the English king, Charles II, established the ROYAL GREENWICH OBSERVATORY and appointed Flamsteed as its director and the first ASTRONOMER ROYAL. However, he had to run this important observatory without any significant financial assistance from the Crown. Nevertheless, he constructed instruments and collected precise astronomical data. In 1712, an impatient SIR ISAAC NEWTON secured a royal command to force publication of the first volume of this data collection without the author's consent. This infuriated Flamsteed, who purchased and burned about 300 copies of the unauthorized book. An extensive three-volume set of Flamsteed's star data was published in 1725 after his death.

Fleming, Williamina Paton (née Stevens) (1857–1911) Scottish-American astronomer who in 1881 joined EDWARD CHARLES PICKERING at the HARVARD COLLEGE OBSERVATORY and devised a classification system for stellar spectra that greatly improved the system used by P. PIETRO ANGELO SECCHI. Her alphabetic system became known as the HARVARD CLASSIFICATION SYSTEM and appeared in the 1890 edition of *The Henry Draper Catalogue of Stellar Spectra.*

Foucault, Jean-Bernard Léon (1819–68) French physicist who investigated the SPEED OF LIGHT in air and water and was the first to demonstrate experimentally that EARTH rotates on its AXIS. He performed this experiment with a Foucault pendulum—a relatively large MASS (about 30 kg) suspended on a very long (about 70 m) wire.

Fowler, William Alfred (1911–95) American astrophysicist who developed the widely accepted theory that nuclear processes in STARS are responsible for all the heavier ELEMENTs in the UNIVERSE beyond HYDROGEN and HELIUM, including those in the human body. He shared the 1983 Nobel Prize in physics for his work on STELLAR EVOLUTION and NUCLEOSYNTHESIS.

Fraunhofer, Joseph von (1787–1826) German optician and physicist who developed the PRISM SPECTROMETER in about 1814 and then used this instrument to discover the dark lines in the SUN's SPECTRUM that now carry his name. In 1823, he observed similar (but different) lines in the spectra of other STARS.

Friedman, Herbert (1916–2000) American astrophysicist who, in 1949, began using SOUNDING ROCKETs in ASTRONOMY—especially for the initial study of the SUN's X-RAY activity. In 1964 as a scientist with the U.S. Naval Research Laboratory (Washington, D.C.), his ROCKET-borne SENSORs detected X rays from the CRAB NEBULA.

Gagarin, Yuri A. (1934–68) The Russian COSMONAUT who became the first human being to travel in OUTER SPACE. He accomplished this feat with his historic one ORBIT of EARTH mission in the *VOSTOK 1* SPACECRAFT on 12 April 1961. A popular hero of the former Soviet Union, he died in an aircraft training flight near Moscow on 27 March 1968.

Galileo Galilei (1564–1642) Italian astronomer, physicist, and mathematician whose innovative use of the TELESCOPE to make astronomical observations ignited the scientific revolution of the 17th century. In 1610, he announced some of his early telescopic findings in the publication *Starry Messenger*—including the discovery of the four major MOONS of JUPITER (now called the GALILEAN SATELLITES). Their behavior like a minature solar system stimulated his enthusiastic support for the HELIOCENTRIC COSMOLOGY of NICHOLAS COPERNICUS. Unfortunately, this scientific work led to a direct clash with church authorities who insisted on retaining the PTOLEMAIC SYSTEM for a number of political and social reasons. By 1632, this conflict earned the fiery Galileo an Inquisition trial at which he was found guilty of heresy (for advocating the COPERNICAN SYSTEM) and confined to house arrest for the remainder of his life.

Galle, Johann Gottfried (1812–1910) German astronomer who was the first person to observe NEPTUNE. Acting upon a request from URBAIN JEAN JOSEPH LEVERRIER, he immediately began a telescopic search at the Berlin Observatory on 23 September 1846. This search soon yielded the new PLANET precisely in the region of the night sky predicted by Leverrier's calculations, which were based on PERTURBATIONS in the ORBIT of URANUS.

Gamow, George (1904–68) Russian-American physicist who, in collaboration in the late 1940s with Ralph Alpher, promoted the BIG BANG theory—a theory that boldly speculated that the UNIVERSE began by means of a huge, ancient explosion. He described this concept in the 1948 paper entitled "The Origin of the Chemical Elements." Gamow based some of this work on the COSMOLOGY concepts previously suggested by GEORGES ÉDOUARD LEMAÎTRE.

Gauss, Carl Friedrich (1777–1855) Brilliant German mathematician whose contributions to CELESTIAL MECHANICS helped many other astronomers efficiently calculate the ORBITS of PLANETS, ASTEROIDS, and COMETS. For example, Gauss's mathematical innovations like the method of least squares enabled URBAIN JEAN JOSEPH LEVERRIER to predict theoretically the position of NEPTUNE in 1846 based on the subtle orbital PERTURBATIONS observed for URANUS.

Giacconi, Ricardo (b. 1931) Italian-American astrophysicist who made pioneering contributions to X-RAY ASTRONOMY. In June 1962, he placed instruments onto a SOUNDING ROCKET and detected the first EXTRASOLAR X-RAY source, Scorpius X-1. He supervised development of NASA's *Uhuru X-Ray SATELLITE* (launched in 1970) and NASA's *HIGH ENERGY ASTRONOMICAL OBSERVATORY 2* (HEAO-2), also called the *EINSTEIN X-ray Observatory,* which was successfully launched in 1978.

Gibson, Edward G. (b. 1936) American ASTRONAUT and SOLAR physicist who served as the science pilot on NASA's *Skylab 4* mission from 16 November 1973 to 8 February 1974. This was the third and final mission to *Skylab*—the first American SPACE STATION placed into LOW EARTH ORBIT.

Gill, Sir David (1843–1914) Scottish astronomer who improved the measured value of the ASTRONOMICAL UNIT (EARTH-SUN distance) by collecting data during his 1877 expedition to Ascension Island in the South Atlantic and who made precise observations of the SOLAR PARALLAX by observing the motion of several ASTEROIDs in 1889.

Giotto di Bondone (1266–1337) Italian artist of the Florentine school who apparently witnessed the 1301 passage of COMET HALLEY and then included the first "scientific" representation of the COMET in his fresco *Adoration of the Magi* found in the Scrovengi Chapel in Padua, Italy. The EUROPEAN SPACE AGENCY named the *GIOTTO* SPACECRAFT in his honor.

Glenn, John Herschel, Jr. (b. 1921) The American ASTRONAUT, U.S. Marine Corps officer, and U.S. senator who was the first American to ORBIT EARTH—a feat that he accomplished on 20 February 1962 as part of NASA's MERCURY PROJECT. Glenn's historic mission, aboard the *Friendship 7* Mercury SPACE CAPSULE, made three orbits of Earth and lasted about five hours. Over three and one-half decades later, he became the oldest human being to travel in OUTER SPACE when he joined

John Herschel Glenn, Jr. (Courtesy of NASA)

the SPACE SHUTTLE *Discovery* crew on its nine-day duration STS-95 orbital mission (from 29 October to 7 November 1998).

Goddard, Robert Hutchings (1882–1945) American physicist and ROCKET scientist who cofounded ASTRONAUTICS early in the 20th century along with (but independent of) KONSTANTIN EDUARDOVICH TSIOLKOVSKY and HERMANN J. OBERTH. Regarded as the Father of Modern Rocketry, he successfully launched the world's first LIQUID-PROPELLANT ROCKET on 16 March 1926 in a snow-covered field in Auburn, Massachusetts. As a brilliant, but reclusive, inventor, he continued his pioneering rocket experiments at a remote desert site near Roswell, New Mexico. In 1960, after recognizing that most modern rockets are really Goddard rockets, the United States government paid his estate 1 million dollars for the use of his numerous patents.

Gold, Thomas (b. 1920) Austrian-American astronomer who, in 1948, while collaborating with SIR FRED HOYLE and SIR HERMANN BONDI, proposed a STEADY-STATE UNIVERSE model in COSMOLOGY. This model, based on the PERFECT COSMOLOGICAL PRINCIPLE, has now been largely abandoned in favor of the BIG BANG theory. In the late 1960s, he suggested that the periodic signals from PULSARs represented emissions from rapidly spinning NEUTRON STARS—a theoretical hypothesis that appears consistent with subsequent observational data.

Goodricke, John (1764–86) English astronomer who, though a deaf-mute, made a significant contribution to 18th-century ASTRONOMY by observing that certain VARIABLE STARs, like Algol, were periodic in nature. Prior to his death from pneumonia at age 21, he further suggested that this periodic behavior might be due to a dark, not visible companion regularly passing in front of (eclipsing) its visible companion—a phenomenon modern astronomers call an eclipsing BINARY (DOUBLE) STAR SYSTEM.

Gordon, Richard F., Jr. (b. 1929) American ASTRONAUT and U.S. Navy officer who occupied the command module pilot seat during NASA's *Apollo 12* mission (14–24 November 1969)—

the second LUNAR-landing mission in which APOLLO PROJECT astronauts ALAN L. BEAN and CHARLES (PETE) CONRAD, JR., walked on the MOON's surface while Gordon remained in lunar ORBIT.

Gould, Benjamin Apthorp (1824–96) American astronomer who founded the *Astronomical Journal* in 1849 and produced an extensive 19th-century STAR catalog for the Southern Hemisphere sky. In 1868, he was invited by the government of Argentina to establish and direct a national observatory at Córdoba. This gave him the opportunity to make detailed observations, resulting in a star catalog comparable to FRIEDRICH WILHELM AUGUST ARGELANDER's work for the northern sky.

Gregory, James (1638–75) Scottish mathematician and astronomer who conceptually described the components and operation of a two-mirror REFLECTING TELESCOPE (a design later called the Gregorian telescope) in his 1663 publication *The Advance of Optics.*

Grissom, Virgil "Gus" I. (1926–67) American ASTRONAUT and U.S. Air Force officer who was the second American to travel in OUTER SPACE—a feat accomplished during NASA's suborbital MERCURY PROJECT/*Liberty Bell 7* flight on 21 July 1961. In March 1965, he served as the command pilot during the first crewed GEMINI PROJECT orbital mission. Gus Grissom, along with his fellow APOLLO PROJECT astronauts EDWARD H. WHITE II and ROGER B. CHAFFEE, died on 27 January 1967 at CAPE CANAVERAL when a flash fire consumed their Apollo SPACECRAFT during a mission simulation and training test at the LAUNCH PAD.

Haise, Fred Wallace, Jr. (b. 1933) American ASTRONAUT who served as the LUNAR EXCURSION MODULE (LEM) pilot for NASA's ill-fated *Apollo 13* mission to the MOON from 11–17 April 1970. Instead of landing as planned in the Fra Mauro region, he helped fellow astronauts JAMES ARTHUR LOVELL, JR., and JOHN LEONARD "JACK" SWIGERT, JR., maneuver an explosion-crippled APOLLO PROJECT SPACECRAFT on a life-saving emergency TRAJECTORY around the Moon and back to safety on EARTH. As the LEM pilot, he was instrumental in converting that two-person lunar LANDER spacecraft into a three-person lifeboat. In 1977, he participated as a pilot-astronaut during the SPACE SHUTTLE approach and landing tests conducted at Edwards Air Force Base in California. These nonpowered, gliding tests served as an important preparation for the shuttle spaceflight program of the 1980s.

Hale, George Ellery (1868–1938) American astronomer who pioneered the field of modern ASTROPHYSICS. In the late 1880s, he invented the spectroheliograph—an instrument that allowed astronomers to photograph the SUN at a particular WAVELENGTH or SPECTRAL LINE. He was also responsible for establishing several major observatories in the United States, including the Yerkes Observatory (Wisconsin), the Mount Wilson Observatory (California), and the PALOMAR OBSERVATORY (California).

Hall, Asaph (1829–1907) American astronomer who discovered the two small MOONS of MARS, PHOBOS and DEIMOS, in 1877 while a staff member at the U.S. NAVAL OBSERVATORY.

Halley, Edmond (aka Edmund) (1656–1742) English mathematician and astronomer who encouraged SIR ISAAC NEWTON to write the *Principia*—one of the most important scientific books ever written. In his own book *A Synopsis of the Astronomy of Comets,* which appeared in 1705, Halley suggested that the COMET he observed in 1682 was actually the same comet that appeared in 1531. He then boldly used NEWTON'S LAWS OF MOTION (as contained in the *Principia,* whose publication he personally financed) to predict that this comet would return again in about 1758. It did (after his death), yet COMET HALLEY carries his name because of the accuracy of his prediction and discovery. He succeeded JOHN FLAMSTEED as the second ASTRONOMER ROYAL (1720–42).

Hawking, Stephen William (b. 1942) English astrophysicist and cosmologist who, despite a disabling neurological disorder, has made major theoretical contributions to the study of GENERAL RELATIVITY and QUANTUM MECHANICS—especially by providing an insight into the unusual physics of BLACK HOLES and mathematical support for the BIG BANG theory. In 1988, he published *A Brief History of Time,* a popular book that introduced millions of readers to modern COSMOLOGY. His concept of HAWKING RADIATION suggests an intimate relationship between GRAVITY, quantum mechanics, and thermodynamics.

Henderson, Thomas (1798–1844) Scottish astronomer who, in the 1830s, made careful PARALLAX measurements of ALPHA CENTAURI from the Cape of Good Hope Observatory (South Africa). As a result of these observations, he reported in 1839 that this STAR (actually a triple star system) was a distance of less than four LIGHT-YEARS away—making it the closest known star (excluding the SUN).

Heraclides of Pontus (c. 388 B.C.E.–315 B.C.E.) Greek astronomer and philosopher who was the first person to suggest that this PLANET had a daily ROTATION on its AXIS from west to east. While believing Earth was the center of the UNIVERSE, he boldly speculated that MERCURY and VENUS might travel in ORBITS around the SUN since neither planet was ever observed far from it. Unfortunately, his revolutionary idea about planets traveling around the Sun clashed with GEOCENTRIC Greek COSMOLOGY—so it remained essentially unnoticed until being rediscovered by NICHOLAS COPERNICUS in the 16th century.

Hero of Alexandria (1st century C.E.: c. 20–c. 80?) Greek mathematician and engineer who invented many clever mechanical devices including the aeolipile—a spinning, steam-powered spherical apparatus that demonstrated the action-reaction principle by which all ROCKET engines work.

Herschel, Caroline (1750–1848) German-born English astronomer who was the sister and assistant of SIR (FREDERICK) WILLIAM HERSCHEL and the first notable woman astronomer. As a skilled observer, she personally discovered eight COMETs between 1786 and 1797. She also observed the ANDROMEDA GALAXY. Consistent with practices of 18th-century ASTRONOMY, she referred to this in 1783 as a NEBULA.

Herschel, Sir (Frederick) William (1738–1822) German-born English astronomer who discovered URANUS in 1781—the first new PLANET since the start of ancient ASTRONOMY and the first found through the use of the TELESCOPE. As a skilled observer, he located more than 2,500 NEBULAS and STAR CLUSTERS as well as over 800 BINARY (DOUBLE) STAR SYSTEMS. His study of the distribution of the observed STARS established a basic understanding of the form of the MILKY WAY GALAXY. In 1800, while investigating the ENERGY content of sunlight (SOLAR RADIATION) with the help of a PRISM and a thermometer, he discovered the existence of (thermal) INFRARED RADIATION—which lies just beyond red LIGHT in a longer WAVELENGTH portion of the ELECTROMAGNETIC (EM) SPECTRUM.

Herschel, Sir John (Frederick William) (1792–1871) English astronomer and the son of SIR (FREDERICK) WILLIAM HERSCHEL, who continued his father's investigations of BINARY (DOUBLE) STAR SYSTEMS and NEBULAS. Between about 1833 and 1838 he performed an extensive survey of the night sky in the Southern Hemisphere from an observatory at the Cape of Good Hope (South Africa).

Hertz, Heinrich Rudolf (1857–94) German physicist who, in 1888, produced and detected RADIO WAVES for the first time. He also demonstrated that ELECTROMAGNETIC RADIATION propagates at the SPEED OF LIGHT. The HERTZ (Hz) is the SI UNIT of FREQUENCY named in his honor.

Hertzsprung, Ejnar (1873–1969) Danish astronomer who, in 1905, showed how the LUMINOSITY of a STAR is related to its COLOR (or SPECTRUM). He contributed to the creation of the famous HERTZSPRUNG-RUSSELL (HR) DIAGRAM that is essential for understanding the theory of STELLAR EVOLUTION.

Hess, Victor Francis (1883–1964) Austrian-American physicist who, between 1911 to 1913, conducted radiation detection measurements while riding in high-altitude balloons that provided him with the initial scientific evidence for the existence of COSMIC RAYS. He shared the 1936 Nobel Prize in physics for this discovery.

Hevelius, Johannes (1611–87) Polish-German astronomer who, in 1647, published a lunar atlas called *Selenographica*—the first detailed description of the NEARSIDE of the MOON. As a skilled observer, he used his large, personally constructed observatory named Stellaburgum located in Danzig (now Gdansk, Poland) to develop a comprehensive catalog of over 1,500 STARS and to discover four COMETS.

Hewish, Anthony (b. 1924) English astronomer who collaborated with SIR MARTIN RYLE in the development of RADIO WAVE-based ASTROPHYSICS. His efforts included the discovery of the PULSAR—for which he shared the 1974 Nobel Prize in physics with Ryle. During a survey of galactic radio waves in August 1967, his graduate student (SUSAN) JOCELYN BELL BURNELL was actually the first person to notice the repetitive signals from this pulsar, but her contributions were inexplicably overlooked by the Nobel awards committee.

Hill, George William (1838–1914) American astronomer and mathematician who made fundamental contributions to CELESTIAL MECHANICS—including pioneering mathematical methods to calculate precisely the MOON's ORBIT with PERTURBATIONS from JUPITER and SATURN.

Hipparchus of Nicaea (c. 190 B.C.E.–120 B.C.E.) Greek astronomer and mathematician who is generally regarded by science historians as the greatest ancient astronomer. Although he embraced GEOCENTRIC COSMOLOGY, he carefully studied the SUN's annual motion to determine the length of a YEAR to an accuracy of about six minutes. His observational legacy includes a STAR catalog completed in about 129 B.C.E. that contained 850 FIXED STARS. He also divided NAKED EYE observations into six MAGNITUDEs, ranging from the faintest (or least visible to the naked eye) to the brightest observable CELESTIAL BODIES in the night sky.

Hohmann, Walter (1880–1945) German engineer who wrote the 1925 book *The Attainability of Celestial Bodies*. In this work, he described the mathematical principles that govern SPACE VEHICLE motion—including the most efficient (that is, minimum-ENERGY) orbit transfer path between two ORBITS in the same geometric plane. This widely used (but time-consuming) orbit transfer technique is now called the HOHMANN TRANSFER ORBIT in his honor.

Horrocks, Jeremiah (1619–41) English astronomer who was the first person to predict and then record a TRANSIT of VENUS across the SUN that occurred on 24 November 1639.

Hoyle, Sir Fred (1915–2001) English astrophysicist and writer who joined with SIR HERMANN BONDI and THOMAS GOLD in 1948 to develop and promote the STEADY-STATE UNIVERSE model. Hoyle first coined the expression BIG BANG theory—essentially as a derogatory remark about this competitive theory in COSMOLOGY. The term stuck and the big bang theory has all but displaced the steady-state universe model in contemporary ASTROPHYSICS. He also collaborated with GEOFFREY RONALD BURBIDGE, (ELEANOR) MARGARET BURBIDGE, and WILLIAM ALFRED FOWLER in the development of the theory of NUCLEOSYNTHESIS. Though often controversial, he popularized ASTRONOMY with several well-known books and was also an accomplished science fiction writer.

Hubble, Edwin Powell (1889–1953) American astronomer who made important observational discoveries in the 1920s and 1930s that completely changed humans' view of the UNIVERSE. In 1923 by using a CEPHEID VARIABLE STAR, he was able to estimate the distance to the ANDROMEDA GALAXY. His results immediately suggested that such spiral NEBULAS were actually large, distant independent stellar systems or ISLAND UNIVERSES. He introduced a classification scheme in 1925 for such nebulas (GALAXIES), calling them either elliptical, spiral, or irregular. This scheme is still used. In 1929, he announced that in an expanding universe, the other galaxies are receding from this one with speeds proportional to their distance—a postulate now known as HUBBLE'S LAW. Hubble's concept of an expanding universe filled with numerous galaxies forms the basis of modern observational COSMOLOGY.

Huggins, Margaret Lindsay (née Murray) (1848–1915) English astronomer who worked with her husband (SIR WILLIAM HUGGINS) to collect pioneering SPECTRA of celestial objects, including the ORION NEBULA.

Huggins, Sir William (1824–1910) English astronomer and spectroscopist who helped revolutionize the observation of CELESTIAL BODIES by performing early SPECTROSCOPY measurements and then comparing the observed stellar SPECTRA with laboratory spectra. In 1868, he collaboratively made the first measurement of a STAR's DOPPLER SHIFT. In 1881 (independent of HENRY DRAPER), he also made a pioneering photograph of the spectrum of a COMET.

Hulse, Russell (b. 1950) American radio astronomer who codiscovered the first binary pulsar—a PULSAR in orbit around another NEUTRON STAR. For this work, he shared the 1993 Nobel Prize in physics with his research professor (Joseph H. Taylor).

Humason, Milton Lassell (1891–1972) American astronomer who assisted EDWIN POWELL HUBBLE in the late 1920s during Hubble's discovery of the recession of the distant GALAXIES and the expansion of the UNIVERSE.

Huygens, Christiaan (1629–95) Dutch astronomer, physicist, and mathematician who discovered TITAN, the largest MOON of SATURN, in 1655. By using a personally constructed TELESCOPE, he was able to discern (~1659) the true (thin disklike) shape of SATURN's RINGS. As a creative and productive individual, he also constructed and patented the first pendulum clock (about 1656) and founded the wave theory of LIGHT.

Innes, Robert Thorburn Ayton (1861–1933) Scottish astronomer who discovered PROXIMA CENTAURI in 1915. Except for the SUN, this STAR is the closest one to Earth (about 4.2 LIGHT-YEARS away) and is the faintest member of the ALPHA CENTAURI triple-star system.

Irwin, James Benson (1930–91) American ASTRONAUT and U.S. Air Force officer who served as the LUNAR EXCURSION MODULE (LEM) pilot during NASA's *Apollo 15* mission to the HADLEY RILLE region of the MOON. As part of this APOLLO PROJECT mission (26 July to 7 August 1971), he became the eighth person to walk on the LUNAR surface.

Jansky, Karl Guthe (1905–50) American radio engineer who started the field of RADIO ASTRONOMY in 1932 by initially detecting and then identifying INTERSTELLAR RADIO WAVES from the direction of the CONSTELLATION Sagittarius—a direction corresponding to the central region of the MILKY WAY GALAXY.

In his honor, the unit describing the strength of an EXTRATERRESTRIAL radio wave signal is called the JANSKY.

Janssen, Pierre Jules César (1824–1907) French astronomer who pioneered the field of SOLAR physics. His precise measurements of the SUN's SPECTRAL LINES in 1868 enabled SIR JOSEPH NORMAN LOCKYER to conclude the presence of a previously unknown ELEMENT (which Lockyer called HELIUM). At the end of the 19th century, Janssen focused his efforts on the new field of ASTROPHOTOGRAPHY and created an extensive collection of over 6,000 photographic IMAGES of the Sun, which he published as a solar atlas in 1904.

Jarvis, Gregory B. (1944–86) American ASTRONAUT who died in the explosion of the SPACE SHUTTLE *Challenger* on 28 January 1986 while serving as a civilian payload specialist on the ill fated STS 51-L mission. He was an electrical engineer for the Hughes Aircraft Company.

Jeans, Sir James Hopwood (1877–1946) English astronomer, physicist, and writer who proposed various theories of STELLAR EVOLUTION and suggested in 1928 that matter was being continuously created in the UNIVERSE. This hypothesis led other scientists to establish a STEADY-STATE UNIVERSE model in COSMOLOGY. From 1928 on, he focused his activities on writing popular books on ASTRONOMY.

Jones, Sir Harold Spencer (1890–1960) English astronomer who made precise measurements of the ASTRONOMICAL UNIT (AU)—that is, the average distance from EARTH to the SUN. He served as the 10th ASTRONOMER ROYAL (1933–55) and wrote a number of popular books on astronomy, including *Life on Other Worlds* (1940).

Kant, Immanuel (1724–1804) German philosopher who was the first to propose the NEBULA hypothesis in the 1755 book *General History of Nature and Theory of the Heavens.* In the nebula hypothesis, he suggested that the SOLAR SYSTEM formed out of a primordial cloud of INTERSTELLAR matter. Kant also introduced the term ISLAND UNIVERSES to describe distant collections of STARS—now called GALAXIES. As a truly brilliant thinker, his works in metaphysics and philosophy had a great influence on Western thinking in the 18th century and beyond.

Kapteyn, Jacobus Cornelius (1851–1922) Dutch astronomer who made significant contributions to ASTROPHOTOGRAPHY and the use of statistical methods to evaluate stellar distributions and motions of stars. By using photographic IMAGES collected by SIR DAVID GILL, Kapteyn finished publishing in 1900 an extensive (three-volume) catalog that contained over 450,000 Southern Hemisphere stars.

Kármán, Theodore von *See* VON KÁRMÁN, THEODORE

Keeler, John Edward (1857–1900) American astronomer who assisted SAMUEL PIERPONT LANGLEY in making measurements of SOLAR RADIATION from Mount Whitney, California, in 1881. By studying the spectral composition of the RINGS OF SATURN, he showed that the rings were composed of collections of discrete PARTICLES as earlier suggested by JAMES CLERK MAXWELL.

Kelvin, Baron William Thomson (aka Sir William Thomson) (1824–1907) Scottish physicist, engineer, and mathematician who made major contributions to the electromagnetic theory of LIGHT and to thermodynamics, including an ABSOLUTE TEMPERATURE scale that now bears his name. The SI UNIT of absolute temperature is called the KELVIN.

Kennedy, John F. (1917–63) The 35th president of the United States who boldly proposed in May 1961 that NASA send ASTRONAUTS to the MOON to demonstrate American space technology superiority over the former Soviet Union during the COLD WAR. He was assassinated on 22 November 1963 and did not live to see the triumphant APOLLO PROJECT LUNAR landings (1969–72)—a magnificent technical accomplishment that his vision and leadership set in motion almost a decade earlier.

Kepler, Johannes (aka Johann) (1571–1630) German astronomer and mathematician who developed three laws of planetary motion that described the ELLIPTICAL ORBITS of the PLANETS around the SUN and provided the empirical basis for the acceptance of NICHOLAS COPERNICUS'S HELIOCENTRIC hypothesis. KEPLER'S LAWS gave ASTRONOMY its modern, mathematical foundation. His publication *The New Star (De Stella Nova)* described the SUPERNOVA in the constellation Ophiuchus that he first observed (with the NAKED EYE) on 9 October 1604.

Kerr, Roy Patrick (b. 1934) New Zealander mathematician who made a major contribution to ASTROPHYSICS in 1963 by applying GENERAL RELATIVITY to describe the properties of a rotating BLACK HOLE—a phenomenon now referred to as a KERR BLACK HOLE.

Khrushchev, Nikita S. (1894–1971) The provocative premier of the former Soviet Union who used early Russian OUTER SPACE achievements to imply Soviet superiority over the United States (and capitalism) during the COLD WAR. With his permission and encouragement, Russian ROCKET engineers, like SERGEI KOROLEV, used powerful military INTERCONTINENTAL BALLISTIC MISSILES as SPACE LAUNCH VEHICLES to place the first ARTIFICIAL SATELLITE successfully into ORBIT around EARTH (*SPUTNIK 1* on 4 October 1957) and to allow the first human to orbit Earth (COSMONAUT YURI A. GAGARIN on 12 April 1961). Despite these Soviet space accomplishments, failing domestic economic programs and a major loss of political prestige during the Cuban Missile Crisis (October 1962) eventually forced him from office in October 1964. Khrushchev's aggressive use of space exploration as a tool of international politics encouraged JOHN F. KENNEDY to initiate the APOLLO PROJECT—ultimately giving all of humanity one of its greatest technical triumphs.

Kirchhoff, Gustav Robert (1824–87) German physicist who cooperated with ROBERT WILHELM BUNSEN in developing the fundamental principles of SPECTROSCOPY. While investigating the phenomenon of BLACKBODY RADIATION, he applied spectroscopy to study the chemical composition of the SUN—especially the production of the Fraunhofer lines (*see* FRAUNHOFER, JOSEPH VON) in the SOLAR SPECTRUM. His pioneering work contributed to the development of astronomical spectroscopy, one of the major tools in modern ASTRONOMY.

Kirkwood, Daniel (1814–95) American astronomer and mathematician who, in 1866, explained the uneven distribution and gaps in the MAIN-BELT ASTEROID population as being the result of orbital resonances with the PLANET JUPITER. Today, these gaps are called the KIRKWOOD GAPS in his honor.

Komarov, Vladimir M. (1927–67) Russian COSMONAUT and Air Force officer who was the first person to make two trips into OUTER SPACE and also the first person to die while engaged in space travel. On 23 April 1967, he ascended into EARTH ORBIT onboard the new Soviet SPACECRAFT, called *SOYUZ 1*. The flight encountered many difficulties. He finally had to execute an emergency REENTRY maneuver on the 18th orbit (24 April). During the final stage reentry over the Kazakh Republic, the recovery parachute became entangled, causing his spacecraft to impact the ground at high speed. He died instantly and was given a hero's state funeral.

Korolev, Sergei (1907–66) The Russian ROCKET engineer who was the driving technical force behind the initial INTERCONTINENTAL BALLISTIC MISSILE (ICBM) program and the early OUTER SPACE exploration projects of the former Soviet Union. In 1954, he started work on the first Soviet ICBM, the *R-7*. This powerful rocket system was capable of carrying a massive PAYLOAD across continental distances. As part of COLD WAR politics, Soviet premier NIKITA S. KHRUSHCHEV allowed Korolev to use this military rocket to place the first ARTIFICIAL SATELLITE *(SPUTNIK 1)* into ORBIT around EARTH on 4 October 1957—an event now generally regarded as the beginning of the Space Age.

Krikalev, Sergei (b. 1958) Russian cosmonaut who flew on the first joint U.S./Russian SPACE SHUTTLE *Discovery* mission (STS-60) in February 1994. A little over four years later, he flew as a crew member on the Space Shuttle *Endeavour* (STS-88) during the inaugural INTERNATIONAL SPACE STATION (ISS) assembly mission (December 1998). This veteran space traveler also served as a member of the ISS *Expedition-1* crew that launched from the BAIKONUR COSMODROME (Kazakhstan) on a Russian Soyuz EXPENDABLE LAUNCH VEHICLE on 31 October 2000. The SPACE STATION's first crew docked with the orbiting international facility (still under assembly) on 2

November. Krikalev and his ISS *Expedition-1* crewmates departed the ISS on 18 March 2001, riding as Earth-bound passengers on the Space Shuttle (STS-102 mission), which landed at NASA'S KENNEDY SPACE CENTER on 21 March.

Kuiper, Gerard Peter (1905–73) Dutch-American astronomer who postulated in 1951 the presence of thousands of icy PLANETESIMALS in an extended region at the edge of the SOLAR SYSTEM beyond the ORBIT of PLUTO—a region now called the KUIPER BELT in his honor. As a skilled SOLAR SYSTEM observer, in 1944 he discovered that SATURN's largest MOON (TITAN) had an ATMOSPHERE; in 1948 Miranda, the fifth-largest moon of URANUS; and in 1949 Nereid, the outermost moon of NEPTUNE. He also served as an adviser to NASA for the early LUNAR missions of the 1960s.

Lacaille, Nicolas Louis de (1713–62) French astronomer who mapped the positions of nearly 10,000 STARS in the Southern Hemisphere from the Cape of Good Hope (South Africa) between 1751–53. This effort was summarized in the book *Star Catalog of the Southern Sky*—published in 1763, the year after he died.

Lagrange, Joseph Louis (aka Giuseppe Luigi Lagrangia) (1736–1813) Italian-French astronomer and mathematician who made significant contributions to CELESTIAL MECHANICS in the 18th century. His book *Analytical Mechanics,* which was published in 1788, is regarded as an excellent compendium on the subject. He is noted for identifying the LAGRANGIAN LIBRATION POINTS. He described them in about 1772 as the five equilibrium points for a small CELESTIAL BODY under the gravitational influence of two larger bodies.

Lalande, Joseph-Jérôme Le Français de (1732–1807) French astronomer who collaborated in 1751 with NICOLAS LOUIS DE LACAILLE in measuring the distance to the MOON. He also published an extensive catalog in 1801, containing about 47,000 stars—including a "star" that actually was the planet Neptune later discovered by JOHANN GOTTFRIED GALLE in 1846.

Landau, Lev Davidovich (1908–68) Russian physicist who, in the early 1930s, speculated about the existence of the NEUTRON

STAR. He was awarded the Nobel Prize in physics in 1962 for his pioneering work in condensed matter, especially involving the properties of HELIUM.

Langley, Samuel Pierpont (1834–1906) American astronomer and aeronautics pioneer who believed that all life and activity on EARTH was made possible by SOLAR RADIATION. In 1878 he invented the bolometer—an instrument sensitive to incident ELECTROMAGNETIC RADIATION and especially suitable for measuring the amount of INFRARED RADIATION (IR). Modern versions of Langley's instrument are placed onto EARTH-OBSERVING SPACECRAFT to provide regional and global measurements of this PLANET's radiant ENERGY budget. He also led an expedition to the top of Mount Whitney (California) in 1881 to analyze the infrared component of the SUN's SPECTRUM.

Laplace, Pierre Simon de Marquis (1749–1827) French mathematician and astronomer who established the mathematical foundations of CELESTIAL MECHANICS. His work provided a complete mechanical interpretation of the SOLAR SYSTEM, including PERTURBATIONS of planetary motions. In 1796 (apparently independent of IMMANUEL KANT), he introduced his own version of the NEBULA hypothesis when he suggested that the SUN and the PLANETs condensed from a primeval INTERSTELLAR cloud of gas.

Lassell, William (1799–1880) English (amateur) astronomer and wealthy brewer who discovered TRITON, the largest MOON of NEPTUNE, on 10 October 1846—just 17 days after JOHANN GOTTFRIED GALLE discovered the PLANET. During other observations of the SOLAR SYSTEM in the 19th century, he codiscovered the eighth SATURNIAN moon, Hyperion (1848), and then discovered two moons of URANUS in 1851, Ariel and Umbriel.

Leavitt, Henrietta Swan (1868–1921) American astronomer who joined the staff at HARVARD UNIVERSITY OBSERVATORY in 1902 and discovered (about 1912) the period-LUMINOSITY relationship for CEPHEID VARIABLE STARs. Her discovery allowed other astronomers, like HARLOW SHAPLEY, to make more accurate estimates of distances in the MILKY WAY

GALAXY and beyond. During her career, she found over 2,400 VARIABLE STARS. Her efforts helped change people's knowledge of the size of the UNIVERSE.

Lemaître, Georges Édouard (1894–1966) Belgian astrophysicist, cosmologist, and priest who suggested, in 1927, that a violent explosion might have started an expanding UNIVERSE. He based this hypothesis on his interpretation of the GENERAL RELATIVITY theory and upon EDWIN POWELL HUBBLE'S contemporary observation of galactic REDSHIFTs—that the universe was indeed expanding. Other physicists, such as GEORGE GAMOW, built upon Lemaître's work and developed it into the widely accepted BIG BANG theory of modern COSMOLOGY. Central to Lemaître's model is the idea of an initial cosmic egg or superdense primeval ATOM that started the universe in a colossal ancient explosion.

Leonov, Alexei Arkhipovich (b. 1934) Russian COSMONAUT who was the first person to perform a SPACE WALK—a tethered EXTRAVEHICULAR ACTIVITY (EVA) on 18 March 1965 outside his EARTH-orbiting *Voskhod 2* SPACECRAFT. In July 1975, he served as the Russian spacecraft commander during the international APOLLO-SOYUZ TEST PROJECT (ASTP).

Leverrier, Urbain Jean Joseph (1811–77) French astronomer and CELESTIAL MECHANICS practitioner whose mathematical predictions in 1846 (independent of JOHN COUCH ADAMS) concerning the possible location of an eighth as yet undetected PLANET allowed JOHANN GOTTFRIED GALLE to discover NEPTUNE quickly by telescopic observation on 23 September 1846. Despite his great success with the predictive discovery of Neptune, he erroneously suggested that observed PERTURBATIONS in the ORBIT or MERCURY were due to another undiscovered planet (called VULCAN), which was much closer to the SUN than Mercury. However, ALBERT EINSTEIN'S GENERAL RELATIVITY theory allowed astronomers in the early 20th century to understand the perturbations of Mercury without resorting to Leverrier's hypothetical, nonexistent planet Vulcan.

Ley, Willy (1906–69) German-American engineer and technical writer who promoted INTERPLANETARY space travel in the United

States following World War II—especially by writing such popular books as *Rockets, Missiles and Space Travel.*

Lindblad, Bertil (1895–1965) Swedish astronomer who studied the dynamics of stellar motions and suggested in the mid-1920s that the MILKY WAY GALAXY was rotating. His pioneering work contributed to the overall theory of galactic motion and structure.

Lippershey, Hans (aka Jan Lippersheim) (c. 1570–1619) Dutch optician who, in 1608, invented and patented a simple, two-lens TELESCOPE. Once the basic concept for this new optical instrument began circulating throughout western Europe, creative persons like GALILEO GALILEI and JOHANNES KEPLER quickly embraced the telescope's role in observational ASTRONOMY and made improved devices of their own design.

Lockyer, Sir Joseph Norman (1836–1920) English physicist who made spectroscopic studies of the SUN (especially SUNSPOTS). In 1868, based on a new line in the SOLAR SPECTRUM (also observed by PIERRE JULES CÉSAR JANSSEN), he concluded that the Sun contained a new ELEMENT, which he called HELIUM. This element was not discovered in Earth's atmosphere until 1895.

Lovell, Sir Alfred Charles Bernard (b. 1913) English radio astronomer who founded and directed the famous Jodrell Bank Experimental Station between 1951 and 1981. It was this RADIO TELESCOPE facility that also recorded the beeping signals from *SPUTNIK 1* and other early Russian SPACECRAFT at the beginning of the Space Age.

Lovell, James Arthur, Jr. (b. 1928) American ASTRONAUT and U.S. Navy officer who was the commander of the near-fatal NASA *APOLLO 13* mission (1970) to the MOON. Prior to that in-flight aborted LUNAR landing mission, he successfully flew into ORBIT around EARTH with FRANK BORMAN during the *GEMINI* 7 mission (1965) and with EDWIN E. "BUZZ" ALDRIN, JR., on the *Gemini 12* mission (1966). Together with Borman and WILLIAM A. ANDERS, Lovell also participated in the historic *Apollo 8* mission (December 1968)—the first human flight to the vicinity of the Moon.

Lowell, Percival (1855–1916) American astronomer who used his own money to establish an astronomical observatory (the Lowell Observatory) near Flagstaff, Arizona—primarily to support his study of MARS and his aggressive search for signs of an intelligent civilization. Lowell had mistakenly interpreted GIOVANNI VIRGINIO SCHIAPARELLI's reports of CANALI on the MARTIAN surface as evidence of large, water-bearing canals built by intelligent beings. He was consumed by this search and wrote *Mars and Its Canals* (1906) and *Mars As the Abode of Life* (1908) to express these ideas. Although his nonscientific (but popular) interpretation of observed surface features on Mars was not accurate, his astronomical instincts were quite correct in another quest. Based on PERTURBATIONS in the ORBIT of NEPTUNE, he predicted, in 1915, the existence of a planet-sized TRANS-NEPTUNIAN OBJECT. The tiny PLANET PLUTO was discovered in 1930 by CLYDE WILLIAM TOMBAUGH while working at the Lowell Observatory.

Shannon W. Lucid
(Courtesy of NASA)

Lucid, Shannon W. (b. 1943) American ASTRONAUT who served as a NASA mission specialist on five SPACE SHUTTLE flights: STS-51G (June 1985), STS-34 (October 1989), STS-43 (August 1991), and STS-58 (October 1993). She also served as "COSMONAUT" engineer 2 on Russia's MIR SPACE STATION (1996), joining the Russian crew from the SPACE SHUTTLE *Atlantis* (STS-76) in March and returning to EARTH after 188 days in ORBIT as a passenger on the Space Shuttle *Atlantis* (STS-79 mission). Dr. Lucid currently holds the record for the most in-orbit hours by any woman space traveler.

Lundmark, Knut Emil (1889–1958) Swedish astronomer who investigated galactic phenomena and suggested that the term *SUPERNOVA* be applied to the brightest NOVAS observed in other GALAXIES.

Luyten, Willem Jacob (1899–1994) Dutch-American astronomer who specialized in the detection and identification of WHITE DWARF STARS.

Lyot, Bernard Ferdinand (1897–1952) French astronomer who invented the coronagraph in 1930. This device allowed astronomers to study the CORONA of the SUN in the absence of a SOLAR ECLIPSE—thereby greatly advancing the field of SOLAR physics.

Mach, Ernst (1838–1916) Austrian physicist who investigated the phenomena associated with high-speed flow.

Magellan, Ferdinand (1480–1521) Portuguese explorer who, in 1519, became the first person to record the two DWARF GALAXIES visible to the NAKED EYE in the Southern Hemisphere. These nearby galaxies are now called the MAGELLANIC CLOUDS in his honor.

Maskelyne, Nevil (1732–1811) English astronomer who served as the fifth ASTRONOMER ROYAL (1765–1811) and made significant contributions to the use of astronomical observations in navigation. In 1769, he studied a TRANSIT of VENUS and used data from this event to calculate the EARTH-SUN distance (the ASTRONOMICAL UNIT) to within 1 percent of its modern value.

Mattingly, Thomas K., II (b. 1936) American ASTRONAUT and U.S. Navy officer who served as the SPACECRAFT pilot during NASA's *APOLLO 16* moon-landing mission (April 1972) while astronauts JOHN W. YOUNG and CHARLES MOSS DUKE, JR., descended to the surface in the LUNAR EXCURSION MODULE (LEM). He also commanded the SPACE SHUTTLE on the STS-4 (1982) and STS-51C (1985) orbital missions.

Maxwell, James Clerk (1831–1879) Scottish theoretical physicist who made many important contributions to physics—including the fundamental concept of ELECTROMAGNETIC RADIATION. He made a major contribution to ASTRONOMY in 1857 by correctly predicting from theoretical principles alone that the RINGS of SATURN must consist of numerous small PARTICLES.

McAuliffe, S. Christa Corrigan (1948–86) American ASTRONAUT and school teacher (NASA Teacher in Space program participant) who died along with the rest of the STS-51L mission crew when the SPACE SHUTTLE *Challenger* exploded during LAUNCH ascent on 28 January 1986.

S. Christa Corrigan McAuliffe (Courtesy of NASA)

McNair, Ronald E. (1950–86) American ASTRONAUT and NASA mission specialist who successfully traveled in OUTER SPACE during the STS-41B mission of the SPACE SHUTTLE *Challenger* (February 1984) but then lost his life along with the rest of the *Challenger's* crew when the Space Shuttle vehicle exploded during LAUNCH ascent on 28 January 1986 at the start of the STS-51L mission.

Messier, Charles (1730–1817) French astronomer who, in about 1781, compiled a list (now called the MESSIER CATALOGUE) of NEBULAS and STAR CLUSTERS. As an avid comet hunter, he personally discovered at least 13 new COMETs and codiscovered perhaps six others. He assembled his list of "unmoving" fuzzy nebulas and star clusters as a tool for those engaged in comet searches, because these celestial objects were often mistaken for comets. He designated each of the approximately 100 entries with a separate number, prefixed by the letter *M*. For example, Messier object M1 is the CRAB NEBULA and M31 is the ANDROMEDA GALAXY. These designations are still used in ASTRONOMY.

Meton (of Athens) (c. 460 B.C.E.–?) Greek astronomer who discovered around 432 B.C.E. that a period of 235 LUNAR months coincides with precisely an interval of 19 YEARs. After each 19-year interval, the phases of the MOON start taking place on the same days of the year. Both the ancient Greek and Jewish CALENDARS used the METONIC CYCLE, and it became the main calendar of the ancient Mediterranean world until replaced by the JULIAN CALENDAR in 46 B.C.E.

Michelson, Albert Abraham (1852–1931) Polish-born, German-American physicist who, in 1878, accurately measured the SPEED OF LIGHT and received the 1907 Nobel Prize in physics for his precise optical measurements. He also collaborated with EDWARD WILLIAMS MORLEY in 1887 to perform an important experiment (now called the Michelson-Morley experiment) that dispelled the prevailing concept of LIGHT traveling through the UNIVERSE by using an invisible ether as the medium. They could not detect this hypothetical, all-pervading ether. Its absence also provided ALBERT EINSTEIN with important empirical evidence upon which to construct his SPECIAL RELATIVITY theory in 1905.

Milne, Edward Arthur (1896–1950) English astrophysicist who studied the dynamic processes and ENERGY transfer mechanisms in stellar atmospheres.

Minkowski, Rudolph Leo (1895–1976) German-American astrophysicist who used SPECTROSCOPY to examine NEBULAS (especially the CRAB NEBULA) and SUPERNOVA remnants. In the mid-1950s, he collaborated with WALTER BAADE in the optical identification of several EXTRAGALACTIC RADIO WAVE sources.

Mitchell, Edgar Dean (b. 1930) American ASTRONAUT and U.S. Navy officer who served as the LUNAR EXCURSION MODULE (LEM) pilot during the *APOLLO 14* MOON-landing mission (January–February 1971). Although this was his first NASA space mission, Mitchell immediately became a member of the exclusive Moon walkers club when he accompanied ALAN B. SHEPARD as they explored the Fra Mauro region of the LUNAR surface.

Mitchell, Maria (1818–89) American astronomer who achieved international recognition for discovering a COMET in 1847. As the first professionally recognized American woman astronomer, she worked as a human computer for the Nautical Almanac Office of the U.S. government from about 1849 to 1868—carefully performing (by hand) precise calculations for the motions of the PLANET VENUS. From 1865 until her death, she served as a professor of ASTRONOMY at the newly opened Vassar (Female) College in Poughkeepsie, New York, and also directed the school's observatory.

Morgan, William Wilson (1906–94) American astronomer who performed detailed investigations of large bluish-white STARS (that is, O-type and B-type SPECTRAL CLASSIFICATION) in the MILKY WAY GALAXY and used these data in 1951 to infer the structure of the two nearest galactic spiral arms.

Morley, Edward Williams (1838–1923) American chemist who collaborated with ALBERT ABRAHAM MICHELSON in conducting the classic either drift experiment of 1887 that led to the collapse of the commonly held (but erroneous) concept that

LIGHT needed this invisible ether medium to propagate through OUTER SPACE.

Newcomb, Simon (1835–1909) Canadian-born American mathematical astronomer who prepared extremely accurate tables for the movement of SOLAR SYSTEM bodies while working for the Nautical Almanac Office of the U.S. NAVAL OBSERVATORY in the 19th century. In 1860, he presented a paper suggesting that the MAIN-BELT ASTEROIDS did not originate from the disintegration of a a single PLANET—as was commonly assumed at the time.

Newton, Sir Isaac (1642–1727) The brilliant, though introverted, English physicist and mathematician whose law of gravitation, three laws of motion, development of calculus, and design of a new type of REFLECTING TELESCOPE identify him as one the greatest scientific minds in human history. Through the patient encouragement and financial support of EDMOND HALLEY, Newton published his great work, *Mathematical Principles of Natural Philosophy* (or *The Principia*), in 1687. This monumental book transformed the practice of physical science and completed the scientific revolution started by NICHOLAS COPERNICUS, JOHANNES KEPLER, and GALILEO GALILEI. *See also* NEWTONIAN TELESCOPE; NEWTON'S LAW OF GRAVITATION; NEWTON'S LAWS OF MOTION.

Oberth, Hermann J. (1894–1989) Transylvanian-born German ROCKET scientist who, like KONSTANTIN TSIOLKOVSKY and ROBERT HUTCHINGS GODDARD, helped establish the field of ASTRONAUTICS and vigorously promoted the concept of space travel throughout his life. His inspirational 1923 publication *The Rocket Into Planetary Space* provided a comprehensive discussion of all the major aspects of space travel. His 1929 award-winning book *Roads to Space Travel* popularized the concept of space travel for technical and nontechnical readers alike.

Olbers, Heinrich Wilhelm (1758–1840) German astronomer who discovered the MAIN-BELT ASTEROIDS PALLAS (1802) and Vesta (1807). He also formulated (in about 1826) the philosophical discussion known as Olber's paradox—namely, the question, Why isn't the night sky (with its infinite number of STARS), as bright as the surface of the SUN? The contemporary explanation uses the concept of an expanding UNIVERSE within which very distant stars become obscure due to extensive REDSHIFT, which weakens their LIGHT and keeps the night sky dark.

Onizuka, Ellison S. (1946–86) American ASTRONAUT and U.S. Air Force officer who served as a NASA mission specialist on the successful Department of Defense-sponsored STS-51C SPACE SHUTTLE *Discovery* mission (January 1985) and then lost his life on 28 January 1986 when the Space Shuttle *Challenger* exploded during LAUNCH ascent at the start of the fatal STS-51L mission.

Oort, Jan Hendrik (1900–92) Dutch astronomer who made pioneering studies of the dimensions and structure of the MILKY WAY GALAXY in the 1920s and proposed, in 1950, that a large swarm of COMETS (now called the OORT CLOUD) circles the SUN at a distance of between 50,000 and 80,000 ASTRONOMICAL UNITS.

Ellison S. Onizuka
(Courtesy of NASA)

Öpik, Ernest Julius (1893–1985) Estonian astronomer with many astronomical interests who gave emphasis to the study of COMETS and METEOROIDS. In 1932, after examining the PERTURBATIONS of cometary ORBITS, he suggested that a huge population of cometary NUCLEI might be at a great distance from the SUN (perhaps at about 60,000 ASTRONOMICAL UNITS). He further speculated that the close passage of a ROGUE STAR might perturb this reservoir of icy CELESTIAL BODIES, causing a few new comets to approach the inner SOLAR SYSTEM. JAN HENDRIK OORT revived this idea in 1950. Öpik also developed a theory of ABLATION to describe how meteoroids burn up and disintegrate in EARTH's ATMOSPHERE due to AERODYNAMIC HEATING.

Payne-Gaposchkin, Cecilia Helena (1900–79) English-American astronomer who, in 1925, proposed that the ELEMENT HYDROGEN was the most abundant element in STARS and in the universe at large. While collaborating with her husband (Russian-American astronomer Sergei Gaposchkin, 1898–1984) in 1938, she published an extensive catalog of VARIABLE STARS.

Peachy, Eleanor Margaret (b. 1922) *See* (ELEANOR) MARGARET BURBIDGE.

Penzias, Arno Allen (b. 1933) German-American physicist who, in the early 1960s, worked with a colleague (ROBERT WOODROW WILSON) at Bell Laboratories to examine natural sources of RADIO WAVE noise that might interfere with COMMUNICATIONS SATELLITES and discovered the COSMIC MICROWAVE BACKGROUND (CMB) in the process. This all-pervading MICROWAVE BACKGROUND RADIATION at the edge of the observable UNIVERSE is the cooled remnant (~3 K) of the BIG BANG explosion. Their discovery provided cosmologists with empirical evidence of a very hot, explosive phase at the beginning of the universe. Penzias and Wilson shared the 1978 Nobel Prize in physics for this work.

Piazzi, Giuseppe (1746–1826) Italian astronomer and monk who discovered the first ASTEROID (or MINOR PLANET) on 1 January 1801 from the observatory that he founded and directed in Palermo, Sicily. He named the asteroid CERES after the ancient Roman goddess of agriculture and the patroness of Sicily. Observation of this asteroid was then lost in the Sun's glare. However, calculations of its ORBIT by CARL FRIEDRICH GAUSS (based on only three of Piazzi's observations) allowed HEINRICH WILHELM OLBERS to relocate the CELESTIAL BODY in 1802. When Piazzi died, only four asteroids were known. However, his discovery pointed out the existence of the MAIN-BELT ASTEROIDS in the gap between MARS and JUPITER. The 1,000th asteroid was discovered in 1923 and named Piazzia in his honor.

Picard, Jean (1620–82) French astronomer and priest whose careful measurement of EARTH's circumference in 1671 (using the length of a degree on the meridian at Paris) provided the first major improvement since ERATOSTHENES OF CYRENE and came quite close to the modern value.

Pickering, Edward Charles (1846–1919) American astronomer and older brother of WILLIAM HENRY PICKERING who, from 1877, served as the director of the HARVARD COLLEGE OBSERVATORY (HCO) for over four decades. He supervised the production of the *Henry Draper Catalogue* (by ANNIE JUMP CANNON,

WILLIAMINA PATON FLEMING, HENRIETTA SWAN LEAVITT, and others), which listed 225,000 STARS according to their spectra as defined by the HARVARD CLASSIFICATION SYSTEM. He promoted ASTROPHOTOGRAPHY and, in 1889, discovered the first spectroscopic BINARY (DOUBLE) STAR SYSTEM—two stars so visually close together that they can be distinguished only by the DOPPLER SHIFT of their SPECTRAL LINES.

Pickering, William Hayward (b. 1910) New Zealander-American physicist and engineer who directed the Caltech Jet Propulsion Laboratory (JPL) from 1954 to 1976 and supervised the development of many of the highly successful U.S. planetary exploration SPACECRAFT including the first American SATELLITE *(EXPLORER 1)*, NASA's *MARINER* spacecraft to VENUS and MARS, NASA's RANGER PROJECT and SURVEYOR PROJECT to the MOON, NASA's VIKING PROJECT to Mars, and the incredible *VOYAGER* missions to the OUTER PLANETS and beyond.

Pickering, William Henry (1858–1938) American astronomer and younger brother of EDWARD CHARLES PICKERING who established an auxiliary astronomical observatory for Harvard at Arequipa, Peru, in 1891. He discovered the ninth MOON of SATURN (called Phoebe) in 1898 at this observatory. He published a photographic atlas of the Moon in 1903 using data collected at Harvard's astronomical station on the island of Jamaica. Unfortunately, he became overzealous in interpreting these LUNAR photographs and reported seeing evidence of vegetation and frost. Influenced by such mistaken evidence, he suspected that life-forms might be on the Moon. He also conducted a photographic search for a PLANET beyond NEPTUNE, and he may have even captured an IMAGE of PLUTO in 1919—but failed to recognize it.

Planck, Max Karl (1858–1947) German physicist who introduced QUANTUM THEORY in 1900—a powerful new theory concerned with the transport of ELECTROMAGNETIC RADIATION in discrete ENERGY packets or QUANTA. His work represents one of the two great pillars of modern physics (the other being ALBERT EINSTEIN'S RELATIVITY). Planck received the 1918 Nobel Prize in physics for this important accomplishment. *See also* PLANCK'S RADIATION LAW.

Pogson, Norman Robert (1829–91) English astronomer who, in 1856, suggested the use of a new, mathematically based scale for describing stellar MAGNITUDES. First, he verified SIR (FREDERICK) WILLIAM HERSCHEL's discovery that a first-magnitude STAR is approximately 100 times brighter than a sixth-magnitude star—the limit of NAKED EYE observing as originally proposed by HIPPARCHUS OF NICAEA and later refined by PTOLEMY. He then suggested that because an interval of five magnitudes corresponds to a factor of 100 in brightness, a one-magnitude difference in brightness should correspond to the fifth root of 100, or 2.512. Pogson's scale is now universally used in modern ASTRONOMY.

Pond, John (1767–1836) English astronomer who was appointed as the sixth ASTRONOMER ROYAL in 1811. He served in this position until 1835, upgrading the ROYAL GREENWICH OBSERVATORY with new instruments and publishing a very accurate STAR catalog in 1833.

Pons, Jean (1761–1831) French astronomer who was an avid comet hunter, discovering (or codiscovering) 37 COMETs between 1801 and 1827—a number that represents a personal observing record in ASTRONOMY and about 75 percent of all comets discovered within that period.

Ptolemy (aka Claudius Ptolemaeus) (c. 100–c. 170 C.E.) Greek astronomer living in Alexandria, Egypt, who wrote (in about 150 C.E.) *Syntaxis (The Great Mathematical Compilation)*—a compendium of astronomical and mathematical knowledge from all the great Greek philosophers and astronomers. His book preserved the Greek GEOCENTRIC COSMOLOGY in which EARTH was considered the unmoving center of the UNIVERSE while the wandering PLANETs and FIXED STARS revolved around it. Arab astronomers translated the book in about 820 C.E., calling it *The Almagest (The Greatest)*. The PTOLEMAIC SYSTEM remained essentially unchallenged in western thinking until NICHOLAS COPERNICUS and the start of the scientific revolution in the 16th century.

Pythagoras (c. 580 B.C.E.–c. 500 B.C.E.) Greek philosopher and mathematician who taught that EARTH was a perfect sphere at the center of the UNIVERSE and that the PLANETs and STARS moved in circles around it. His thoughts influenced Greek COSMOLOGY and western thinking for centuries. However, they arose primarily from his mystical belief that the circle was a perfect form rather than via careful observation. Although NICHOLAS COPERNICUS successfully challenged GEOCENTRIC cosmology in the 16th century, JOHANNES KEPLER displaced the concept of circular ORBITS for the planets early in the 17th century.

Reber, Grote (b. 1911) American radio engineer who built the world's first RADIO TELESCOPE in his backyard in 1937 and for several years thereafter was the world's only practicing radio astronomer. His steerable, approximately 10 m diameter, parabolic dish antenna detected many radio sources, including the RADIO WAVE signals in the CONSTELLATION Sagittarius from the direction of the center of the MILKY WAY GALAXY— confirming KARL GUTHE JANSKY's discovery in the early 1930s.

Rees, Sir Martin John (b. 1942) English astrophysicist who investigated the nature of the DARK MATTER of the UNIVERSE and was appointed as current (15th) ASTRONOMER ROYAL in 1995.

Reinmuth, Karl (1892–1979) German astronomer who discovered the 1.4 km diameter ASTEROID that came within 10 million km (about 0.07 ASTRONOMICAL UNIT) of EARTH in 1932. Called Apollo, this asteroid now gives its name to the APOLLO GROUP of NEAR-EARTH ASTEROIDS whose ORBITS cross Earth's orbit around the SUN.

Resnik, Judith A. (1949–86) American ASTRONAUT and NASA mission specialist who successfully traveled into ORBIT around EARTH during the SPACE SHUTTLE *Discovery's* STS-41D mission (1984) but then lost her life when the Space Shuttle *Challenger* exploded during LAUNCH ascent on 28 January 1986 at the start of the STS-51L mission.

Richer, Jean (1630–96) French astronomer who led a scientific expedition to Cayenne, French Guiana, in 1671 that provided careful measurements of MARS at the same time GIOVANNI DOMENICO CASSINI was making similar observations in France.

These simultaneous observations supported the first adequate PARALLAX of Mars and enabled Cassini to calculate the distance from EARTH to Mars, leading to improved dimensions for the SOLAR SYSTEM that closely approached modern values. On this expedition, Richer also made geophysical observations with a pendulum that eventually allowed SIR ISAAC NEWTON to discover the OBLATENESS of Earth.

Ride, Sally K. (b. 1951) American ASTRONAUT and NASA mission specialist who was the first American woman to travel in OUTER SPACE. She did this as a crew member on the STS-7 mission of the SPACE SHUTTLE *Challenger,* which lifted off from the KENNEDY SPACE CENTER on 18 June 1983. She flew again in space as a crew member on the STS-41G mission of Space Shuttle *Challenger* (1984). She also served as a member of the presidential commission that investigated the *CHALLENGER* ACCIDENT (January 1986).

Roche, Edouard Albert (1820–83) French mathematician who calculated how close a natural SATELLITE can orbit around its parent PLANET (PRIMARY BODY) before the influence of GRAVITY creates tidal FORCEs that rip the MOON apart, creating a RING of debris. By assuming the planet and its satellite have the same DENSITY, he calculated that this limit (called the ROCHE LIMIT) occurs when the moon is at a distance of approximately 2.5 times the radius of the planet or less.

Roemer, Olaus Christensen (aka Ole Römer) (1644–1710) Danish astronomer who, while working at the PARIS OBSERVATORY with GIOVANNI DOMENICO CASSINI, noticed time discrepancies for successive predicted ECLIPSES of the MOONS of JUPITER and correctly concluded that LIGHT must have a finite VELOCITY. He then attempted to calculate the SPEED OF LIGHT in 1675. However, his numerical results (about 227,000 km/s) were lower than the currently accepted value (299,792 km/s) mainly because of the inaccuracies in accepted SOLAR SYSTEM distances at that time.

Roosa, Stuart Allen (1933–94) American ASTRONAUT and U.S. Air Force officer who served as the Apollo spacecraft pilot during the *Apollo 14* MOON-landing mission (1971). He remained in LUNAR ORBIT as fellow APOLLO PROJECT astronauts ALAN B.

SHEPARD and EDGAR DEAN MITCHELL explored the lunar
surface in the Fra Mauro region.

Rosse, third earl of (aka William Parsons) (1800–67) Irish
astronomer who used his personal wealth to construct a
massive 1.8 m diameter (72 in) REFLECTING TELESCOPE on his
family castle grounds in Ireland. Although located in a
geographic region poorly suited for observation, the earl
engaged in ASTRONOMY as a hobby and used his giant
TELESCOPE (the largest in the world at the time) to detect and
study "fuzzy spiral NEBULAS" (many of which were later
identified as SPIRAL GALAXIES). He gave the CRAB NEBULA its
name in 1848 and also made detailed observations of the
ORION NEBULA.

Rossi, Bruno Benedetto (1905–93) Italian-American physicist and
astronomer who investigated the fundamental nature of
COSMIC RAYS in the 1930s and then collaborated with RICARDO
GIACCONI and others in the 1962 discovery of X-RAY sources
outside the SOLAR SYSTEM using SENSORS carried into OUTER
SPACE by SOUNDING ROCKETS. NASA named a scientific
SATELLITE launched in 1995 the *Rossi X-ray Timing Explorer
(RXTE)* in his honor.

Russell, Henry Norris (1877–1957) American astronomer who
collaborated in 1913 with EJNAR HERTZSPRUNG in the
development of the HERTZSPRUNG-RUSSELL (HR) DIAGRAM that
is of fundamental importance in understanding the theory of
STELLAR EVOLUTION. In 1928, he performed a detailed
analysis of the SUN'S SPECTRUM—showing hydrogen as its
major constituent but also noting the presence of other
ELEMENTS and their relative abundances.

Ryle, Sir Martin (1918–84) English radio astronomer who established
a center for RADIO ASTRONOMY at Cambridge University after
World War II and developed (in about 1960) the technique of
APERTURE SYNTHESIS. He shared the 1974 Nobel Prize in
physics with ANTHONY HEWISH for the discovery of the
PULSAR and their pioneering techniques in radio astronomy,
including the use of aperture synthesis. He also served from
1972 to 1982 as the 12th ASTRONOMER ROYAL.

Sagan, Carl Edward (1934–96) American astronomer and science writer who investigated the origin of life on EARTH and the possibility of EXTRATERRESTRIAL LIFE. In the early 1960s, he suggested that a RUNAWAY GREENHOUSE effect could be operating on VENUS. In the 1970s and beyond, he used his collection of popular books and a television series *(Cosmos)* to communicate science effectively, especially ASTRONOMY and ASTROPHYSICS, to the general public.

Sänger, Eugen (1905–64) Austrian ASTRONAUTICS pioneer who envisioned ROCKET planes and reusable space transportation systems in the 1920s and 1930s—a technical vision partially fulfilled by NASA's SPACE SHUTTLE half a century later. During World War II, he directed a rocket research program for the German Air Force and focused his attention on a long-range, winged rocket bomber that could travel intercontinental distances by skipping in and out of EARTH's ATMOSPHERE.

Scheiner, Christoph (1573–1650) German mathematician, astronomer, and Jesuit priest who, independently of GALILEO GALILEI, designed and used his own TELESCOPE to observe the SUN, discovering SUNSPOTs in 1611. By attempting to preserve ARISTOTLE's hypothesis of the immutable (unchanging) heavens, Scheiner interpreted the sunspots as being small SATELLITEs that encircled the Sun as opposed to changing, moving features of the Sun itself. This interpretation stirred up a great controversy with Galileo, who vigorously endorsed the Copernican revolution in ASTRONOMY.

Schiaparelli image on a Hungarian stamp (Courtesy of author)

Schiaparelli, Giovanni Virginio (1835–1910) Italian astronomer who carefully observed MARS in the 1870s and made a detailed map of its surface, including some straight markings that he described as CANALI, meaning channels (in Italian). Unfortunately, when translated into English as *canals,* some astronomers (like PERCIVAL LOWELL) completely misunderstood Schiaparrelli's meaning and launched a frantic observational search for canals—as the presumed artifacts of an intelligent alien civilization. He also worked on the relationship between METEOR showers and the passage and/or disintegration of COMETs.

Schirra, Walter M. (b. 1923) American ASTRONAUT and U.S. Navy officer who was selected as one of NASA's original seven astronauts and is the only one to have traveled into OUTER SPACE in all three 1960s programs: the MERCURY PROJECT, the GEMINI PROJECT, and the APOLLO PROJECT.

Schmidt, Maarten (b. 1929) Dutch-American astronomer who discovered the first QUASAR in 1963. The quasar had an enormous REDSHIFT in its SPECTRUM, indicating the very distant object was traveling away from EARTH at more than 15 percent of the SPEED OF LIGHT.

Schmitt, Harrison H. (b. 1935) American ASTRONAUT and former U.S. senator who was a member of the *APOLLO 17* LUNAR-landing mission (December 1972). He and EUGENE A. CERNAN became the last human beings to walk on the MOON in the 20th century. They explored the Tauraus-Littrow region of the lunar surface.

Schriever, Bernard (b. 1913) American U.S. Air Force officer and engineer who supervised the rapid development of the ATLAS, THOR, and TITAN BALLISTIC MISSILES during the COLD WAR. He created and applied an innovative systems engineering approach that saved a great deal of development time. His ROCKETs not only served the United States in the area of national defense, they also supported the SPACE LAUNCH VEHICLE needs of NASA and the civilian space exploration community.

Schwabe, Sammuel Heinrich (1789–1875) German pharmacist and amateur astronomer who made systematic observations of the SUN for many years while searching for a hypothetical PLANET (VULCAN) that traveled around the Sun inside the ORBIT of MERCURY. Instead of this fictitious planet, he discovered that SUNSPOTs have a cycle of about 11 years or so. He announced his findings in 1843 and eventually received recognition for this discovery in the 1860s.

Schwarzschild, Karl (1873–1916) German astronomer who applied RELATIVITY theory to very high-DENSITY objects and point masses (SINGULARITIES). In 1916, he introduced the concept of the SCHWARZSCHILD RADIUS—the zone (or EVENT HORIZON) around a superdense, gravitationally collapsing STAR from which nothing, not even LIGHT, can escape. His work marks the start of BLACK HOLE ASTROPHYSICS.

Scobee, Francis R. (1939–86) American ASTRONAUT and U.S. Air Force officer who successfully flew into OUTER SPACE as the pilot of the SPACE SHUTTLE *Challenger* on the STS-41C mission (April 1984) and then as commander during the STS-51L mission was killed on 28 January 1986 when the Space Shuttle *Challenger* exploded during LAUNCH ascent.

Scott, David R. (b. 1932) American ASTRONAUT and U.S. Air Force officer who participated in the GEMINI PROJECT and APOLLO PROJECT and traveled to the MOON during the *Apollo 15* mission (1971). In this successful LUNAR landing mission, he and astronaut JAMES BENSON IRWIN used an electric-battery powered LUNAR ROVER vehicle for the first time.

Secchi, Pietro Angelo (1818–78) Italian astronomer and Jesuit priest who was the first to apply SPECTROSCOPY to ASTRONOMY systematically. By 1863, he completed the first major spectroscopic survey of the STARS and published a catalog containing the SPECTRA of more than 4,000 stars. After examining these data, he proposed (in 1867) that such stellar spectra can be divided into four basic classes—the Secchi classification system. His pioneering work eventually evolved into the HARVARD CLASSIFICATION SYSTEM and to an improved understanding of STELLAR EVOLUTION. He also pursued advances in ASTROPHOTOGRAPHY by photographing SOLAR ECLIPSES to assist in the study of solar phenomena such as PROMINENCES.

Seyfert, Carl Keenan (1911–60) American astronomer who, in 1943, discovered a special group of SPIRAL GALAXIES now known as SEYFERT GALAXIES.

Shapley, Harlow (1885–1972) American astronomer who used a detailed study of VARIABLE STARS (especially CEPHEID VARIABLES) in about 1914 to establish more accurate dimensions for the MILKY WAY GALAXY and to discover that the SUN was actually two-thirds of the way out in the rim of this SPIRAL GALAXY—some 30,000 LIGHT-YEARS from its center. Up until then, astronomers thought the Sun was located near the center of the galaxy. In 1921, he engaged in a public debate with HEBER DOUST CURTIS concerning the nature of distant spiral NEBULAS, which Shapley originally believed were either part of this galaxy or very close neighbors. He also studied the MAGELLANIC CLOUDS and CLUSTERS OF GALAXIES.

Shepard, Alan B., Jr. (1923–99) American astronaut and U.S. Navy officer who was the first American to travel into OUTER SPACE on 5 May 1961 on the MERCURY PROJECT-Redstone ROCKET suborbital mission. As one of the original seven Mercury Project astronauts, he rode for about 15 minutes in his tiny *Freedom 7* SPACECRAFT on a BALLISTIC TRAJECTORY and was recovered about 450 km downrange from CAPE CANAVERAL in the Atlantic Ocean. He made a second, much longer journey into space in 1971 as the commander of the *APOLLO 14* MOON-landing mission. Together with EDGAR DEAN MITCHELL, he explored the LUNAR surface in the Fra Mauro region (February 1971).

Alan B. Shepard, Jr.
(Courtesy of NASA)

Slayton, Deke, Jr. (1924–93) American ASTRONAUT who was selected as one of the original seven MERCURY PROJECT astronauts and who traveled into OUTER SPACE as part of the first cooperative international RENDEZVOUS and DOCKING mission, the APOLLO-SOYUZ TEST PROJECT (July 1975).

Slipher, Earl Carl (1883–1964) American astronomer and brother of VESTO MELVIN SLIPHER who contributed to advances in ASTROPHOTOGRAPHY, especially innovative techniques to produce high-quality IMAGES of MARS, JUPITER, and SATURN.

Slipher, Vesto Melvin (1875–1969) American astronomer and brother of EARL CARL SLIPHER who began spectroscopic studies in 1912 of the LIGHT from spiral nebulas (now recognized as GALAXIES) and observed DOPPLER SHIFT (REDSHIFT) phenomena—suggesting that these objects were receding from EARTH at very high speed. His work provided the foundation upon which EDWIN POWELL HUBBLE and others developed the concept of an expanding UNIVERSE.

Smith, Michael J. (1945–1986) American ASTRONAUT and U.S. Navy officer who served as the pilot on the STS-51L mission of the SPACE SHUTTLE *Challenger*—a fatal mission in which the launch vehicle exploded shortly after LIFTOFF, claiming the lives of Smith and the other six crew members on 28 January 1986.

Stafford, Thomas P. (b. 1930) American ASTRONAUT and U.S. Air Force officer who flew into OUTER SPACE as the pilot on NASA's *GEMINI 6* mission (December 1965), as the commander on the *Gemini 9* mission (June 1966), as the commander of the *APOLLO 10* mission (May 1969), and finally as U.S. commander of the APOLLO-SOYUZ TEST PROJECT (July 1975). The *Apollo 10* crew traveled to the MOON and demonstrated all the operational steps for a landing mission while in LUNAR ORBIT except the actual physical landing on the lunar surface— that mission and historic moment were assigned by NASA to the crew of *Apollo 11*.

Stefan, Josef (1835–93) Austrian physicist who, in about 1879, experimentally demonstrated that the ENERGY radiated per unit time by a BLACKBODY was proportional to the fourth power of the body's ABSOLUTE TEMPERATURE. In 1884, LUDWIG BOLTZMANN provided the theoretical foundations for this relationship. They collaborated on the formulation of the STEFAN-BOLTZMANN LAW—a physical principle of great importance to astronomers and astrophysicists.

Struve, Friedrich Georg Wilhelm (1793–1864) German-Russian astronomer and father of OTTO WILHELM STRUVE who set up the Pulkovo Observatory near St. Petersburg, Russia, in the mid-1830s and spent many years investigating and cataloging BINARY (DOUBLE) STAR SYSTEMS. In about 1839, he made one of the early attempts to quantify INTERSTELLAR distances by measuring the PARALLAX of the STAR Vega.

Struve, Otto (1897–1963) Russian-American astronomer and great-grandson of FRIEDRICH GEORG WILHELM STRUVE who investigated STELLAR EVOLUTION and was a strong proponent for the existence of EXTRASOLAR PLANETs, especially around SUNLIKE STARS. As the childless son of Ludwig Struve, his death in 1963 ended the famous four-generation Struve family dynasty in ASTRONOMY.

Struve, Otto Wilhelm (1819–1905) Russian-born German astronomer and son of FRIEDRICH GEORG WILHELM STRUVE who succeeded his father as director of the Pulkovo Observatory and continued the study of BINARY (DOUBLE) STAR SYSTEMS, adding some 500 new binary systems to the list. He had two sons who were also astronomers. Hermann Struve (1854–1920) became director of the Berlin Observatory. Ludwig Struve (1858–1920) became a professor of ASTRONOMY at Kharkov University (Ukraine).

Sullivan, Kathryn D. (b. 1951) American ASTRONAUT and NASA mission specialist who became the first American woman to perform an EXTRAVEHICULAR ACTIVITY (EVA). This space walk occurred during the STS-41G mission of the SPACE SHUTTLE *Challenger* in October 1984. She traveled again in OUTER SPACE as part of the STS-31 mission (1990) during which the Space Shuttle *Discovery* crew deployed the *HUBBLE SPACE TELESCOPE* and as part of the STS-45 mission (1992) during which Space Shuttle *Atlantis* performed EARTH-observing experiments.

Swigert, John Leonard "Jack," Jr. (1931–82) American ASTRONAUT who flew as part of NASA's *APOLLO 13* mission to the MOON (April 1970). An explosion within the Apollo SPACECRAFT service module some 55 hours into the translunar flight forced the crew to reconfigure the LUNAR EXCURSION MODULE (LEM) into a lifeboat. Any attempt at landing in the programmed Fra Mauro area was aborted. Although a spaceflight rookie, his skillful response to this serious in-flight emergency helped fellow astronauts JAMES ARTHUR LOVELL, JR., and FRED WALLACE HAISE, JR., survive this perilous journey and return safely to EARTH.

Tereshkova, Valentina (b. 1937) Russian COSMONAUT who was the first woman to travel in OUTER SPACE. She accomplished this feat on 16 June 1963 by riding the *VOSTOK 6* SPACECRAFT into ORBIT. During her mission, she completed 48 orbits of EARTH. Upon her return, she was awarded the Order of Lenin and made a hero of the (former) Soviet Union.

Thomson, William *See* BARON WILLIAM THOMSON KELVIN.

Titius, Johann Daniel (1729–96) German astronomer who, in 1766, was the first to notice an EMPIRICAL relationship describing the distances of the six known PLANETs from the SUN. JOHANN ELERT BODE later popularized this empirical relationship which has become alternately called Bode's law or the Titius-Bode law.

Titov, Gherman S. (1935–2000) Russian COSMONAUT who was the second person to travel in ORBIT around EARTH. In August 1961, his *VOSTOK 2* SPACECRAFT made 17 orbits of the PLANET, during which he became the first of many space travelers to experience SPACE SICKNESS.

Tombaugh, Clyde William (1906–97) American astronomer who discovered the PLANET PLUTO on 18 February 1930 while working as an assistant at the Lowell Observatory. He used the blinking comparator (an innovative approach to ASTROPHOTOGRAPHY based on the difference in photographic IMAGES taken a few days apart) to detect the elusive trans-Neptunian planet whose existence had been predicted by PERCIVAL LOWELL. He continued to use the comparative technique for the next decade or so to find many new ASTEROIDS, STAR CLUSTERS, and CLUSTERS OF GALAXIES.

Tsiolkovsky, Konstantin Eduardovich (1857–1935) Russian space travel pioneer who is regarded as one of the three founding fathers of ASTRONAUTICS—the other two technical visionaries being ROBERT HUTCHINGS GODDARD and HERMANN J. OBERTH. Tsiolkovsky was a nearly deaf schoolteacher in an obscure rural town within czarist Russia. Yet, despite the physical handicap and remote location, his writings accurately projected future space technologies. In 1895, he published the book *Dreams of Earth and Sky,* in which he discussed the concept of an ARTIFICIAL SATELLITE orbiting EARTH. Many of the most important principles of astronautics appeared in his seminal 1903 work *Exploration of Space by Reactive Devices.* This book linked the use of the ROCKET to space travel and even introduced a design for a LIQUID-PROPELLANT ROCKET ENGINE, using LIQUID HYDROGEN and LIQUID OXYGEN as its chemical PROPELLANTs. His 1924 work *Cosmic Rocket Trains* introduced the concept of the MULTISTAGE ROCKET. Although he never personally constructed any of the rockets proposed in his visionary books, these works inspired many future rocket scientists, including SERGEI KOROLEV, whose powerful rockets did place the world's first artificial satellite *(SPUTNIK 1)* into ORBIT in 1957.

Ulugh, Beg (aka Muhammed Targai) (1394–1449) Mongol astronomer, mathematician, and prince (the grandson of the conqueror Timur) who, in about 1420, began developing a great observatory in the city of Samarkand (modern Uzbekistan). By using large instruments, he and his staff carefully observed the positions of the SUN, MOON, PLANETs, and 994 FIXED STARS, publishing an accurate STAR catalog in 1437 called the *Zij-i Sultani*—the first improved listing of stellar positions and MAGNITUDES since PTOLEMY. Following the death, in 1447, of his father the king, he briefly served as ruler over the region but was assassinated by his own son ('Abd al-Latif) in 1449. With his politically inspired murder died Mongol ASTRONOMY and the greatest astronomer of the time.

Urey, Harold Clayton (1893–1981) American chemist and early astrobiologist who investigated the possible origins of life on EARTH from a chemical perspective and summarized some of his concepts in the 1952 book *The Planets: Their Origin and Development.* In 1953, together with his student Stanley Miller, he performed a classic EXOBIOLOGY experiment (the Urey-Miller experiment) in which gaseous mixtures, simulating Earth's primitive ATMOSPHERE, were subjected to energy sources (such as ULTRAVIOLET RADIATION and lightning discharges). Amino acids (life-forming organic compounds) formed. He had previously been awarded the Nobel Prize in chemistry in 1934 for his discovery of DEUTERIUM. He played a key research role in the American atomic bomb development program (called the Manhattan Project) during World War II.

Van Allen, James Alfred (b. 1914) American physicist and pioneering space scientist whose instruments on the first U.S. SATELLITE *(EXPLORER 1)* detected EARTH'S TRAPPED RADIATION BELTS— zones of magnetically trapped atomic PARTICLES now called the VAN ALLEN RADIATION BELTS in his honor. Following World War II, he performed space environment-related research through the use of SOUNDING ROCKETs. He assisted NASA in the development of scientific instrument PAYLOADS for many of the American satellites that were launched in the early 1960s.

Verne, Jules (1828–1905) French writer and technical visionary who created modern science fiction and its dream of space travel with his classic 1865 novel *From the Earth to the Moon,* an acknowledged source of youthful inspiration for KONSTANTIN EDUARDOVICH TSIOLKOVSKY, ROBERT HUTCHINGS GODDARD, HERMANN J. OBERTH, and WERNHER VON BRAUN.

Vogel, Hermann Carl (1841–1907) German astronomer who performed spectroscopic analyses of STARS that supported the study of STELLAR EVOLUTION and used DOPPLER SHIFT measurements to obtain their radial velocities. In the course of this work in the late 1880s, he was the first to detect spectroscopic BINARY (DOUBLE) STAR SYSTEMS.

Wernher Von Braun (Courtesy of NASA/ Marshall Space Flight Center)

Von Braun, Wernher (1912–77) German-American ROCKET engineer and space travel advocate who developed the V-2 rocket during World War II for the German Army and then assisted the COLD WAR era U.S. space program in the development of both military rockets and civilian LAUNCH VEHICLES. Inspired by HERMANN J. OBERTH's vision of rockets for INTERPLANETARY travel, von Braun devoted his professional life to the development of ever more powerful LIQUID-PROPELLANT ROCKET ENGINES. In the mid-1950s, a professional friendship with WALT DISNEY allowed von Braun to communicate his dream of space travel to millions of Americans. He fulfilled a significant portion of these dreams by developing the mighty *Saturn V* launch vehicle that successfully sent the first human beings to the MOON during NASA's APOLLO PROJECT.

Von Kármán, Theodore (1881–1963) Hungarian-American mathematician and research engineer who pioneered the application of advanced mathematics in aerodynamics and ASTRONAUTICS. In 1944, he cofounded Caltech's Jet Propulsion Laboratory and initiated research on both solid- and liquid-propellant ROCKETS. This laboratory would eventually create some of NASA's most successful deep-space exploration SPACECRAFT. In 1963, President JOHN F. KENNEDY awarded him the first National Medal of Science.

Von Neumann, John (1903–57) Hungarian-born, German-American mathematician who made significant contributions to nuclear physics, game theory, and computer science. He was selected by the U.S. Air Force in 1953 to chair a special panel of experts who evaluated the American strategic BALLISTIC MISSILE program in the face of an anticipated Soviet ballistic missile threat. As a result of the recommendations from von Neumann's panel, President DWIGHT D. EISENHOWER gave strategic ballistic missile development the highest national priority. General BERNARD SCHRIEVER was tasked with creating an operational ATLAS INTERCONTINENTAL BALLISTIC MISSILE (ICBM) as quickly as possible.

Weizsäcker, Carl Friedrich Baron (b. 1912) German theoretical physicist who, in 1938, (independent of the work of HANS ALBRECHT BETHE) suggested that STARs derive their ENERGY from a chain of thermonuclear FUSION reactions, primarily by joining HYDROGEN into HELIUM in a process called the CARBON CYCLE. Reviving (in part) some of the 18th-century work of IMMANUEL KANT and PIERRE SIMON DE MARQUIS LAPLACE, in 1944 he developed a 20th-century version of the NEBULA hypothesis in an attempt to explain how the SOLAR SYSTEM might have formed from an ancient cloud of INTERSTELLAR gas and dust. One of the consequences of his modern version of this hypothesis is that stars with PLANETs are a normal part of STELLAR EVOLUTION.

Whipple, Fred Lawrence (b. 1906) American astronomer who proposed in 1949 the dirty snowball hypothesis for COMET NUCLEI, a hypothesis confirmed when the *GIOTTO* SPACECRAFT encountered COMET HALLEY at close range in March 1986.

White, Edward H., II. (1930–67) American ASTRONAUT and U.S. Air Force officer who performed the first EXTRAVEHICULAR ACTIVITY (EVA) from an orbiting American SPACECRAFT (*GEMINI 4* mission on 7 June 1965) and then died in the flash fire that consumed the interior of the APOLLO PROJECT spacecraft during a training test on the LAUNCH PAD at CAPE CANAVERAL on 26 January 1967. His fellow Apollo astronauts VIRGIL "GUS" I. GRISSOM and ROGER B. CHAFFEE also perished in this fatal accident.

Wildt, Rupert (1905–76) German-American astronomer who performed detailed spectroscopic studies of the ATMOSPHERES of JUPITER and SATURN in the 1930s that suggested the presence of methane and ammonia.

Wilson, Robert Woodrow (b. 1936) American physicist who collaborated with ARNO ALLEN PENZIAS at Bell Laboratories in the mid-1960s and detected the COSMIC MICROWAVE BACKGROUND (CMB) radiation, a discovery providing EMPIRICAL evidence supporting the BIG BANG theory and for which he and Penzias were awarded the 1978 Nobel Prize for physics.

Wolf, Johann Rudolph (1816–93) Swiss astronomer and mathematician who counted SUNSPOTs and sunspot groups to confirm SAMMUEL HEINRICH SCHWABE's work and discovered that the period of the sunspot cycle was approximately 11 years. He was also one of several observers who noticed the relationship between sunspot number and SOLAR ACTIVITY. He became director of the Bern Observatory in 1847. In 1849, he formalized his system for describing solar activity by counting sunspots and sunspot groups, a system formerly referred to as Wolf's sunspot numbers and now called the relative sunspot number.

Wolf, Maximillian (Max) Franz Joseph Cornelius (1863–1932) German astronomer who pioneered the use of ASTROPHO-TOGRAPHY to search for ASTEROIDS. He discovered more than 200 MINOR PLANETs, including asteroid Achilles in 1906, the first of the TROJAN GROUP of asteroids that move around the SUN in JUPITER's ORBIT 60 degrees ahead of and 60 degrees behind the GIANT PLANET.

Woolley, Sir Richard van der Riet (1906–86) English astronomer who investigated stellar and SOLAR ATMOSPHERES, GLOBULAR CLUSTERS, and STELLAR EVOLUTION. He also served as the 11th ASTRONOMER ROYAL from 1956 to 1971.

Worden, Alfred Merrill (b. 1932) American ASTRONAUT and U.S. Air Force officer who served as the command module pilot during NASA's *Apollo 15* LUNAR landing mission (July–August 1971). While astronauts DAVID R. SCOTT and JAMES BENSON IRWIN explored the MOON's surface in the HADLEY RILLE area, he orbited overhead in the APOLLO PROJECT SPACECRAFT.

Young, John W. (b. 1930) American ASTRONAUT and U.S. Navy officer who traveled in OUTER SPACE on six separate NASA missions, starting with the first crewed GEMINI PROJECT flight *(Gemini 3),* which he flew with VIRGIL "GUS" I. GRISSOM on 23 March 1965. After the *Gemini 10* flight (July 1966), he traveled to the MOON as part of the *Apollo 10* mission (May 1969), a key rehearsal mission that orbited the Moon and completed all appropriate APOLLO PROJECT operations just short of landing on the surface. Then, in April 1972, he served as the commander of the *Apollo 16* Moon-landing mission and walked on the LUNAR surface with CHARLES MOSS DUKE, JR. As a tribute to his astronaut skills, NASA selected him to command the SPACE SHUTTLE *Columbia* on the inaugural mission (STS-1) of the SPACE TRANSPORTATION SYSTEM (April 1981). He returned to space for the sixth time as the commander of the Space Shuttle *Columbia* during the STS-9 mission (1983), which was the first flight of the EUROPEAN SPACE AGENCY's *Spacelab.*

Zucchi, Niccolo (1586–1670) Italian astronomer, instrument maker, and Jesuit priest who made high-quality early TELESCOPES in the early 17th century, including one of the earliest known REFLECTING TELESCOPES in about 1616. He personally provided JOHANNES KEPLER with one of his telescopes and reported observing colored belts and spots on JUPITER in 1630 (most likely its GREAT RED SPOT) and spots on Mars (about 1640).

Zwicky, Fritz (1898–1974) Swiss-American astronomer who, while collaborating with WALTER BAADE on SUPERNOVA phenomena in 1934, postulated the creation of a NEUTRON STAR as a result of a supernova explosion. He published an extensive catalog of GALAXIES and CLUSTERS OF GALAXIES in the 1960s.

SECTION THREE
CHRONOLOGY

c. 3000 B.C.E. (to perhaps 1000 B.C.E.) ● STONEHENGE erected on the Salisbury Plain of southern England (possible use: ancient astronomical CALENDAR for prediction of summer SOLSTICE)

c. 1300 B.C.E ● Egyptian astronomers recognize all the PLANETs visible to the NAKED EYE (MERCURY, VENUS, MARS, JUPITER, and SATURN); they also identify over 40 STAR CONSTELLATIONS

c. 500 B.C.E. ● Babylonians devise ZODIAC that is later adopted and embellished by Greeks and used by other early peoples

c. 432 B.C.E. ● The early Greek astronomer METON (OF ATHENS) discovers that 235 LUNAR months makes up about 19 years

c. 375 B.C.E. ● The early Greek mathematician and astronomer EUDOXUS OF CNIDUS starts codifying the ANCIENT CONSTELLATIONS from tales of Greek mythology

c. 366 B.C.E. ● The ancient Greek astronomer and mathematician EUDOXUS OF CNIDUS constructs a NAKED-EYE astronomical observatory. Eudoxus also postulates that EARTH is the center of the UNIVERSE and develops a system of 27 concentric spheres (or shells) to explain and predict the motion of the visible PLANETs (MERCURY, VENUS, MARS, JUPITER, and SATURN), the MOON, and the SUN. His GEOCENTRIC model represents the first attempt at a coherent theory of COSMOLOGY. With ARISTOTLE's endorsement, this early Greek cosmological model eventually leads to the dominance of the PTOLEMAIC SYSTEM in Western culture for almost two millennia

c. 350 B.C.E. ● Greek astronomer HERACLIDES OF PONTUS suggests EARTH spins daily on an AXIS of ROTATION; also teaches that the PLANETs MERCURY and VENUS move around the SUN.

● In his work *Concerning the Heavens,* the great Greek philosopher ARISTOTLE suggests Earth is a sphere and makes an attempt to estimate its circumference. He also endorses and modifies the GEOCENTRIC model of EUDOXUS OF CNIDUS. Because of Aristotle's intellectual prestige, his concentric crystal-sphere model of the heavens

(COSMOLOGY) with Earth at the center of the UNIVERSE essentially remains unchallenged in Western civilization until NICHOLAS COPERNICUS in the 16th century

c. 275 B.C.E. ● The Greek astronomer ARISTARCHUS OF SAMOS suggests an astronomical model of the UNIVERSE (SOLAR SYSTEM) that anticipates the modern HELIOCENTRIC theory proposed by NICHOLAS COPERNICUS. However, these "correct" thoughts that Aristarchus presented in his work *On the Size and Distances of the Sun and the Moon* are essentially ignored in favor of the GEOCENTRIC model proposed by EUDOXUS OF CNIDUS and endorsed by ARISTOTLE

c. 225 B.C.E. ● The Greek astronomer ERATOSTHENES OF CYRENE uses mathematical techniques to estimate the circumference of EARTH

c. 129 B.C.E. ● The Greek astronomer HIPPARCHUS OF NICAEA completes a catalog of 850 STARs that remains important until the 17th century

0 C.E. ● According to Christian tradition, the Star of Bethlehem guides three wise men (thought to be Middle Eastern astronomers) to the nativity

c. 60 C.E. ● The Greek engineer and mathematician HERO OF ALEXANDRIA creates the aeoliphile, a toylike device that demonstrates the action-reaction principle that is the basis of operation of all ROCKET engines

Steam

Hollow Sphere

Hero's aeoliphile

c. 150 ● Greek astronomer PTOLEMY writes *Syntaxis* (later called the *ALMAGEST* by Arab astronomers and scholars)—an important book that summarizes all the astronomical knowledge of the ancient astronomers, including the GEOCENTRIC model of the UNIVERSE that dominates Western science for more than one and one-half millennia

820 ● Arab astronomers and mathematicians establish a school of ASTRONOMY in Baghdad and translate PTOLEMY's work into Arabic after which it became known as the *Great Work* or *al-Majisti* (the *ALMAGEST* by Medieval scholars)

850 ● The Chinese begin to use gunpowder for festive fireworks, including a ROCKET-like device

964 ● Arab astronomer ABD AL-RAHMAN AL-SUFI publishes his *Book of the Fixed Stars,* a STAR catalog with Arabic star names; includes earliest known reference to the ANDROMEDA GALAXY

1232 ● The Chinese army uses fire arrows (crude gunpowder ROCKETs) to repel Mongol invaders at the battle of Kai-fung-fu. This is the first reported use of the rocket in warfare

Battle of Kai-fung-fu

1280–90 ● The Arab historian Al-Hasan al Rammah writes *The Book of Fighting on Horseback and War Strategies,* in which he gives instructions for making both gunpowder and ROCKETs

1379 ● ROCKETs appear in western Europe; used in the siege of Chioggia (near Venice), Italy

1420 ● The Italian military engineer Joanes de Fontana writes *Book of War Machines,* a speculative work that suggests military applications of gunpowder ROCKETs, including a rocket-propelled battering ram and a rocket-propelled torpedo

1424 ● Mongolian ruler and astronomer BEG ULUGH constructs a great observatory at Samarkand

1429 ● The French army uses gunpowder ROCKETs to defend the city of Orléans. During this period, arsenals throughout Europe begin to test various types of gunpowder rockets as an alternative to early cannons

c. 1500 ● According to early ROCKET lore, a Chinese official named Wan-Hu (or Wan-Hoo) attempts to use an innovative rocket-propelled kite assembly to fly through the air. As he sat in the pilot's chair, his servants lit the assembly's 47 gunpowder (black powder) rockets. Unfortunately, this early rocket test pilot disappeared in a bright flash and explosion

1519 ● The Portuguese explorer FERDINAND MAGELLAN becomes the first to record observing the two irregular DWARF GALAXIES visible to the NAKED EYE in the Southern

Hemisphere that now bear his name as the MAGELLANIC CLOUDS

1543 ● The Polish astronomer NICHOLAS COPERNICUS changes history and initiates the Scientific Revolution with his book *On the Revolution of Celestial Orbs (De Revolutionibus Orbium Coelestium)*. This important book, published while Copernicus lay on his deathbed, proposed a sun-centered (HELIOCENTRIC) model of the UNIVERSE in contrast to the long-standing EARTH-centered (GEOCENTRIC) model advocated by PTOLEMY and many of the early Greek astronomers

1572 ● Danish astronomer TYCHO BRAHE discovers a SUPERNOVA that appears in the CONSTELLATION of Cassiopeia. He describes his precise (pretelescope) observations in *De Nova Stella* (1573). The dynamic nature of this "brilliant star" causes Brahe and other astronomers to question the long-cherished hypothesis of ARISTOTLE concerning the unchanging nature of the heavens

1576 ● With the generous support of Danish King Frederick II, TYCHO BRAHE starts construction of a great NAKED-EYE astronomical observatory on the island of Hven in the Baltic Sea. For the next two decades, Brahe's Uraniborg (Castle of the Sky) serves as the world's center for ASTRONOMY. The precise data from this observatory eventually allow JOHANNES KEPLER to develop his important laws of planetary motion

1577 ● The observation of a COMET's elongated TRAJECTORY causes TYCHO BRAHE to question ARISTOTLE's GEOCENTRIC, crystal spheres model of planetary motion. He proposes his own Tychonic system—a modified version of Ptolemaic COSMOLOGY in which all the PLANETs (except EARTH) revolve around the SUN and the Sun with its entire assembly of planets and comets revolves around a stationary Earth

1601 ● Following the death of TYCHO BRAHE, German Emperor Rudolf II appoints JOHANNES KEPLER to succeed him as the imperial mathematician in Prague. Kepler thus acquires Brahe's collection of precise, pretelescopic astronomical observations and uses them to help develop his three laws of planetary motion

1602–04 ● The German astronomer DAVID FABRICIUS makes precise, preTELESCOPE observations of the ORBIT of MARS. These data greatly assist JOHANNES KEPLER in the development of his three laws of planetary motion

1608 ● The Dutch optician HANS LIPPERSHEY develops a crude TELESCOPE

1609 ● The German astronomer JOHANNES KEPLER publishes *New Astronomy* in which he modifies NICHOLAS COPERNICUS'S HELIOCENTRIC model of the UNIVERSE by announcing that the PLANETs have ELLIPTICAL ORBITS rather than circular ones. KEPLER'S LAWS of planetary motion help put an end to more than 2,000 years of GEOCENTRIC Greek ASTRONOMY

1610 ● On 7 January 1610, GALILEO GALILEI uses his TELESCOPE to gaze at JUPITER and discovers the PLANET's four major MOONS (CALLISTO, EUROPA, IO, and GANYMEDE). He announces this and other astronomical observations in his book *Starry Messenger (Sidereus Nuncius)*. Discovery of Jupiter's moons encourages Galileo to advocate the HELIOCENTRIC theory of NICHOLAS COPERNICUS vigorously and brings him into direct conflict with church authorities

1619 ● JOHANNES KEPLER publishes *The Harmony of the World* in which he presents his third law of planetary motion

1639 ● On 24 November (JULIAN CALENDAR), the English astronomer JEREMIAH HORROCKS makes the first observation of a TRANSIT of VENUS

1642 ● GALILEO GALILEI dies while under house arrest near Florence, Italy, for his clashes with church authorities concerning the HELIOCENTRIC theory of NICHOLAS COPERNICUS

1647 ● The Polish-German astronomer JOHANNES HEVELIUS publishes his work, *Selenographia,* in which he provides a detailed description of features on the surface (NEARSIDE) of the MOON

1655 • Dutch astronomer CHRISTIAAN HUYGENS discovers TITAN, the largest MOON of SATURN

1668 • GIOVANNI DOMENICO CASSINI publishes tables that describe the motions of JUPITER's major SATELLITES

• SIR ISAAC NEWTON constructs the first REFLECTING TELESCOPE

1675 • GIOVANNI DOMENICO CASSINI discovers the division of SATURN'S RINGS, a feature that now carries his name

• English King Charles II establishes the ROYAL GREENWICH OBSERVATORY and appoints JOHN FLAMSTEED as its director and also the first ASTRONOMER ROYAL

1680 • Russian Czar Peter the Great sets up a facility to manufacture ROCKETs in Moscow. The facility later moves to Saint Petersburg and provides the Czarist army with a variety of gunpowder rockets for bombardment, signaling, and nocturnal battlefield illumination

1687 • Financed and encouraged by EDMOND HALLEY, SIR ISAAC NEWTON publishes his great work, *The Principia (Philosophiae Naturalis Principia Mathematica)*. This book provides the mathematical foundations for understanding the motion of almost everything in the UNIVERSE, including the orbital motion of PLANETs and the TRAJECTORIES of ROCKET-propelled vehicles

1728 • The English astronomer JAMES BRADLEY discovers the optical phenomenon of ABERRATION OF STARLIGHT. He explains this phenomenon as a combination of EARTH's motion around the SUN and LIGHT having a finite speed

1740 • The Swedish astronomer ANDERS CELSIUS assumes responsibility for the construction and management of the Uppsala Observatory. In 1742, he suggests a temperature scale based on 0° as the boiling point of water and 100° as

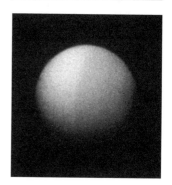

Saturn's largest moon, Titan, discovered in 1655 (NASA/JPL)

the melting point of ice. Following his death in 1744, his colleagues at the Uppsala Observatory continue to use this new temperature scale but invert it. Their action creates the familiar CELSIUS TEMPERATURE SCALE with 0° as the melting point of ice and 100° as the boiling point of water

1748 ● The French physicist PIERRE BOUGUER invents the HELIOMETER—an instrument that measures the LIGHT of the SUN and other luminous bodies

1755 ● The German philosopher IMMANUEL KANT introduces the NEBULA hypothesis, suggesting that the SOLAR SYSTEM may have formed out of an ancient cloud of INTERSTELLAR material. He also uses the term ISLAND UNIVERSES to describe distant DISKlike collections of STARS

1758 ● The English optician JOHN DOLLAND and his son make the first achromatic (that is, without CHROMATIC ABERRATION) TELESCOPE

1766 ● The German astronomers JOHANN ELERT BODE and JOHANN DANIEL TITIUS discover an EMPIRICAL rule (eventually called Bode's law) that describes the apparently proportional distances of the six known PLANETs from the SUN. Their work stimulates the search by other astronomers to find a "missing" planet between MARS and JUPITER, eventually leading to the discovery of MAIN-BELT ASTEROIDS

1772 ● The Italian-French mathematician JOSEPH LEWIS LAGRANGE describes the points in the plane of two objects in ORBIT around their common CENTER OF GRAVITY at which their combined FORCES of GRAVITATION are zero. Today, such points in space are called LAGRANGIAN LIBRATION POINTS

1781 ● German-born English astronomer SIR (FREDERICK) WILLIAM HERSCHEL discovers URANUS, the first PLANET to be found through the use of the TELESCOPE. Herschel originally called this new planet Georgium Sidus (George's Star) to honor the English king, George III. However, astronomers from around the world insisted that a more traditional name from Greco-Roman mythology be used. Herschel bowed to peer pressure and eventually chose the name Uranus, as suggested by the German astronomer JOHANN ELERT BODE

1780s ● The Indian ruler Hyder Ally of Mysore creates a ROCKET corps within his army. Hyder's son, Tippo Sultan, successfully uses rockets against the British in a series of battles in India between 1782 and 1799

1782 ● The English astronomer JOHN GOODRICKE is the first to recognize that the periodic behavior of the VARIABLE STAR Algol (Beta Persei) is actually that of an eclipsing BINARY (DOUBLE) STAR SYSTEM

1786 ● The German-born English astronomer CAROLINE HERSCHEL becomes the world's first (recognized) woman astronomer. While working with her brother SIR (FREDERICK) WILLIAM HERSCHEL, she discovers eight COMETs during the period 1786–97

1796 ● The French mathematician and astronomer PIERRE SIMON DE MARQUIS LAPLACE formalizes the nebular hypothesis of planetary formation—suggesting that the SOLAR SYSTEM originated from a massive cloud of gas and that, over time, the FORCES of GRAVITATION helped the center of this cloud to collapse to form the SUN and smaller remnant clumps of matter to form the PLANETs

1801 ● The Italian astronomer GIUSEPPE PIAZZI discovers the first ASTEROID (MINOR PLANET), CERES, on 1 January

1802 ● The German astronomer HEINRICH WILHELM OLBERS discovers the second ASTEROID (Pallas) and then continues his search, finding the brightest MINOR PLANET (Vesta) in 1807

1804 ● SIR WILLIAM CONGREVE writes *A Concise Account of the Origin and Progress of the Rocket System* and documents the British military's experience in India. He then starts the development of a series of British military (black-powder) ROCKETs

1807 ● The British use about 25,000 of SIR WILLIAM CONGREVE's improved military (black-powder) ROCKETs to bombard Copenhagen during the Napoleonic Wars

1809 ● The brilliant German mathematician, astronomer, and physicist CARL FRIEDRICH GAUSS publishes a major work on CELESTIAL MECHANICS that revolutionizes the calculation of PERTURBATIONS in planetary ORBITs. His work paves the way for other 19th-century astronomers to anticipate mathematically and then discover NEPTUNE (in 1846), using perturbations in the orbit of URANUS

1812 ● British forces use SIR WILLIAM CONGREVE's military ROCKETs against American troops during the War of 1812. British rocket bombardment of Fort William McHenry inspires Francis Scott Key to add "the rockets' red glare" verse in the *Star Spangled Banner*

1814–1817 ● The German physicist JOSEPH VON FRAUNHOFER develops the PRISM SPECTROMETER into a precision instrument. He subsequently discovers the dark lines in the SUN'S SPECTRUM that now bear his name

1840 ● The American astronomers HENRY DRAPER and JOHN WILLIAM DRAPER (his son) photograph the MOON and start the field of ASTROPHOTOGRAPHY

1842 ● The Austrian physicist CHRISTIAN JOHANN DOPPLER describes the DOPPLER SHIFT in a scientific paper. This phenomenon is experimentally verified in 1845 through an interesting experiment in which a locomotive pulls an open railroad car carrying several trumpeters

1846 ● On 23 September, the German astronomer JOHANN GOTTFRIED GALLE discovers NEPTUNE in the location theoretically predicted and calculated by URBAIN JEAN JOSEPH LEVERRIER

1847 ● American astronomer MARIA MITCHELL makes the first telescopic discovery of a COMET

1850 ● American astronomer WILLIAM CRANCH BOND helps expand the field of ASTROPHOTOGRAPHY by successfully photographing the PLANET JUPITER

1851 ● French physicist JEAN-BERNARD LÉON FOUCAULT, using a 65m long pendulum, demonstrates the ROTATION of EARTH to a large crowd in a Parisian church

　　 ● The English astronomer WILLIAM LASSELL discovers the two SATELLITES of URANUS, named Ariel and Umbriel

1852 ● French physicist JEAN-BERNARD LÉON FOUCAULT constructs the first GYRO

1864 ● Italian astronomer GIOVANNI BATTISTA DONATI collects the SPECTRUM of Comet Tempel

1865 ● The French science fiction writer JULES VERNE publishes his famous story, *From the Earth to the Moon (De la Terre a la Lune).* This story interests many people in the concept of space travel, including three young readers who go on to become the founders of ASTRONAUTICS: ROBERT HUTCHINGS GODDARD, HERMANN J. OBERTH, and KONSTANTIN EDUARDOVICH TSIOLKOVSKY

1866 ● American astronomer DANIEL KIRKWOOD explains the unequal distribution of ASTEROIDs found in the main asteroid belt as a CELESTIAL MECHANICS resonance phenomenon, involving the GRAVITATION of JUPITER. In his honor, these empty spaces or gaps are now called the KIRKWOOD GAPS. *See* MAIN-BELT ASTEROID

1868 ● During a total ECLIPSE of the SUN, the French astronomer PIERRE JULES CÉSAR JANSSEN makes important spectroscopic observations that reveal the gaseous nature of SOLAR PROMINENCES

　　 ● The English astronomer SIR WILLIAM HUGGINS becomes the first to observe the gaseous spectrum of a NOVA

　　 ● While observing the SUN, the English astronomer SIR JOSEPH NORMAN LOCKYER postulates the existence of an unknown element that he names HELIUM (meaning "the SUN's ELEMENT")—but not until 1895 is helium is detected on EARTH

1869 ● American clergyman and writer Edward Everett Hale writes *The Brick Moon*—a story that is the first fictional account of a human-crewed SPACE STATION

1874 ● The Scottish astronomer SIR DAVID GILL makes the first of several very precise measurements of the distance from EARTH to the SUN while on an expedition to Mauritius (Indian Ocean). He then visits Ascension Island (South Atlantic Ocean) in 1877 to perform similar measurements

1877 ● While a staff member at the U.S. NAVAL OBSERVATORY in Washington, D.C., the American astronomer ASAPH HALL discovers and names the two tiny MARTIAN MOONs, Deimos and Phobos

1881 ● The American astronomer and aeronautical engineer SAMUEL PIERPOINT LANGLEY leads an expedition to Mount Whitney in the Sierra Nevada to examine how incoming SOLAR RADIATION is absorbed by EARTH's ATMOSPHERE

1897 ● English author H. G. Wells writes the science fiction story *War of the Worlds*—the classic tale about EXTRATERRESTRIAL invaders from MARS

1903 ● The Russian technical visionary KONSTANTIN EDUARDOVICH TSIOLKOVSKY becomes the first to link the ROCKET and space travel when he publishes *Exploration of Space with Reactive Devices*

1906 ● American astronomer GEORGE ELLERY HALE sets up the first tower TELESCOPE for SOLAR research on Mount Wilson in California

1918 ● American physicist ROBERT HUTCHINGS GODDARD writes *The Ultimate Migration*—a far-reaching technology piece within which he postulates an atomic-powered space ark to carry human beings away from a dying SUN. Fearing professional ridicule, Goddard hides the visionary manuscript and it remains unpublished until November 1972—many years after his death in 1945

1919 ● American ROCKET pioneer ROBERT HUTCHINGS GODDARD publishes the Smithsonian monograph *A Method of Reaching Extreme Altitudes*. This important work presents all the fundamental principles of modern rocketry. Unfortunately, members of the press completely miss the true significance of his technical contribution and decide to sensationlize his comments about possibly reaching the MOON with a small, rocket-propelled package. For such "wild fantasy," newspaper reporters dubbed Goddard with the unflattering title of "Moon man"

● On 29 May, a British expedition led by the English astronomer SIR ARTHUR EDDINGTON observes a total SOLAR ECLIPSE and provides scientific verification of ALBERT EINSTEIN'S GENERAL RELATIVITY theory by detecting the slight deflection of a beam of LIGHT from a STAR near the edge of the "darkened" SUN. This important experiment helps usher in a new era in humans' understanding of the physical UNIVERSE

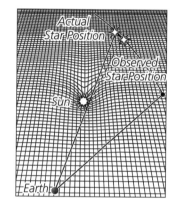

Sun's gravity bending starlight

1922 ● American astronomer WILLIAM WALLACE CAMPBELL measures the deflection of a beam of starlight that just skims the SUN's edge during another SOLAR ECLIPSE. This work provides additional scientific evidence in favor of ALBERT EINSTEIN'S GENERAL RELATIVITY theory

1923 ● Independently of ROBERT HUTCHINGS GODDARD and KONSTANTIN EDUARDOVICH TSIOLKOVSKY, the German space travel visionary HERMANN J. OBERTH publishes the inspirational book *The Rocket Into Planetary Space (Die Rakete zu den Planetenräumen)*

● American astronomer EDWIN POWELL HUBBLE uses CEPHEID VARIABLE STARS as astronomical distance indicators and shows that the distance to the spiral ANDROMEDA GALAXY is well beyond the MILKY WAY GALAXY. As a result of Hubble's work, astronomers begin to postulate that an expanding universe contains many such GALAXIES or ISLAND UNIVERSES

1924 ● The German engineer WALTER HOHMANN writes *The Attainability of Celestial Bodies (Die Erreichbarkeit der Himmelskörper)*—an important work that details the mathematical principles of ROCKET and SPACECRAFT motion. He includes a description of the most efficient (that is, minimum ENERGY) ORBIT transfer path between two coplanar orbits—a frequently used space operations maneuver now called the HOHMANN TRANSFER ORBIT

1926 ● On 16 March in a snow-covered farm field in Auburn, Massachusetts, American physicist ROBERT HUTCHINGS GODDARD makes space technology history by successfully firing the world's first LIQUID-PROPELLANT ROCKET. Although his primitive gasoline (fuel) and LIQUID OXYGEN (OXIDIXER) device burns for only two and one-half seconds and lands about 60 m away, it represents the technical ancestor of all modern liquid-propellant rocket engines

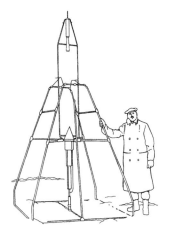

Goddard's first liquid rocket (Courtesy of NASA)

● In April, the first issue of *Amazing Stories* appears. The publication becomes the world's first magazine dedicated exclusively to science fiction. Through science fact and fiction, the modern ROCKET and space travel become firmly connected. As a result of this union, the visionary high-technology dream for many people in the 1930s (and beyond) becomes that of INTERPLANETARY travel

1927 ● Belgian astrophysicist and cosmologist GEORGES ÉDOUARD LEMAÎTRE publishes a major paper introducing a COSMOLOGY model that features an ancient explosion that starts an expanding UNIVERSE. His work leads to the development of the BIG BANG theory

1929 ● American astronomer EDWIN POWELL HUBBLE announces that his measurements of galactic REDSHIFT values indicate that other GALAXIES are receding from this MILKY WAY GALAXY with speeds that increase in proportion to their distance from humans' home galaxy. This observation becomes known as HUBBLE'S LAW

● British scientist and writer JOHN DESMOND BERNAL speculates about large human settlements in OUTER SPACE (later called BERNAL SPHERES) in his book *The World, The Flesh, and the Devil*

● German space travel visionary HERMANN J. OBERTH writes the award-winning book *Roads To Space Travel (Wege zur Raumschiffahrt)* that helps popularize the notion of space travel among nontechnical audiences

1930 ● On 18 February, American astronomer CLYDE WILLIAM TOMBAUGH discovers the PLANET PLUTO

1932 ● In December, American radio engineer KARL GUTHE JANSKY announces his discovery of a stellar radio source in the CONSTELLATION of Sagittarius. His paper is regarded as the birth of RADIO ASTRONOMY

● Following the discovery of the NEUTRON by the English physicist Sir James Chadwick, Russian physicist LEV DAVIDOVICH LANDAU suggests the existence of NEUTRON STARS

1933 ● P. E. Cleator founds the British Interplanetary Society (BIS), which becomes one of the world's most respected space travel advocacy organizations

1935 ● KONSTANTIN EDUARDOVICH TSIOLKOVSKY publishes his last book, *On the Moon,* in which he strongly advocates the SPACESHIP as the means of LUNAR and INTERPLANETARY travel

1936 ● P. E. Cleator, founder of the British Interplanetary Society, writes *Rockets Through Space,* the first serious treatment of ASTRONAUTICS in the United Kingdom. However, several established British scientific publications ridicule his book as the premature speculation of an unscientific imagination

1937 ● American radio engineer GROTE REBER builds the first RADIO TELESCOPE—a modest device located in his backyard. For several years following this construction, Reber is the world's only radio astronomer

1939–1945 ● Throughout World War II, combating nations use of ROCKETs and GUIDED MISSILEs of all sizes and shapes. Of these, the most significant with respect to space exploration is the development of the LIQUID-PROPELLANT V-2 ROCKET by the German Army at Peenemünde under WERNHER VON BRAUN

1942 ● On 3 October, the German A-4 ROCKET (later renamed VENGEANCE WEAPON TWO or V-2 Rocket) completes its first successful flight from the Peenemünde test site on the Baltic Sea. This is the birth date of the modern military BALLISTIC MISSILE

1944 ● In September, the German Army begins a BALLISTIC MISSILE offensive by launching hundreds of unstoppable V-2 ROCKETs (each carrying a one-ton high-explosive WARHEAD) against London and southern England

1945 ● After recognizing the war was lost, the German rocket scientist WERNHER VON BRAUN and key members of his staff surrender to American forces near Reutte, Germany, in early May. Within months, U.S. intelligence teams, under Operation Paperclip, interrogate German ROCKET personnel and sort through carloads of captured documents and equipment. Many of these German scientists and engineers join von Braun in the United States to continue their rocket work. Hundreds of captured V-2 rockets are also disassembled and shipped back to the United States

● Similarly, on 5 May, the Soviet Army captures the German rocket facility at Peenemünde and hauls away any remaining equipment and personnel. In the closing days of World War II in Europe, captured German rocket technology and personnel help set the stage for the great MISSILE and space race of the COLD WAR

● On 16 July, the United States explodes the world's first nuclear weapon. The test shot, code named Trinity, occurs in a remote portion of southern New Mexico and changes the nature of warfare forever. As part of the Cold War confrontation between the United States and the former Soviet Union, the nuclear-armed BALLISTIC MISSILE will become the most powerful weapon ever developed by the human race

● In October, British engineer and writer SIR ARTHUR C. CLARKE suggests the use of SATELLITES in GEOSTATIONARY ORBIT to support global communications. His article, "Extra-Terrestrial Relays" in *Wireless World,* represents the birth of the COMMUNICATIONS SATELLITE concept—a use of space technology that vigorously promotes the information revolution

1946 ● On 16 April, the U.S. Army launches the first American-adapted, captured German V-2 ROCKET from the White Sands Proving Ground in southern New Mexico

● Between July and August, the Russian rocket engineer SERGEI KOROLEV develops a stretched-out version of the German V-2 rocket. As part of his engineering improvements, he increases the rocket engine's THRUST and lengthens the vehicle's PROPELLANT tanks

1947 ● On 30 October, Russian ROCKET engineers successfully launch a modified German V-2 rocket from a desert LAUNCH SITE near a place called KAPUSTIN YAR. This rocket impacts about 320 km DOWNRANGE from the launch site

German V-2 rocket (Courtesy of U.S. Army)

1948 ● The September issue of the *Journal of the British Interplanetary Society (JBIS)* starts a four-part series of technical papers by L. R. Shepherd and A. V. Cleaver that explores the feasibility of applying nuclear ENERGY to space travel, including the concepts of NUCLEAR-ELECTRIC PROPULSION and the NUCLEAR ROCKET

1949 ● On 29 August, the former Soviet Union detonates its first nuclear device at a secret test site in the Kazakh Desert. Code named First Lightning (Pervaya Molniya), this successful test breaks the nuclear weapon monopoly enjoyed by the United States. It plunges the world into a massive nuclear arms race that includes the accelerated development of strategic BALLISTIC MISSILES capable of traveling thousands of kilometers. Because they are well behind the United States in nuclear weapons technology, the leaders of the former Soviet Union decide to develop powerful, high-THRUST ROCKETs to carry their heavier, more primitive-design nuclear weapons. That decision gives the Soviet Union a major LAUNCH VEHICLE advantage when both superpowers decide to race into OUTER SPACE (starting in 1957) as part of a global demonstration of national power

1950 ● On 24 July, the United States successfully LAUNCHes a modified German V-2 ROCKET with an American-designed WAC Corporal second-stage rocket from the U.S. Air Force's newly established long-range proving ground at CAPE CANAVERAL, Florida. The hybrid, MULTISTAGE ROCKET (called the *Bumper 8)* inaugurates the incredible sequence of military MISSILE and SPACE VEHICLE launches to take place from Cape Canaveral—the world's most famous LAUNCH SITE.

● In November, English technical visionary SIR ARTHUR C. CLARKE publishes "Electromagnetic Launching As a Major Contribution To Space-Flight." Clarke's article suggests mining the MOON and launching the mined LUNAR material into OUTER SPACE with an electromagnetic catapult

1951 ● Cinema audiences are shocked by the science fiction movie *The Day The Earth Stood Still.* This classic story involves the arrival of a powerful, humanlike EXTRATERRESTRIAL and his robot companion who come to warn the governments of the world about the foolish nature of their nuclear arms race. It is the first major science fiction story to portray powerful space aliens as friendly, intelligent creatures who come to help EARTH

● Dutch-American astronomer GERARD PETER KUIPER suggests the existence of a large population of small, icy PLANETESIMALS beyond the ORBIT of PLUTO—a collection of frozen CELESTIAL BODIES now known as the KUIPER BELT

1952 ● *Collier's* magazine helps stimulate a surge of American interest in space travel by publishing a beautifully illustrated series of technical articles written by space experts such as WERNHER VON BRAUN and WILLY LEY. The first of the famous eight-part series appears on 22 March and is boldly titled "Man Will Conquer Space Soon." The magazine also hires the most influential space artist, Chesley Bonestell, to provide stunning color illustrations. Subsequent articles in the series introduce millions of American readers to the concept of a SPACE STATION, a mission to the MOON, and an expedition to MARS.

● Wernher Von Braun publishes *The Mars Project (Das Marsprojekt),* the first serious technical study regarding a human-crewed expedition to Mars. His visionary proposal involves a convoy of 10 SPACESHIPs with a total combined crew of 70 ASTRONAUTs to explore the RED PLANET for about one year and then return to EARTH

Von Braun's space station concept (Courtesy of NASA)

1953 ● In August, the Soviet Union detonates its first thermonuclear weapon (a hydrogen bomb). This is a technological feat that intensifies the superpower nuclear arms race and increases emphasis on the emerging role of strategic, nuclear-armed, BALLISTIC MISSILES

● In October, the U.S. Air Force forms a special panel of experts, headed by JOHN VON NEUMANN, to evaluate the American strategic BALLISTIC MISSILE program. In 1954, this panel recommends a major reorganization of the American ballistic missile effort

1954 ● Following the recommendations of the JOHN VON NEUMANN panel, President DWIGHT D. EISENHOWER gives strategic BALLISTIC MISSILE development the highest national priority. The COLD WAR missile race explodes on the world stage as the fear of a strategic ballistic missile gap sweeps through the American government. CAPE CANAVERAL becomes the famous proving ground for such important ballistic missiles as the THOR, ATLAS, TITAN, Minuteman, and Polaris. Once developed, many of these powerful military ballistic missiles also serve the United States as SPACE LAUNCH VEHICLES. Air Force General BERNARD SCHRIEVER oversees the time-critical development of the Atlas ballistic missile—an astonishing feat of engineering and technical management

1955 ● WALT DISNEY (the American entertainment visionary) promotes space travel by producing an inspiring three-part television series that includes appearances by noted space experts like WERNHER VON BRAUN. The first episode, "Man In Space," airs on 9 March and popularizes the dream of space travel for millions of American television viewers. This show, along with its companion episodes, "Man and the Moon" and "Mars and Beyond," make von Braun and the term *rocket scientist* a popular household phrase

1957 ● On 4 October, Russian engineer SERGEI KOROLEV with permission from Soviet Premier NIKITA S. KHRUSHCHEV uses a powerful military ROCKET to place *SPUTNIK 1* (the world's first ARTIFICIAL SATELLITE) successfully into ORBIT around EARTH. News of the Soviet success sends a political and technical shock wave across the United States. The LAUNCH of *Sputnik 1* marks the beginning of the Space Age. It also is the start of the great space race of the COLD WAR—a period when people measured national strength and global prestige by accomplishments (or failures) in OUTER SPACE.

● On 3 November, the former Soviet Union launches *Sputnik 2*—the world's second ARTIFICIAL SATELLITE. It is a massive SPACECRAFT (for the time) that carries a live dog named Laika, which is euthanized at the end of the mission

● The highly publicized attempt by the United States to launch its first SATELLITE with a newly designed civilian rocket ends in complete disaster on 6 December. The *Vanguard* rocket explodes after rising only a few centimeters above its LAUNCH PAD at *CAPE CANAVERAL*. Soviet successes with *Sputnik 1* and *Sputnik 2* and the dramatic failure of the *Vanguard* rocket heighten American anxiety. The exploration and use of OUTER SPACE becomes a highly visible instrument of COLD WAR politics

Mongolian stamp honoring *Sputnik 2* and Laika (Courtesy of author)

1958 ● On January 31, the United States successfully launches *EXPLORER 1*—the first American SATELLITE in ORBIT around EARTH. A hastily formed team from the U.S. Army Ballistic Missile Agency (ABMA) and Caltech's Jet Propulsion Laboratory (JPL), led by WERNHER VON BRAUN, accomplishes what amounts to a national prestige rescue mission. The team uses a military BALLISTIC MISSILE as the LAUNCH VEHICLE. With instruments supplied by JAMES ALFRED VAN ALLEN of the State University of Iowa, *Explorer 1* discovers EARTH'S TRAPPED RADIATION BELTS—now called the VAN ALLEN RADIATION BELTS in his honor

● The National Aeronautics and Space Administration (NASA) becomes the official civilian space agency for the United States government on 1 October. On 7 October, the newly created NASA announces the start of the MERCURY PROJECT—apioneering program to put the first American ASTRONAUTs into orbit around Earth

● In mid-December, an entire ATLAS ROCKET lifts off from CAPE CANAVERAL and goes into orbit around Earth. The military BALLISTIC MISSILE'S PAYLOAD compartment carries Project SCORE (*S*ignal *C*ommunications *O*rbit *R*elay *E*xperiment)—a prerecorded Christmas season message from President DWIGHT D. EISENHOWER. This is the first time the human voice is broadcast back to Earth from OUTER SPACE

Pickering, Van Allen, and Von Braun hoist an *Explorer 1* model in celebration (Courtesy of NASA)

1959 ● On 2 January, the former Soviet Union sends a 360 kg SPACECRAFT (called *Lunik 1*) toward the MOON. Although it misses hitting the Moon by between 5,000 and 7,000 km, it is the first human-made object to escape EARTH'S GRAVITY and go into ORBIT around the SUN

● In mid-September, the former Soviet Union launches *Lunik 2*. The 390 kg spacecraft successfully impacts on the Moon and becomes the first human-made object to land (crash) on another world. *Lunik 2* carries Soviet emblems and banners to the LUNAR surface

● The September issue of *Nature* contains the article by Philip Morrison and G. Cocconi, "Searching for Interstellar Communications"—marking the start of SETI (SEARCH FOR EXTRATERRESTRIAL INTELLIGENCE)

● On 4 October, the former Soviet Union sends *Lunik 3* on a mission around the Moon. The spacecraft successfully circumnavigates the Moon and takes the first images of the lunar FARSIDE. Because of the SYNCHRONOUS ROTATION of the Moon around Earth, only the NEARSIDE of the lunar surface is visible to observers on Earth

1960 ● The United States launches the *Pioneer 5* SPACECRAFT on 11 March into ORBIT around the SUN. The modest-sized (42 kg) spherical American SPACE PROBE reports conditions in INTERPLANETARY space between EARTH and VENUS over a distance of about 37 million km

● On 1 April, NASA successfully launches the world's first METEOROLOGICAL SATELLITE, called TIROS (*T*elevision and *I*nfra*r*ed *O*bservation *S*atellite). The IMAGES of cloud patterns from OUTER SPACE create a revolution in weather forecasting

● The U.S. Navy places the world's first experimental NAVIGATION SATELLITE (called *TRANSIT 1B*) into EARTH ORBIT on 13 April. The SPACECRAFT serves as a space-based beacon, providing RADIO WAVE signals that allow military users to determine their location at sea more precisely

● On 24 May, the U.S. Air Force launches a MIDAS (*Mi*ssile *D*efense *A*larm *S*ystem) satellite from CAPE CANAVERAL. This event inaugurates an important American program of military SURVEILLANCE SATELLITES intended to detect enemy INTERCONTINENTAL BALLISTIC MISSILE (ICBM) launches by observing the characteristic INFRARED RADIATION (heat) signature of a ROCKET's EXHAUST PLUME. Essentially unknown to the general public for decades because of the classified nature of their mission, the emerging family of missile surveillance satellites provides U.S. government authorities with a reliable early-warning system concerning a surprise enemy (Soviet) ICBM attack. Surveillance satellites help support the national policy of strategic nuclear deterrence throughout the COLD WAR and prevent an accidental nuclear conflict

● The U.S. Air Force successfully launches the *Discoverer 13* spacecraft from VANDENBERG AIR FORCE BASE on 10 August. This spacecraft is actually part of a highly classified Air Force and Central Intelligence Agency (CIA) RECONNAISSANCE SATELLITE program called Corona. Started under special executive order from President DWIGHT D. EISENHOWER, the joint agency SPY SATELLITE program begins to provide important photographic IMAGES of denied areas of the world from OUTER SPACE. On 18 August, *Discoverer 14* (also called *Corona XIV)* provides the U.S. intelligence community its first SATELLITE-acquired images of the former Soviet Union. The era of satellite reconnaissance is born. Data collected by the spy satellites of the NATIONAL RECONNAISSANCE OFFICE (NRO) contribute significantly to U.S. national security and help preserve global stability during many politically troubled times

● On 12 August, NASA successfully launches the *Echo 1* experimental spacecraft. This large (30.5 m diameter) inflatable, metallized balloon becomes the world's first passive COMMUNICATIONS SATELLITE. At the dawn of space-based TELECOMMUNICATIONS, engineers bounce RADIO WAVE signals off the large, inflated satellite between the United States and the United Kingdom

● The former Soviet Union launches *Sputnik 5* into orbit around Earth on 19 August. This large spacecraft is actually a test vehicle for the new *VOSTOK* spacecraft that will soon carry COSMONAUTs into outer space. *Sputnik 5* carries two dogs, Strelka and Belka. When the spacecraft's recovery capsule functions properly the next day, these two dogs become the first living creatures to return to Earth successfully from an orbital flight

1961 ● On 31 January, NASA launches a Redstone ROCKET with a MERCURY PROJECT SPACE CAPSULE on a suborbital flight

Shepard departs Cape Canaveral on a Redstone rocket (Courtesy of NASA)

from CAPE CANAVERAL. The passenger ASTROCHIMP Ham is recovered DOWNRANGE in the Atlantic Ocean after reaching an ALTITUDE of 250 km. This successful primate space mission is a key step in sending American ASTRONAUTS safely into OUTER SPACE

● The former Soviet Union achieves a major space exploration milestone by successfully launching the first human being into ORBIT around EARTH on 12 April. COSMONAUT YURI A. GAGARIN travels into OUTER SPACE in the *VOSTOK 1* SPACECRAFT and becomes the first person to observe Earth directly from an orbiting SPACE VEHICLE

● On 5 May, NASA uses a Redstone rocket to send ASTRONAUT ALAN B. SHEPARD, JR., on his historic 15-minute suborbital flight into outer space from CAPE CANAVERAL. While riding inside his MERCURY PROJECT *Freedom 7* SPACE CAPSULE, Shepard reaches an ALTITUDE of 186 km and becomes the first American to travel in space

● President JOHN F. KENNEDY addresses a joint session of the U.S. Congress on 25 May. In an inspiring speech concerning many urgent national needs, the newly elected president creates a major space challenge for the United States when he declares, "I believe that this nation should commit itself to achieving the goal, before this decade is out, of landing a man on the MOON and returning him safely to Earth." This is a daring decision to use a highly visible civilian (NASA) space mission to support national political objectives. Because of his visionary leadership, when American ASTRONAUTS NEIL A. ARMSTRONG and EDWIN E. "BUZZ" ALDRIN, JR., step onto the LUNAR surface for the first time on 20 July 1969, the United States is recognized around the world as the undisputed winner of the COLD WAR space race

● On 29 June, the United States launches the *Transit 4A* NAVIGATION SATELLITE into orbit around Earth. The spacecraft uses a RADIOISOTOPE THERMOELECTRIC GENERATOR (RTG) to provide supplementary electric power.

The mission represents the first successful use of a nuclear power supply in outer space

● On 21 July, astronaut VIRGIL "GUS" I. GRISSOM becomes the second American to travel in outer space. A Redstone rocket successfully hurls his NASA MERCURY PROJECT *Liberty Bell 7* space capsule on a 15-minute suborbital flight from CAPE CANAVERAL

● The former Soviet Union launches the *Vostok 2* spacecraft into orbit on 6 August. It carries COSMONAUT GHERMAN S. TITOV—the second person to orbit Earth in a spacecraft successfully. About 10 hours into the flight, Titov also becomes the first of many space travelers to suffer from the temporary discomfort of SPACE SICKNESS, or space adaptation syndrome (SAS)

1962 ● On 20 February, ASTRONAUT JOHN HERSCHEL GLENN, JR., becomes the first American to ORBIT EARTH in a SPACECRAFT. An ATLAS ROCKET launches the NASA MERCURY PROJECT *Friendship 7* SPACE CAPSULE from CAPE CANAVERAL. After completing three orbits, Glenn's capsule safely splashes down in the Atlantic Ocean

● NASA launches *Telstar 1* on 10 July—the world's first commercially constructed and funded active COMMUNICATIONS SATELLITE. Despite its relatively low operational orbit (about 950 km by 5,600 km), the pioneering American Telephone and Telegraph (AT&T) SATELLITE triggers a revolution in international television broadcasting and TELECOMMUNICATIONS services

● In late August, NASA sends the *MARINER 2* Spacecraft to VENUS from CAPE CANAVERAL. *Mariner 2* passes within 35,000 km of the PLANET on 14 December 1962—thereby becoming the world's first successful INTERPLANETARY SPACE PROBE. The spacecraft observes very high surface temperatures (~430°C). These data shatter pre–space age visions about Venus being a lush, tropical planetary twin of EARTH

● During October, the placement of nuclear-armed Soviet offensive BALLISTIC MISSILES in Fidel Castro's Cuba precipitates the Cuban Missile Crisis. This dangerous superpower confrontation brings the world perilously close to nuclear warfare. Fortunately, the crisis dissolves when the Soviet ballistic missiles are withdrawn by Premier NIKITA S. KHRUSHCHEV after much skillful political maneuvering by President JOHN F. KENNEDY and his national security advisers

1963 ● Soviet COSMONAUT VALENTINA TERESHKOVA becomes the first woman to travel in OUTER SPACE. On 16 June, she ascends into space onboard the *VOSTOK 6* SPACECRAFT and then returns to EARTH after a flight of almost three days. While in ORBIT, she flies the *Vostok 6* within 5 km of the *Vostok 5* spacecraft, piloted by Cosmonaut Valery Bykovskiy. During their proximity flight, the two cosmonauts communicate with each other by radio

● On 26 July, NASA successfully launches the *Syncom 2* SATELLITE from CAPE CANAVERAL. This SPACECRAFT is the first COMMUNICATIONS SATELLITE to operate in high ALTITUDE (figure eight) SYNCHRONOUS ORBIT and helps fulfill the vision of SIR ARTHUR C. CLARKE. About a year later, *Syncom 3* will achieve a true GEOSYNCHRONOUS ORBIT above the EQUATOR. Both of these experimental NASA satellites clearly demonstrate the feasibility and great value of placing communications satellites into geostationary orbits. The age of instantaneous global communications is born

● In October, President JOHN F. KENNEDY signs the Limited Test Ban Treaty for the United States, and the important new treaty enters force on 10 October. Within a week (on 16 October), the U.S. Air Force successfully launches the first pair of Vela nuclear detonation detection satellites from CAPE CANAVERAL. These spacecraft orbit Earth at a very high altitude and continuously monitor the PLANET and outer space for nuclear detonations in violation of the Test Ban Treaty. From that time on, American spacecraft carrying nuclear detonation detection instruments continuously provide the United States with a reliable technical ability to monitor Earth's ATMOSPHERE and outer space for nuclear treaty violations

- Dutch-American astronomer MAARTEN SCHMIDT discovers the first QUASAR

1964
- Following a series of heartbreaking failures, NASA successfully launches the *RANGER 7* SPACECRAFT to the MOON from CAPE CANAVERAL on 24 July. About 68 hours after LIFTOFF, the robot SPACE PROBE transmits over 4,000 high-resolution television IMAGES before crashing into the lunar surface in a region known as the Sea of Clouds. The *Ranger 7, 8,* and *9* spacecraft help prepare the way for the lunar landing missions by the APOLLO PROJECT ASTRONAUTs (1969–72)

- In August, the International Telecommunications Satellite Organization (INTELSAT) is formed—its mission to develop a global COMMUNICATIONS SATELLITE system

- On 12 October, the former Soviet Union launches the first three-person crew into OUTER SPACE when the specially configured two-person *Voskhod 1* spacecraft is used to carry COSMONAUTs VLADIMIR M. KOMAROV, Boris Yegorov, and Konstantin Feoktistov into ORBIT around EARTH

- On 28 November, NASA's *MARINER 4* spacecraft departs CAPE CANAVERAL on its historic journey as the first spacecraft from Earth to visit MARS. It successfully encounters the RED PLANET on 14 July 1965 at a FLYBY distance of about 9,800 km. *Mariner 4*'s close-up images reveal a barren, desertlike world and quickly dispel any pre–space age notions about the existence of ancient MARTIAN cities or a giant network of artificial canals

1965
- German-American physicist ARNO ALLEN PENZIAS and American physicist ROBERT WOODROW WILSON detect the faint COSMIC MICROWAVE BACKGROUND considered by cosmologists to be the lingering signal from the BIG BANG explosion

● On 18 March, the former Soviet Union launches the *Voskhod 2* SPACECRAFT, carrying COSMONAUTs Pavel Belyayev and Alekesey Leonov. During this mission, Leonov becomes the first person to leave the confines of an orbiting spacecraft and perform an EXTRAVEHICULAR ACTIVITY (EVA). While tethered, he conducts this historic 10-minute SPACE WALK. Then he encounters some significant difficulties when he tries to get back into the *Voskhod's* AIRLOCK with a bloated and cumbersome SPACESUIT. Their REENTRY proves equally challenging, as they land in an isolated portion of a snowy forest with only wolves to greet them. Rescuers arrive the next day, and the entire cosmonaut rescue group departs the improvised camp site on skis

● A TITAN II ROCKET carries ASTRONAUT VIRGIL "GUS" I. GRISSOM and JOHN W. YOUNG into ORBIT on 23 March from CAPE CANAVERAL inside a two-person GEMINI PROJECT SPACECRAFT. NASA's *Gemini 3* flight is the first crewed mission for the new spacecraft and marks the beginning of more sophisticated space activities by American crews in preparation for the APOLLO PROJECT LUNAR missions

● On 6 April, NASA places the *Intelsat-1 (Early Bird)* COMMUNICATIONS SATELLITE into orbit from CAPE CANAVERAL. It is the first commercial communications satellite placed into GEOSYNCHRONOUS ORBIT

● The former Soviet Union launches its first communications satellite on 23 April. Designed to facilitate TELECOM-MUNICATIONS across locations at high northern latitudes, the *Molniya 1A* spacecraft uses a special highly elliptical, 12-hour orbit (about 500 km by 12,000 km). This type of orbit is now called a MOLNIYA ORBIT

● On 3 June, NASA launches the *Gemini 4* mission from Cape Canaveral with astronauts James McDivitt and EDWARD H. WHITE II as the crew. During this NASA GEMINI PROJECT mission, astronaut White conducts the first American SPACE WALK, spending about 21 minutes on a tether outside the spacecraft

CHRONOLOGY   **1965**

● In December, NASA expands the scope of the GEMINI PROJECT activities. *Gemini 7* lifts off on 4 December, carrying astronauts FRANK BORMAN and JAMES ARTHUR LOVELL, JR., into OUTER SPACE for an almost 14-day mission. On 15 December, they are joined in orbit by the *Gemini 6* spacecraft, carrying astronauts WALTER M. SCHIRRA and THOMAS P. STAFFORD. Once in orbit, the *Gemini 6* spacecraft comes within two meters of the *Gemini 7* "target spacecraft," thereby accomplishing the first successful orbital RENDEZVOUS operation

1966 ● The former Soviet Union sends the *LUNA 9* SPACECRAFT to the MOON on 31 January. The 100 kg MASS spherical spacecraft SOFT LANDS in the Ocean of Storms region on 3 February, rolls to a stop, opens four petal-like covers, and then transmits the first panoramic television images from the Moon's surface

● On March 16, NASA launches the *Gemini 8* mission from CAPE CANAVERAL using a TITAN II ROCKET. Astronauts NEIL A. ARMSTRONG and DAVID R. SCOTT guide their spacecraft to a Gemini AGENA target vehicle (GATV), accomplishing the first successful RENDEZVOUS and DOCKING operation between a crewed CHASER SPACECRAFT and an uncrewed target vehicle. However, after an initial period of stable flight, the docked spacecraft begin to tumble erratically. Only quick, corrective action by the astronauts prevents a major space disaster. They make an emergency REENTRY and are recovered in a contingency landing zone in the Pacific Ocean

● The former Soviet Union launches the *LUNA 10* to the Moon on 31 March. This massive (1,500 kg) spacecraft becomes the first human-made object to achieve ORBIT around the Moon

● On May 30, NASA sends the *SURVEYOR 1* LANDER spacecraft to the Moon. The versatile robot spacecraft successfully makes a SOFT LANDING (1 June) in the Ocean of Storms. It then transmits over 10,000 IMAGES from the LUNAR surface and performs numerous soil mechanics experiments in preparation for the APOLLO PROJECT human-landing missions

- In mid-August, NASA sends the *LUNAR ORBITER 1* spacecraft to the Moon from CAPE CANAVERAL. It is the first of five successful missions to collect detailed IMAGES of the Moon from lunar orbit. At the end of each mapping mission, the ORBITER spacecraft is intentionally crashed into the Moon to prevent interference with future orbital activities

- On 12 September, NASA launches the *Gemini 11* mission from CAPE CANAVERAL. The GEMINI PROJECT spacecraft with astronauts CHARLES (PETE) CONRAD, JR., and RICHARD F. GORDON, JR., quickly accomplishes RENDEZVOUS and DOCKING with an AGENA target vehicle. The astronauts then use the Agena's restartable rocket engine to propel themselves (in this docked configuration) to a record-setting ALTITUDE of 1,370 km—the highest ever flown by an EARTH-orbiting, human-crewed spacecraft

1967 On 27 January, disaster strikes NASA's APOLLO PROJECT. While inside their *Apollo 1* SPACECRAFT during a training exercise on LAUNCH PAD 34 at CAPE CANAVERAL, ASTRONAUTS VIRGIL "GUS" I. GRISSOM, EDWARD H. WHITE, JR., and ROGER B. CHAFFEE are killed when a flash fire sweeps through their SPACE CAPSULE. The MOON landing program is delayed by 18 months, while major design and safety changes are made in the Apollo spacecraft

- On 23 April, tragedy also strikes the Russian space program when the Soviets launch COSMONAUT VLADIMIR M. KOMAROV in the new SOYUZ (union) SPACECRAFT. Following an orbital mission plagued with difficulties, Komarov is killed (on 24 April) during emergency REENTRY operations when the spacecraft's parachute fails to deploy properly and the vehicle impacts the ground at high speed

- On 27 and 30 October (respectively), the former Soviet Union launches two uncrewed SOYUZ-type spacecraft, called *COSMOS 186* and *188*. On 30 October, the orbiting spacecraft accomplish the first automatic RENDEZVOUS and DOCKING operation. The craft then separate and are recovered on 31 October and 2 November

1968 ● NASA launches the *Apollo 7* mission into ORBIT around EARTH. ASTRONAUTS WALTER M. SCHIRRA, DON F. EISELE, and R. WALTER CUNNINGHAM perform a variety of orbital operations with the redesigned APOLLO PROJECT SPACECRAFT

● On 21 December, NASA's *Apollo 8* spacecraft (Command and Service Module only) departs Launch Complex 39 at the KENNEDY SPACE CENTER during the first flight of the mighty *Saturn V* LAUNCH VEHICLE with a human crew as part of the PAYLOAD. Astronauts FRANK BORMAN, JAMES ARTHUR LOVELL, JR., and WILLIAM A. ANDERS become the first people to leave Earth's gravitational influence. They go into orbit around the MOON and capture IMAGES of an incredibly beautiful Earth "rising" above the starkly barren LUNAR horizon—pictures that inspire millions and animate an emerging environmental movement. After 10 orbits around the Moon, the first lunar astronauts return safely to Earth (27 December)

1969 ● In a full dress rehearsal for the first MOON landing, NASA's *Apollo 10* mission departs the KENNEDY SPACE CENTER on 18 May. ASTRONAUT EUGENE A. CERNAN, JOHN W. YOUNG, and THOMAS P. STAFFORD successfully demonstrate the complete APOLLO PROJECT mission profile and evaluate the performance of the LUNAR EXCURSION MODULE (LEM) down to within 15 km of the LUNAR surface

● The entire world watches as NASA's *Apollo 11* mission leaves for the Moon on 16 July from the KENNEDY SPACE CENTER. Astronauts NEIL A. ARMSTRONG, MICHAEL COLLINS, and EDWIN E. "BUZZ" ALDRIN, JR., make a long-held dream of humanity a reality. On 20 July, American Neil Armstrong cautiously descends the steps of the lunar excursion module's ladder and steps onto the lunar surface, exclaiming: "One small step for a man, one giant leap for mankind!" He and Buzz Aldrin become the first two people

First Moon landing, 20 July 1969 (Courtesy of NASA)

to walk on another world. Many people regard the APOLLO PROJECT lunar landings as the greatest technical accomplishment in all human history

- On 14 November, NASA's *Apollo 12* mission lifts off from the KENNEDY SPACE CENTER. Astronauts CHARLES (PETE) CONRAD, JR., RICHARD F. GORDON, JR., and ALAN L. BEAN continue the scientific objectives of the APOLLO PROJECT. Conrad and Bean become the third and fourth "Moon walkers," collecting samples from a larger area and deploying the Apollo Lunar Surface Experiment Package (ALSEP)

1970 - NASA's *Apollo 13* mission leaves for the MOON on 11 April. Suddenly, on 13 April, a life-threatening explosion occurs in the service module portion of the Apollo SPACECRAFT. ASTRONAUTS JAMES ARTHUR LOVELL, JR., JOHN LEONARD SWIGERT, and FRED WALLACE HAISE, JR., must use their LUNAR EXCURSION MODULE (LEM) as a lifeboat. While an anxious world waits and listens, the crew rides their disabled spacecraft around the Moon. With critical supplies running low, they limp back to EARTH on a free-return TRAJECTORY. At just the right moment on 17 April, they abandon the LEM *Aquarius* and board the APOLLO PROJECT spacecraft (command module) for a successful atmospheric REENTRY and recovery in the Pacific Ocean

- The former Soviet Union launches its *VENERA 7* mission to VENUS on 17 August. When the spacecraft arrives on 15 December, a PROBE is parachuted into the dense VENUSIAN ATMOSPHERE. Subsequent analysis of the data from this hardy instrumented capsule confirms that it has landed on the surface. It records an ambient temperature of approximately 475°C and an atmospheric pressure that is 90 times the pressure found at sea level on Earth. It is the first successful transmission of scientific data from the surface of another PLANET

- On 12 September, the former Soviet Union sends the *LUNA 16* robot spacecraft to the Moon. Russian engineers use

TELEOPERATION of the spacecraft's drill to collect about 100 g of LUNAR dust, which is then placed into a sample return canister. On 21 September, the canister leaves the lunar surface on a TRAJECTORY back to Earth. Three days later, it lands in Russia. *Luna 16* is the first robot spacecraft to return a sample of material from another world. In similar missions, *Luna 20* (February 1972) and *Luna 24* (August 1976) return automatically collected lunar samples, while *Luna 18* (September 1971) and *Luna 23* (October 1974) land on the Moon but fail to return samples for various technical reasons

- On November 10, the former Soviet Union launches another interesting robot mission to the Moon. *Luna 17* lands in the Sea of Rains. On command from Earth, a 750 kg robot ROVER (called *LUNAKHOD 1*) rolls down an extended ramp and begins exploring the lunar surface. Russian engineers use teleoperation to control this eight-wheeled rover as it travels for months across the lunar surface—transmitting more than 20,000 television IMAGES and performing more than 500 soil tests at various locations. The mission represents the first successful use of a mobile, remotely controlled (teleoperated) robot vehicle to explore another planetary body. The *Luna 23* mission in January 1973 successfully deploys another rover *(Lunakhod 2)*. However, the technical significance of these machine missions is all but ignored in the global glare of NASA's APOLLO PROJECT and its incredible human triumphs

1971 • NASA sends the *Apollo 14* mission to the MOON on 31 January. While ASTRONAUT STUART ALLEN ROOSA orbits overhead in the APOLLO PROJECT SPACECRAFT, astronauts ALAN B. SHEPARD, JR., and EDGAR DEAN MITCHELL descend to the lunar surface. After departing the LUNAR EXCURSION MODULE (LEM), they become the fifth and sixth Moon walkers

- On 19 April, the Soviet Union launches the first SPACE STATION (called *SALYUT 1*). It remains initially uninhabited because the three-COSMONAUT crew of the *SOYUZ 10* mission (launched on 22 April) attempts to dock with the station but cannot go on board

- At the end of May, NASA launches the *MARINER 9* spacecraft to MARS with an ATLAS-CENTAUR ROCKET. The spacecraft successfully enters ORBIT around the PLANET on 13 November 1971 and provides numerous IMAGES of the MARTIAN surface

- The second Russian spaceflight tragedy occurs in late June. The fatal accident takes place as the crew separates from *SALYUT 1* to return to EARTH in their *SOYUZ 11* spacecraft after spending 22 days on board the space station. While not wearing pressure suits for REENTRY, cosmonauts Victor Patseyev, Vladislav Volkov, and Georgi Dobrovolsky suffocate when a pressure valve malfunctions and the air rushes out of their *Soyuz 11* spacecraft. On 30 June, a startled recovery team on Earth finds all three men dead inside their spacecraft

- On July 26, NASA launches the *Apollo 15* mission to the Moon. Astronaut JAMES BENSON IRWIN remains in the APOLLO PROJECT spacecraft orbiting the Moon, while astronauts DAVID R. SCOTT and ALFRED MERRILL WORDEN become the seventh and eighth Moon walkers. They also are the first persons to drive a motor vehicle (electric powered) on another world, using their LUNAR ROVER to scoot across the surface

1972 - In early January, President Richard M. Nixon approves NASA's SPACE SHUTTLE program. This decision shapes the major portion of NASA's program for the next three decades

- On 2 March, an ATLAS-CENTAUR LAUNCH VEHICLE successfully sends NASA's *PIONEER 10* spacecraft from CAPE CANAVERAL on its historic mission. This far-traveling robot spacecraft becomes the first to transit the MAIN-BELT ASTEROIDS, the first to encounter JUPITER (3 December 1973), and then by crossing the ORBIT of NEPTUNE on 13 June 1983 (which at the time was the farthest planet from the Sun), the first human-made object ever to leave the planetary boundaries of the SOLAR SYSTEM. Now on an INTERSTELLAR trajectory, *Pioneer 10* (and its twin, *Pioneer 11*) carries a special plaque, greeting any intelligent alien civilization that might find it drifting through interstellar space millions of years from now

- On 16 April, NASA launches the *Apollo 16* mission—the fifth human-landing mission to the MOON. While ASTRONAUT THOMAS K. MATTINGLY II orbits the Moon in the APOLLO PROJECT SPACECRAFT, astronauts JOHN W. YOUNG and CHARLES MOSS DUKE, JR., become the ninth and tenth Moon walkers. They also use the battery-powered LUNAR ROVER to travel across the Moon's surface in the Descartes LUNAR HIGHLANDS

- NASA places a new type of SATELLITE, called the *Earth Resources Technology Satellite-1 (ERTS-1),* into a SUN-SYNCHRONOUS ORBIT from VANDENBERG AIR FORCE BASE on 23 July. Renamed *LANDSAT-1,* it is the first civilian spacecraft to provide relatively high-resolution, multispectral IMAGES of EARTH's surface, creating a revolution in the way people look at their home PLANET. Over the next three decades, a technically evolving family of LANDSAT spacecraft helps scientists study the EARTH SYSTEM

- On 7 December, NASA's *Apollo 17* mission, the last expedition to the Moon in the 20th century, departs from the KENNEDY SPACE CENTER, propelled by a mighty *Saturn V* ROCKET. While astronaut RONALD E. EVANS remains in lunar orbit, fellow astronauts EUGENE A. CERNAN and HARRISON H. SCHMITT become the 11th and 12th members of the exclusive Moon walkers club. By using a lunar rover, they explore the Taurus-Littrow region. Their safe return to Earth on 19 December brings to a close one of the epic periods of human exploration

1973 - In early April, while propelled by an ATLAS-CENTAUR ROCKET, NASA's *PIONEER 11* spacecraft departs on an INTERPLANETARY journey from CAPE CANAVERAL. The spacecraft encounters JUPITER (2 December 1974) and then uses a GRAVITY ASSIST maneuver to establish a FLYBY TRAJECTORY to SATURN. It is the first spacecraft to view Saturn at close range (encountered on 1 September 1979) and then follows a path into INTERSTELLAR space

- On 14 May, NASA launches *SKYLAB*—the first American SPACE STATION. A giant *Saturn V* ROCKET is used to place this large facility into ORBIT. The first crew of three American ASTRONAUTS arrives on 25 May and makes the emergency repairs necessary to save the station, which suffered damage during LAUNCH. Astronauts CHARLES (PETE) CONRAD, JR., Paul J. Weitz, and Joseph P. Kerwin stay on board for 28 days. They are replaced by Astronauts ALAN L. BEAN, Jack R. Lousma, and Owen K. Garriott, who arrive on 28 July and live in space for about 59 days. The final *Skylab* crew (Astronauts Gerald P. Carr, William R. Pogue, and EDWARD G. GIBSON) arrive on 11 November and reside in the station until 8 February 1974—setting a spaceflight endurance record (for the time) of 84 days. *Skylab* is then abandoned

- In early November, NASA launches the *MARINER 10* spacecraft from CAPE CANAVERAL. It encounters VENUS (5 February 1974) and uses a gravity assist maneuver to become the first and only spacecraft to investigate MERCURY at close range

1974 - On 30 May, NASA launches the *Applications Technology Satellite (ATS)-6,* which demonstrates the use of a large antenna structure on a GEOSTATIONARY ORBIT COMMUNICATIONS SATELLITE to transmit good-quality television signals to small, inexpensive ground receivers

1975 - In June, the former Soviet Union sends twin *VENERA 8* and *9* SPACECRAFT to VENUS. *Venera 9* (launched 8 June) goes into ORBIT around Venus on 22 October. It releases a LANDER that reaches the surface and transmits the first television images of the PLANET's infernolike landscape. *Venera 10* (launched 10 June) follows a similar mission profile

- In July, the United States and the former Soviet Union conduct the first cooperative international RENDEZVOUS and DOCKING mission, called the APOLLO-SOYUZ TEST PROJECT (ASTP). On 15 July, the Russians launch the *SOYUZ 19* spacecraft with COSMONAUTS ALEXEI ARKHIPOVICH LEONOV and Valerie N. Kubasov onboard. Several hours later, NASA launches the *Apollo 18* spacecraft with ASTRONAUTS THOMAS P. STAFFORD, Vance D. Brand, and DEKE SLAYTON, JR., on board

● In late August and early September, NASA launches the twin *Viking 1* (20 August) and *Viking 2* (9 September) ORBITER/LANDER combination spacecraft to the RED PLANET from CAPE CANAVERAL. After arriving at MARS in 1976, all VIKING PROJECT spacecraft (two landers and two orbiters) perform exceptionally well—but the detailed search for microscopic ALIEN LIFE-FORMS on the MARTIAN surface remains inconclusive

● On 16 October, NASA launches the *Geostationary Operational Environmental Satellite* (GOES-1) for the U.S. National Oceanic and Atmospheric Administration (NOAA). It is the first in a long series of operational METEOROLOGICAL SATELLITES that monitor weather conditions on a hemispheric scale—providing warnings about hurricanes and other severe weather patterns

1976 ● NASA launches the first *la*ser *geo*dynamics *s*atellite (LAGEOS) on 6 May into a precise orbit around EARTH from VANDENBERG AIR FORCE BASE. The heavy but small (60 cm diameter), golf ball-shaped SPACECRAFT has its surface completely covered with mirrorlike retroreflectors. This joint NASA-Italian Space Agency project demonstrates the use of ground-to-satellite laser-ranging systems in the study of solid Earth dynamics, an important part of EARTH SYSTEM science

1977 ● On 20 August, NASA sends the *VOYAGER 2* SPACECRAFT from CAPE CANAVERAL on an epic grand tour mission during which it encounters all four JOVIAN PLANETs and then departs the SOLAR SYSTEM on an INTERSTELLAR TRAJECTORY. By using GRAVITY ASSIST maneuvers, *Voyager 2* visits JUPITER (9 July 1979), SATURN (25 August 1981), URANUS (24 January 1986), and NEPTUNE (25 August 1989). The resilient, far-traveling spacecraft (and its twin *Voyager 1*) also carries a special interstellar message from EARTH—a digital record entitled *The Sounds of Earth*

- On 5 September, NASA sends the *VOYAGER 1* spacecraft from CAPE CANAVERAL on its fast trajectory journey to Jupiter (5 March 1979), Saturn (12 March 1980), and beyond the solar system

- In late September, the former Soviet Union launches the *SALYUT 6* SPACE STATION—a second-generation design with several important improvements, including an additional DOCKING port and the use of automated *PROGRESS* resupply spacecraft

- NASA launches *Meteosat-1* from CAPE CANAVERAL on 22 November for the EUROPEAN SPACE AGENCY (ESA). Upon reaching GEOSTATIONARY ORBIT, it becomes Europe's first WEATHER SATELLITE

1978 - In May, the British Interplanetary Society releases its Project Daedalus report—a conceptual study about a one-way robot SPACECRAFT mission to BARNARD'S STAR at the end of the 21st century

- NASA successfully launches the PIONEER VENUS MISSION ORBITER *(Pioneer 12)* from CAPE CANAVERAL on 20 May. After arriving 4 December, it becomes the first American spacecraft to ORBIT VENUS. It uses its radar mapping system (from 1978–92) to image extensively the hidden surface of the cloud-enshrouded PLANET

- American astronomer James Christy discovers PLUTO's large MOON CHARON on 22 June

- In early August, NASA launches the PIONEER VENUS Multiprobe *(Pioneer 13),* which encounters the planet on 9 December and releases four PROBES into the VENUSIAN ATMOSPHERE

- During a visit to NASA's KENNEDY SPACE CENTER on 1 October, President Jimmy Carter publicly mentions that American RECONNAISSANCE SATELLITES have made immense contributions to international security

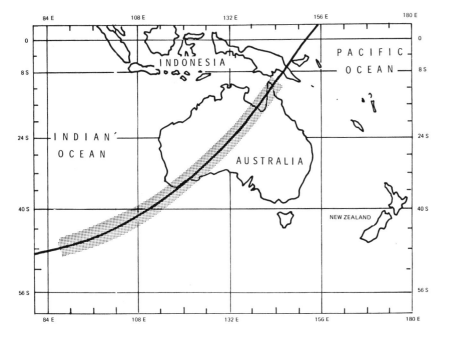

***Skylab* debris footprint
(Courtesy of NASA)**

1979 ● When unable to maintain its own ORBIT, the abandoned NASA *SKYLAB* becomes a dangerous orbiting derelict that eventually decays in a dramatic fiery plunge through EARTH's ATMOSPHERE on 11 July—a REENTRY that leaves debris fragments scattered over remote regions of western Australia

● On 24 December, the EUROPEAN SPACE AGENCY (ESA) successfully launches the first *ARIANE 1* ROCKET from the Guiana Space Center in Kourou, French Guiana

1980 ● India's Space Research Organization (ISRO) successfully places a modest 35 kg test SATELLITE (called *Rohini*) into LOW EARTH ORBIT on 1 July. The LAUNCH VEHICLE is a four-stage, SOLID-PROPELLANT ROCKET manufactured in India. The SLV-3 (Standard Launch Vehicle-3) gives India independent national access to OUTER SPACE

1981 ● On 12 April, NASA launches the SPACE SHUTTLE *Columbia* on its maiden orbital flight from Complex 39-A at the KENNEDY SPACE CENTER. ASTRONAUTS JOHN W. YOUNG and

ROBERT L. CRIPPEN thoroughly test the new AEROSPACE VEHICLE. Upon REENTRY, it becomes the first SPACECRAFT to return to EARTH by gliding through the ATMOSPHERE and landing like an airplane. Unlike all previous one-time use SPACE VEHICLES, *Columbia* is prepared for other missions in OUTER SPACE

- In Autumn, the former Soviet Union sends the *VENERA 13* and *14* SPACECRAFT to VENUS. *Venera 13* departs on 30 October, and its LANDER touches down on the VENUSIAN surface on 1 March 1982. *Venera 14* lifts off on 4 November 1981, and its capsule lands on Venus on 5 March 1982. Both hardly robot landers successfully return color IMAGES of the infernolike surface of Venus and perform the first soil-sampling experiments on that PLANET

1982 - On 11 November, NASA launches the SPACE SHUTTLE *Columbia* with a crew of four ASTRONAUTs on the first operational flight (called STS-5) of the U.S. SPACE TRANSPORTATION SYSTEM

1983 - An expendable DELTA ROCKET places the Infrared Astronomy Satellite (IRAS) into a POLAR ORBIT from VANDENBERG AIR FORCE BASE on 25 January. The international scientific SPACECRAFT (NASA-United Kingdom-Netherlands) completes the first comprehensive (all-sky) INFRARED RADIATION survey of the UNIVERSE

- The first flight of the *Challenger* occurs on 4 April when NASA launches the STS-6 SPACE SHUTTLE mission. During the mission, ASTRONAUTs Donald Peterson and Story Musgrave put on their SPACESUITs and perform the first EXTRAVEHICULAR ACTIVITY (EVA) from an orbiting Shuttle.

- On 18 June, NASA launches the Space Shuttle *Challenger* (STS-7 mission) with astronaut SALLY K. RIDE—the first American woman to travel in OUTER SPACE

- In late August, the Space Shuttle *Challenger* (STS-8 mission) flies into space with an astronaut crew of five, including GUION S. BLUFORD, JR.—first African American to orbit EARTH

1984 ● In his State of the Union Address on 25 January, President Ronald Reagan calls for a permanent American SPACE STATION. However, his vision must wait until December 1998, when a combined ASTRONAUT and COSMONAUT crew assembles the first two components of the INTERNATIONAL SPACE STATION as part of the STS-88 SPACE SHUTTLE mission

● The former Soviet Union launches the *SOYUZ T-12* SPACECRAFT on 17 July. The spacecraft carries three cosmonauts, including Svetlana Savistskaya. While the *Soyuz T-12* docks with the *SALYUT-7* space station, she performs a series of experiments in OUTER SPACE during an EXTRAVEHICULAR ACTIVITY (EVA)—becoming the first female space walker

● During the STS 41-G space shuttle mission (launched 5 October), astronaut KATHRYN D. SULLIVAN becomes the first American woman to perform an EXTRAVEHICULAR ACTIVITY (EVA)

1985 ● On 12 April, NASA's SPACE SHUTTLE *Discovery* carries U.S. Senator "Jake" Garn into orbit as a member of the ASTRONAUT crew of the STS 51-D mission

1986 ● On 24 January, NASA's *VOYAGER 2* SPACECRAFT encounters URANUS

● On 28 January, the SPACE SHUTTLE *Challenger* lifts off from the NASA KENNEDY SPACE CENTER on its final voyage. At just under 74 seconds into the STS 51-L mission, a deadly explosion occurs, killing the crew and destroying the vehicle. Led by President Ronald Reagan, the United States mourns the seven ASTRONAUTs lost in the *CHALLENGER* ACCIDENT

● In late February, the former Soviet Union launches the first segment of a third-generation SPACE STATION—the modular orbiting complex called *MIR* (peace)

● In March, an international armada of spacecraft encounter COMET HALLEY. The EUROPEAN SPACE AGENCY'S *GIOTTO* SPACECRAFT makes and survives the most hazardous FLYBY—streaking within 610 km of the COMET'S NUCLEUS on 14 March at a relative velocity of 68 km/s

1987 ● In late February, SUPERNOVA 1987A is observed in photographic IMAGES of the LARGE MAGELLANIC CLOUD by Canadian astronomer Ian Shelton. It is the first supernova visible to the NAKED EYE since the one discovered by JOHANNES KEPLER in 1604

1988 ● On 19 September, the State of Israel uses a *Shavit* (comet) three-stage ROCKET to place the country's first SATELLITE (called *Ofeq 1*) into an unusal east-to-west ORBIT—one that is opposite to the direction of EARTH'S ROTATION but necessary because of launch safety restrictions

● As the *Discovery* successfully lifts off on the STS-26 mission, NASA returns the SPACE SHUTTLE to service following a 32-month hiatus after the *CHALLENGER* ACCIDENT

1989 ● During the STS-30 mission in early May, the ASTRONAUT crew of the SPACE SHUTTLE *Atlantis* deploys NASA'S *MAGELLAN* SPACECRAFT and sends it on an INTERPLANETARY TRAJECTORY to VENUS

● On 25 August, the *VOYAGER 2* spacecraft encounters NEPTUNE

● During the STS-34 mission in mid-October, the space shuttle *Atlantis* deploys NASA's *GALILEO* spacecraft for its long INTERPLANETARY journey to JUPITER

● In mid-November, NASA launches the *COSMIC BACKGROUND EXPLORER (COBE)* into a POLAR ORBIT from VANDENBERG AIR FORCE BASE. The spacecraft carefully measures the COSMIC MICROWAVE BACKGROUND, helping scientists answer key questions about the BIG BANG explosion

1990 ● NASA officially begins the Voyager Interstellar Mission (VIM) on 1 January. This is an extended mission in which both VOYAGER SPACECRAFT search for the HELIOPAUSE

- During the STS-31 mission in late April, the ASTRONAUT crew of the SPACE SHUTTLE *Discovery* deploys the NASA *HUBBLE* SPACE TELESCOPE (HST) into ORBIT around EARTH. Subsequent shuttle missions (in 1993, 1997, 1999, and 2002) will repair design flaws and perform maintenance on this orbiting optical observatory

1991 ● In early April, NASA uses the SPACE SHUTTLE *Atlantis* to deploy the *COMPTON GAMMA RAY OBSERVATORY* (GRO)—a major EARTH-orbiting astrophysical observatory that investigates the UNIVERSE in the GAMMA RAY portion of the ELECTROMAGNETIC (EM) SPECTRUM

- On its way to JUPITER, NASA's *GALILEO* SPACECRAFT passes within 1,600 km of the ASTEROID 951 Gaspra. The ENCOUNTER provides the first close-up IMAGES of a MAIN-BELT ASTEROID. Gaspra is a type-S (silicaceous) asteroid about $19 \times 12 \times 11$ km in size

1992 ● On 11 February, the NATIONAL SPACE DEVELOPMENT AGENCY OF JAPAN (NASDA) successfully launches that country's first EARTH-OBSERVING SPACECRAFT from the Tanegashima Space Center using a Japanese-manufactured H-1 ROCKET

- NASA successfully launches the *Mars Observer (MO)* SPACECRAFT from CAPE CANAVERAL on 25 September. For unknown reasons, all contact with the SPACECRAFT is lost in late August 1993—just a day or so before it is to go into ORBIT around MARS

1993 ● While coasting to JUPITER, NASA's *GALILEO* SPACECRAFT encounters Ida (a MAIN-BELT ASTEROID) on 28 August at a distance of 2,400 km. *Galileo's* imagery reveals that Ida has a tiny SATELLITE of its own, named DACTYL

- In early December during the STS-61 mission, the ASTRONAUT crew of the SPACE SHUTTLE *Endeavour* perform a complicated in-orbit repair of NASA's *HUBBLE* SPACE TELESCOPE—thereby restoring the orbiting observatory to its planned scientific capabilities

1994 ● In late January, a joint Department of Defense and NASA advanced technology demonstration SPACECRAFT, called Clementine, lifts off for the MOON from VANDENBERG AIR FORCE BASE. Some of the spacecraft's data suggest that the Moon may actually possess significant quantities of water ice in its permanently shadowed polar regions

● On 3 February, SPACE SHUTTLE *Discovery* lifts off from the NASA KENNEDY SPACE CENTER The six-person crew of the STS-60 mission includes COSMONAUT SERGEI KRIKALEV— the first Russian to travel into OUTER SPACE using an American LAUNCH VEHICLE

● In March, the U.S. Air Force successfully launches the final SATELLITE in the GLOBAL POSITIONING SYSTEM (GPS), and the NAVIGATION SATELLITE system becomes fully operational. GPS revolutionizes navigation on land, at sea, and in the air for numerous military and civilian users

● In mid-July, fragments of COMET Shoemaker-Levy 9 slam into the PLANET JUPITER, causing observable disturbances in the Jovian ATMOSPHERE that persist for many months

1995 ● In February during NASA's STS-63 mission, the SPACE SHUTTLE *Discovery* approaches (encounters) the Russian MIR space station as a prelude to the development of the INTERNATIONAL SPACE STATION (ISS). ASTRONAUT EILEEN MARIE COLLINS serves as the first female shuttle pilot

● On 14 March, the Russians launch the *SOYUZ TM-21* (UNION) spacecraft to the *Mir* space station from the BAIKONUR COSMODROME. The crew of three includes American astronaut Norman Thagard—the first American to travel into OUTER SPACE on a Russian ROCKET and the first to stay on the *Mir* space station. The *Soyuz TM-21* COSMONAUTs also relieve the previous *Mir* crew, including cosmonaut Valeriy Polyakov, who returns to EARTH on 22 March after setting a world record for remaining in space for 438 days

● In late June, NASA's space shuttle *Atlantis* docks with the Russian *Mir* space station for the first time. During this shuttle mission (STS-71), *Atlantis* delivers the *Mir 19* crew (cosmonauts Anatoly Solovyev and Nikolai Budarin) to the Russian space station and then returns the *Mir 18* crew back to Earth—including American astronaut Norman Thagard, who has just spent 115 days in space on board the *Mir.* The shuttle-*Mir* docking program is the first phase of the INTERNATIONAL SPACE STATION (ISS). A total of nine shuttle-*Mir* docking missions will occur between 1995–98

● In early December, NASA's GALILEO spacecraft arrives at JUPITER and starts its multiyear scientific mission by successfully deploying a PROBE into the Jovian ATMOSPHERE

1996 ● In late March, the SPACE SHUTTLE *Atlantis* delivers ASTRONAUT SHANNON W. LUCID, who becomes a COSMONAUT researcher and the first American woman to live on the Russian MIR space station

● In the summer, a NASA research team from the Johnson Space Center announces that they have found evidence within a MARTIAN METEORITE (called ALH84001) that "strongly suggests primitive life may have existed on MARS more than 3.6 billion years ago." The Martian microfossil hypothesis touches off a great deal of technical debate

● In mid-September, the space shuttle *Atlantis* docks with the Russian *Mir* space station and returns astronaut Shanon W. Lucid to EARTH after she spends 188 days on board the *Mir*— setting a new U.S. and world spaceflight record for a woman

● NASA launches the *MARS GLOBAL SURVEYOR* (MGS) on 7 November and then the *MARS PATHFINDER* on 4 December from CAPE CANAVERAL

1997 ● In February, the ASTRONAUT crew of the SPACE SHUTTLE *Discovery* (STS-82 mission) successfully accomplishes the second *HUBBLE* SPACE TELESCOPE (HST)-servicing mission

1998 ● In early January, NASA sends the *LUNAR PROSPECTOR* to the MOON from CAPE CANAVERAL. Data from this ORBITER SPACECRAFT reinforces previous hints that the LUNAR polar regions may contain large reserves of water ice in a mixture of frozen dust lying at the frigid bottom of some permanently shadowed CRATERS

● On 29 October, the SPACE SHUTTLE *Discovery* (STS-95 mission) lifts off from the NASA KENNEDY SPACE CENTER. Its crew includes ASTRONAUT (U.S. Senator) JOHN HERSCHEL GLENN, JR.—who returns to OUTER SPACE after 36 years and becomes the oldest person (at age 77) to experience spaceflight in the 20th century

● In early December, the Space Shuttle *Endeavour* ascends from the NASA KENNEDY SPACE CENTER on the first assembly mission of the INTERNATIONAL SPACE STATION (ISS). During the STS-88 shuttle mission, *Endeavour* performs a RENDEZVOUS with the previously launched Russian-built *Zarya* (sunrise) module. An international crew connects this module with the American-built *Unity* module carried in the shuttle's CARGO BAY

1999 ● In July, ASTRONAUT EILEEN MARIE COLLINS serves as the first female SPACE SHUTTLE commander (STS-93 mission) as the *Columbia* carries NASA'S *CHANDRA X-RAY OBSERVATORY* (CXRO) into ORBIT

● After a successful launch from CAPE CANAVERAL (11 December 1998) and an uneventful INTERPLANETARY trip, NASA loses all contact with the *Mars Climate Orbiter (MCO)* as it at approaches the RED PLANET on 23 September. A TRAJECTORY calculation error has most likely caused the ORBITER SPACECRAFT to approach the MARTIAN ATMOSPHERE too steeply and burn up

● On 3 December, NASA also loses all contact with the *Mars Polar Lander* mission just prior to its arrival at MARS. Although successfully launched from CAPE CANAVERAL on 3 January, it is the second NASA mission to fail upon arrival at the RED PLANET in 1999

2000 ● Aware of the growing SPACE (ORBITAL) DEBRIS problem, NASA SPACECRAFT controllers intentionally de-orbit the massive *COMPTON GAMMA RAY OBSERVATORY* on 4 June at the end of its useful mission. This insures that any pieces surviving atmospheric REENTRY will fall safely into a remote part of the Pacific Ocean

2001 ● NASA launches the 2001 *MARS ODYSSEY* mission to the RED PLANET in early April—the SPACECRAFT orbits the PLANET in October

● On 23 March, Russian officials intentionally de-orbit the decommissioned *MIR* SPACE STATION, which has become a large space derelict. Their action assures that any components surviving atmospheric REENTRY will fall harmlessly into a remote part of the Pacific Ocean

2002 ● In February, NASA's 2001 *MARS ODYSSEY* SPACECRAFT starts using its collection of scientific instruments to begin a detailed study of the RED PLANET

● In March, a seven-member ASTRONAUT crew flew NASA's SPACE SHUTTLE *Columbia* into ORBIT around EARTH as part of the STS-109 mission to successfully service and upgrade the *HUBBLE* SPACE TELESCOPE

SECTION FOUR
CHARTS & TABLES

Special units for astronomical investigations

Astronomical unit *(AU):* The mean distance from Earth to the Sun—approximately 1.495979×10^{11} m

Light-year *(ly):* The distance light travels in 1 year's time—approximately 9.46055×10^{15} m

Parsec *(pa):* The parallax shift of 1 second of arc (3.26 light-years)—approximately 3.085768×10^{16} m

Speed of light *(c):* 2.9979×10^8 m/s

Source: NASA.

International System (SI) units and their conversion factors

Quantity	Name of unit	Symbol	Conversion factor
distance	meter	m	1 km = 0.621 mi.
			1 m = 3.28 ft.
			1 cm = 0.394 in.
			1 mm = 0.039 in.
			1 μm = 3.9×10^{-5} in. = 104 Å
			1 nm = 10 Å
mass	kilogram	kg	1 tonne = 1.102 tons
			1 kg = 2.20 lb.
			1 g = 0.0022 lb. = 0.035 oz.
			1 mg = 2.20×10^{-6} lb. = 3.5×10^{-5} oz.
time	second	s	1 yr. = 3.156×10^7 s
			1 day = 8.64×10^4 s
			1 hr. = 3,600 s
temperature	kelvin	K	273 K = 0°C = 32°F
			373 K = 100°C = 212°F
area	square meter	m^2	1 m^2 = 10^4 cm^2 = 10.8 $ft.^2$
volume	cubic meter	m^3	1 m^3 = 10^6 cm^3 = 35 $ft.^3$
frequency	hertz	Hz	1 Hz = 1 cycle/s
			1 kHz = 1,000 cycles/s
			1 MHz = 10^6 cycles/s
density	kilogram per cubic meter	kg/m^3	1 kg/m^3 = 0.001 g/cm^3
			1 g/cm^3 = density of water
speed, velocity	meter per second	m/s	1 m/s = 3.28 ft./s
			1 km/s = 2,240 mi./hr.
force	newton	N	1 N = 10^5 dynes = 0.224 lbf
pressure	newton per square meter	N/m^2	1 N/m^2 = 1.45×10^{-4} lb./$in.^2$
energy	joule	J	1 J = 0.239 cal
photon energy	electronvolt	eV	1 eV = 1.60×10^{-19} J; 1 J = 10^7 erg
power	watt	W	1 W = 1 J/s
atomic mass	atomic mass unit	amu	1 amu = 1.66×10^{-27} kg
wavelength of light	angstrom	Å	1 Å = 0.1 nm = 10^{-10} m
acceleration of gravity	g	g	1 g = 9.8 m/s^2

Source: NASA.

Common metric/English conversion factors (for space technology activities)

	Multiply	By	To obtain
length	inches	2.54	centimeters
	centimeters	0.3937	inches
	feet	0.3048	meters
	meters	3.281	feet
	miles	1.6093	kilometers
	kilometers	0.6214	miles
	kilometers	0.54	nautical miles
	nautical miles	1.852	kilometers
	kilometers	3281	feet
	feet	0.0003048	kilometers
weight and mass	ounces	28.350	grams
	grams	0.0353	ounces
	pounds	0.4536	kilograms
	kilograms	2.205	pounds
	tons	0.9072	metric tons
	metric tons	1.102	tons
liquid measure	fluid ounces	0.0296	liters
	gallons	3.7854	liters
	liters	0.2642	gallons
	liters	33.8140	fluid ounces
temperature	degrees Farenheit plus 459.67	0.5555	kelvins
	degrees Celsius plus 273.16	1.0	kelvins
	kelvins	1.80	degrees Fahrenheit minus 459.67
	kelvins	1.0	degrees Celsius minus 273.16
	degrees Fahrenheit minus 32	0.5555	degrees Celsius
	degrees Celsius	1.80	degrees Fahrenheit plus 32
thrust (force)	pounds force	4.448	newtons
pressure	newtons	0.225	pounds
	millimeters mercury	133.32	pascals (newtons per square meter)
	pounds per square inch	6.895	kilopascals (1,000 pascals)
	pascals	0.0075	millimeters mercury at 0° C
	kilopascals	0.1450	pounds per square inch

Source: NASA.

Recommended SI unit prefixes

Prefix	Abbreviation	Factor by which unit is multiplied
tera-	T	10^{12}
giga-	G	10^{9}
mega-	M	10^{6}
kilo-	k	10^{3}
hecto-	h	10^{2}
centi-	c	10^{-2}
milli-	m	10^{-3}
micro-	μ	10^{-6}
nano-	n	10^{-9}
pico-	p	10^{-12}

Source: NASA.

Greek alphabet

A	α	alpha	N	ν	nu
B	β	beta	Ξ	ξ	xi
Γ	γ	gamma	O	o	omicron
Δ	δ	delta	Π	π	pi
E	ε	epsilon	P	ρ	rho
Z	ζ	zeta	Σ	σ	sigma
H	η	eta	T	τ	tau
Θ	θ	theta	Y	υ	upsilon
I	ι	iota	Φ	φ	phi
K	κ	kappa	X	χ	chi
Λ	λ	lambda	Ψ	ψ	psi
M	μ	mu	Ω	ω	omega

Stellar spectral classes

Type	Description	Typical surface temperature (K)	Remarks/Examples
O	very hot, large blue stars (hottest)	28,000–40,000	ultraviolet stars; very short lifetimes (3–6 million years)
B	large, hot blue stars	11,000–28,000	Rigel
A	blue-white, white stars	7,500–11,000	Vega, Sirius, Altair
F	white stars	6,000–7,500	Canopis, Polaris
G	yellow stars	5,000–6,000	the Sun
K	orange-red stars	3,500–5,000	Arcturus, Aldebaran
M	red stars (coolest)	<3,500	Antares, Betelgeuse

Source: NASA.

Constellations of the Zodiac

Spring Signs
♈ Aries   the Ram
♉ Taurus   the Bull
♊ Gemini   the Twins

Summer Signs
♋ Cancer   the Crab
♌ Leo   the Lion
♍ Virgo   the Virgin

Autumn Signs
♎ Libra   the Scales
♏ Scorpio   the Scorpion
♐ Sagittarius the Archer

Winter Signs
♑ Capricorn   the Goat
♒ Aquarius   the Water Bearer
♓ Pisces   the Fishes

Keplerian Elements

a   Semi-Major Axis—gives the size of the orbit

e   Eccentricity—gives the shape of the orbit

i   Inclination Angle—gives the angle of the orbit plane to the central body's equator

Ω   Right Ascension of the Ascending Node—which gives the rotation of the orbit plane from reference axis

ω   Argument of Perigee—gives the rotation of the orbit in its plane

θ   True Anomaly—gives the location of the satellite on the orbit

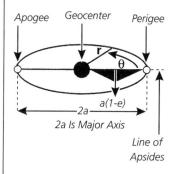

2a Is Major Axis

Source: NASA.

Constellations of the Zodiac – Keplerian Elements

CHARTS & TABLES

NASA space shuttle launches (1981–2000)

Year	Launches
1981	STS-1, STS-2
1982	STS-3, STS-4, STS-5
1983	STS-6, STS-7, STS-8, STS-9
1984	41-B, 41-C, 41-D, 41-G, 51-A
1985	51-C, 51-D, 51-B, 51-G, 51-F, 51-I, 51-J, 61-A, 61-B
1986	61-C, 51-L (*Challenger* accident)
1987	No launches
1988	STS-26, STS-27
1989	STS-29, STS-30, STS-28, STS-34, STS-33
1990	STS-32, STS-36, STS-31, STS-41, STS-38, STS-35
1991	STS-37, STS-39, STS-40, STS-43, STS-48, STS-44
1992	STS-42, STS-45, STS-49, STS-50, STS-46, STS-47, STS-52, STS-53
1993	STS-54, STS-56, STS-55, STS-57, STS-51, STS-58, STS-61
1994	STS-60, STS-62, STS-59, STS-65, STS-64, STS-68 STS-66
1995	STS-63, STS-67, STS-71, STS-70, STS-69, STS-73, STS-74
1996	STS-72, STS-75, STS-76, STS-77, STS-78, STS-79, STS-80
1997	STS-81, STS-82, STS-83, STS-84, STS-94, STS-85 STS-86, STS-87
1998	STS-89, STS-90, STS-91, STS-95, STS-88
1999	STS-96, STS-93, STS-103
2000	STS-99, STS-101, STS-106, STS-92

Source: NASA (STS flights as of October 31, 2000.)

Major U.S. launch vehicles that supported space exploration in the 20th century

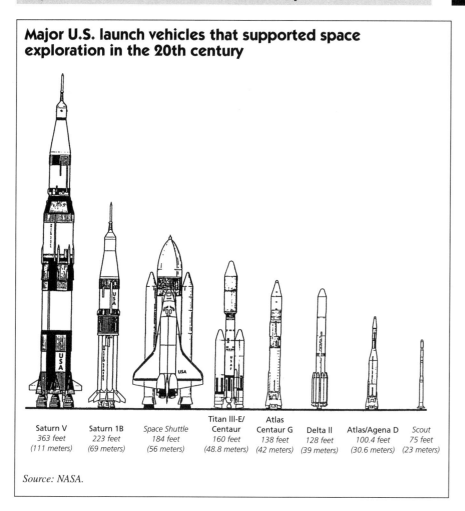

Saturn V	Saturn 1B	Space Shuttle	Titan III-E/ Centaur	Atlas Centaur G	Delta II	Atlas/Agena D	Scout
363 feet	223 feet	184 feet	160 feet	138 feet	128 feet	100.4 feet	75 feet
(111 meters)	(69 meters)	(56 meters)	(48.8 meters)	(42 meters)	(39 meters)	(30.6 meters)	(23 meters)

Source: NASA.

Characteristics of some of the world's launch vehicles

Country	Launch Vehicle	Stages	First launch	Performance
China	Long March 2 (CZ-2C)	2 hypergolic, optional solid upper stage	1975	3,175 kg to LEO
	Long March 2F	2 hypergolic, 4 hypergolic strap-on rockets	1992	8,800 kg to LEO
	Long March 3	2 hypergolic, 1 cryogenic	1984	5,000 kg to LEO
	Long March 3A	2 hypergolic, 1 cryogenic	1994	8,500 kg to LEO
	Long March 4	3 hypergolic	1988	4,000 kg to LEO
Europe	Ariane 40	2 hypergolic, 1 cryogenic	1990	4,625 kg to LEO
(ESA/France)	Ariane 42P	2 hypergolic, 1 cryogenic, 2 strap-on solid rockets	1990	6,025 kg to LEO
	Ariane 42L	2 hypergolic, 1 cryogenic 2 hypergolic strap-on rockets	1993	3,550 kg to GTO
	Ariane 5	2 large solid boosters, cryogenic core, hypergolic upper stage	1996	18,000 kg to LEO, 6,800 kg to GTO
India	Polar Space Launch Vehicle (PSLV)	2 solid stages, 2 hypergolic, 6 strap-on solid rockets	1993	3,000 kg to LEO
Israel	Shavit	3 solid-rocket stages	1988	160 kg to LEO
Japan	M-3SII	3 solid-rocket stages, 2 strap-on solid rockets	1985	770 kg to LEO
	H-2	2 cryogenic, 2 strap-on solid rockets	1994	10,000 kg to LEO, 4,000 kg to GTO
Russia	Soyuz	2 cryogenic, 4 cryogenic strap-on rockets	1963	6,900 kg to LEO
	Rokot	3 hypergolic	1994	1,850 kg to LEO
	Tsyklon	3 hypergolic	1977	3,625 kg to LEO
	Proton (D-I)	3 hypergolic	1968	20,950 kg to LEO
	Energia	cryogenic core, 4 cryogenic strap-on rockets, optional cryogenic upper stages	1987	105,200 kg to LEO
USA	Atlas I	1-1/2 cryogenic lower stage, 1 cryogenic upper stage	1990	5,580 kg to LEO, 2,250 GTO
	Atlas II	1-1/2 cryogenic lower stage, 1 cryogenic upper stage	1991	6,530 kg to LEO, 2,800 kg to GTO
	Atlas IIAS	1-1/2 cryogenic lower stage, 1 cryogenic upper stage, 4 strap-on solid rockets	1993	8,640 kg to LEO, 4,000 kg to GTO
	Delta II	1 cryogenic, 1 hypergolic, 1 solid stage, 9 strap-on solid	1990	5,050 kg to LEO, 1,820 kg to GTO
	Lockheed Martin Launch Vehicle (LMLV 1)	2 solid stages	1995	815 kg to LEO
	Pegasus (aircraft-launched)	3 solid stages	1990	290 kg to LEO
	Space shuttle	2 large solid-rocket boosters, cryogenic core	1981	25,000 kg to LEO
	Taurus	4 solid stages	1994	1,300 to LEO
	Titan 4	2 hypergolic stages, 2 large strap-on solid rockets, variety of upper stages	1989	18,100 to LEO

LEO, low Earth orbit; GTO, geostationary transfer orbit.

Source: NASA, DoD, OTA (U.S. Congress), and others.

International Space Station

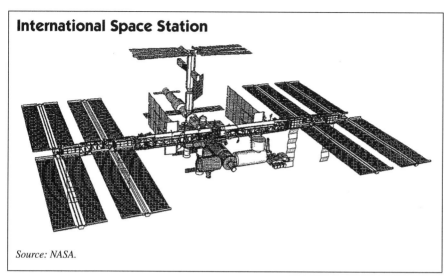

Source: NASA.

A synoptic view of Earth's magnetosphere

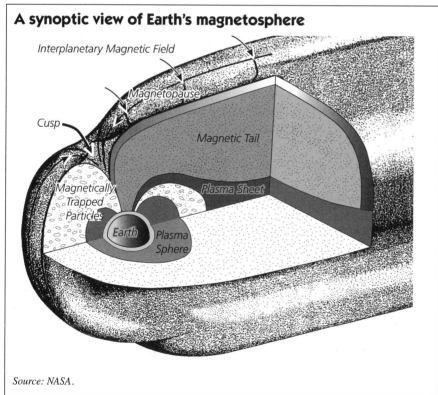

Interplanetary Magnetic Field

Magnetopause

Cusp

Magnetic Tail

Magnetically Trapped Particles

Plasma Sheet

Earth

Plasma Sphere

Source: NASA.

Apollo Project summary

Spacecraft name	Crew	Date	Flight time (hours, minutes, seconds)	Revolutions	Remarks
Apollo 7	Walter H. Schirra Donn Eisele Walter Cunningham	10/11–22/68	260:8:45	163	First crewed Apollo flight demonstrated the spacecraft, crew, and support elements. All performed as required.
Apollo 8	Frank Borman James A. Lovell, Jr. William Anders	12/21–27/68	147:00:41	10 rev. of Moon	History's first crewed flight to the vicinity of another celestial body.
Apollo 9	James A. McDivitt David R. Scott Russell L. Schweikart	3/3–13/69	241:00:53	151	First all-up crewed Apollo flight (with *Saturn V* and command, service, and lunar modules). First Apollo extravehicular activity. First docking of command service module with lunar module (LM).
Apollo 10	Thomas P. Stafford John W. Young Eugene A. Cernan	5/18–26/69	192:03:23	31 rev. of Moon	Apollo LM descended to within 14.5 km of Moon and later rejoined command service module. First rehearsal in lunar environment.
Apollo 11	Neil A. Armstrong Michael Collins Edwin E. Aldrin, Jr.	7/16–24/69	195:18:35	30 rev. of Moon	First landing of a person on the Moon. Total stay time: 21 hr., 36 min.
Apollo 12	Charles Conrad, Jr. Richard F. Gordon, Jr. Alan L. Bean	11/14–24/69	244:36:25	45 rev. of Moon	Second crewed exploration of the Moon. Total stay time: 31 hr., 31 min.
Apollo 13	James A. Lovell, Jr. John L. Swigert, Jr. Fred W. Haise, Jr.	4/11–17/70	142:54:41	—	Mission aborted because of service module oxygen tank failure.
Apollo 14	Alan B. Shepard, Jr. Stuart A. Roosa Edgar D. Mitchell	1/31–2/9/71	216:01:59	34 rev. of Moon	First crewed landing in and exploration of lunar highlands. Total stay time: 33 hr., 31 min.
Apollo 15	David R. Scott Alfred M. Worden James B. Irwin	7/26–8/7/71	295:11:53	74 rev. of Moon	First use of lunar roving vehicle. Total stay time: 66 hr., 55 min.
Apollo 16	John W. Young Thomas K. Mattingly II Charles M. Duke, Jr.	4/16–27/72	265:51:05	64 rev. of Moon	First use of remote-controlled television camera to record liftoff of the lunar module ascent stage from the lunar surface. Total stay time: 71 hr., 2 min.
Apollo 17	Eugene A. Cernan Ronald E. Evans Harrison H. Schmitt	12/7–19/72	301:51:59	75 rev. of Moon	Last crewed lunar landing and exploration of the Moon in the Apollo Program returned 110 kg of lunar samples to Earth. Total stay time: 75 hr.

Source: NASA.

Components of the natural space radiation environment

Galactic cosmic rays

Typically 85% protons, 13% alpha particles, 2% heavier nuclei

Integrated yearly fluence:

  1×10^8 protons/cm^2 (approximately)

Integrated yearly radiation dose:

  4 to 10 rads (approximately)

Geomagnetically trapped radiation

Primarily electrons and protons

Radiation dose depends on orbital altitude

Crewed flights below 300 km altitude avoid Van Allen belts

Solar particle events

Occur sporadically; not predictable

Energetic protons and alpha particles

Solar-flare events may last for hours to days

Dose very dependent on orbital altitude and amount of shielding

Skylab mission summary (1973–74)

Mission	Dates	Crew	Mission duration	Remarks
Skylab 1	Launched May 14, 1973	No crew	Reentered atmosphere July 11, 1979	90 metric-ton space station visited by three astronaut crews
Skylab 2	May 25, 1973–June 22, 1973	Charles Conrad, Jr. Paul J. Weitz Joseph P. Kerwin, M.D.	28 days, 49 min.	Repaired *Skylab;* 392 hr. experiments; 3 EVAs
Skylab 3	July 28, 1973–September 25, 1973	Alan L Bean Jack R. Lousma Owen K. Garriott, Ph.D.	59 days, 11 hr.	Performed maintenance; 1,081 hr. experiments; 3 EVAs
Skylab 4	November 16, 1973–February 8, 1974	Gerald P. Carr William R. Pogue Edward G. Gibson, Ph.D.	84 days, 1 hr.	Observed Comet Kohoutek; 4 EVAs; 1,563 hr. experiments

Russian space station experience

First-generation space stations (1964–77)

Name	Type	Launched	Remarks
Salyut-1	civilian	1971	first space station
no crew	civilian	1972	failure
Salyut-2	military	1973	first Almaz station; failure
Cosmos 557	civilian	1973	failure
Salyut-3	military	1974–75	Almaz station
Salyut-4	civilian	1974–77	
Salyut-5	military	1976–77	Last Almaz station

Second-generation space stations (1977–85)

Salyut-6	civilian	1977–82	highly successful
Salyut-7	civilian	1982–91	last staffed in 1986

Third-generation space stations (1986–2001)

Mir	civilian	1986–2001	first permanent space station

Source: NASA.

The Planets

Mercury	☿
Venus	♀
Earth	⊕
Mars	♂
Jupiter	♃
Saturn	♄
Uranus	♅
Neptune	♆
Pluto	♇

Dynamic and physical properties of planet Earth

radius	
equatorial	6,378 km
polar	6,357 km
mass	5.98×10^{24} kg
density (average)	5.52 g/cm^3
surface area	5.1×10^{14} m^2
volume	1.08×10^{21} m^3
distance from the Sun (average)	1.496×10^8 km (1 AU)
eccentricity	0.01673
orbital period (sidereal)	365.256 days
period of rotation (sidereal)	23.934 hours
inclination of equator	23.45°
mean orbital velocity	29.78 km/s
acceleration of gravity, g (sea level)	9.807 m/s^2
solar flux at Earth (above atmosphere)	$1,371 \pm 5$ watts/m^2
number of natural satellites	1 (the Moon)

Source: NASA and other geophysical data sources.

Physical and astrophysical properties of the Moon

diameter (equatorial)	3,476 km
mass	7.350×10^{22} kg
mass (Earth's mass = 1.0)	0.0123
average density	3.34 g/cm^3
mean distance from Earth (center to center)	384,400 km
surface gravity (equatorial)	1.62 m/s^2
escape velocity	2.38 km/s
orbital eccentricity (mean)	0.0549
inclination of orbital plane (to ecliptic)	5° 09′
sidereal month (rotation period)	27.322 days
albedo (mean)	0.07
mean visual magnitude (at full)	−12.7
surface area	37.9×10^6 km^2
volume	2.20×10^{10} km^3
atmospheric density (at night on surface)	2×10^5 molecules/cm^3
surface temperature	102 K–384 K

Source: NASA.

Physical and dynamic properties of Jupiter

diameter (equatorial)	142,982 km
mass	1.9×10^{27} kg
density (mean)	1.32 g/cm^3
surface gravity (equatorial)	23.1 m/s^2
escape velocity	59.5 km/s
albedo (visual geometric)	0.52
atmosphere	hydrogen (~89%), helium (~11%), also ammonia, methane, and water
natural satellites	16
rings	3 (1 main, 2 minor)
period of rotation (a Jovian day)	0.413 day
average distance from Sun	7.78×10^8 km (5.20 AU) (43.25 light-min.)
eccentricity	0.048
period of revolution around Sun (a Jovian year)	11.86 years
mean orbital velocity	13.1 km/s
magnetosphere	yes (intense)
radiation belts	yes (intense)
mean atmospheric temperature (at cloud tops)	~129 K
solar flux at planet (at top of atmosphere)	50.6 W/m^2 (at 5.2 AU)

Source: NASA.

Physical and dynamic data for Mars

diameter (equatorial)	6,794 km
mass	6.42×10^{23} kg
density (mean)	3.9 g/cm^3
surface gravity	3.73 m/s^2
escape velocity	5.0 km/s
albedo (geometric)	0.15
atmosphere (main components by volume)	
carbon dioxide (CO_2)	95.32%
nitrogen (N_2)	2.7%
argon (Ar)	1.6%
oxygen (O_2)	0.13%
carbon monoxide (CO)	0.07%
water vapor (H_2O)	0.03% (variable)
natural satellites	2 (Phobos and Deimos)
period of rotation (a Martian day)	1.026 days
average distance from Sun	2.28×10^8 km (1.523 AU)
eccentricity	0.093
period of revolution around Sun (a Martian year)	687 days
mean orbital velocity	24.1 km/s
solar flux at planet (at top of atmosphere)	590 W/m^2 (at 1.52 AU)

Source: NASA.

Physical and dynamic properties of the planet Mercury

diameter (mean equatorial)	4,878 km
mass	3.30×10^{23} kg
mean density	5.44 g/cm^3
acceleration of gravity (at the surface)	3.70 m/s^2
escape velocity	4.25 km/s
normal albedo (averaged over visible spectrum)	0.12
surface temperature extremes	~100 K to 700 K (–173° C to 427° C)
atmosphere	negligible (transitory wisp)
number of natural satellites	none
magnetic field	yes (but weak)
flux of solar radiation	
aphelion (~0.467 AU)	6,290 W/m^2
perihelion (~0.31 AU)	14,490 W/m^2
eccentricity	0.2056 (most elliptical planetary orbit, except Pluto's)
semimajor axis	5.79×10^7 km (0.387 AU)
perihelion distance	4.60×10^7 km (0.308 AU)
aphelion distance	6.98×10^7 km (0.467 AU)
orbital inclination	7.00°
mean orbital velocity	47.87 km/s
sidereal day (a Mercurean day)	58.646 Earth days
sidereal year (a Mercurean year)	87.969 Earth days

Source: NASA.

Physical and dynamic properties of Neptune

diameter (equatorial)	49,532 km
mass	1.02×10^{26} kg
density (mean)	1.64 g/cm^3
surface gravity (equatorial)	11 m/s^2 (approximate)
escape velocity (equatorial)	23.5 km/s (approximate)
albedo (visual geometric)	0.4 (approximate)
atmosphere	hydrogen (~80%), helium (~18.5%), methane (~1.5%)
temperature (blackbody)	33.3 K
natural satellites	8
rings	6 (Galle, LeVerrier, Lassell, Arago, Unnamed, Adams [arcs])
period of rotation (a Neptunian day)	0.6715 day (16 hr., 7 min.)
average distance from Sun	4.5×10^9 km (30.06 AU)
eccentricity	0.0086
period of revolution around Sun (a Neptunian year)	164.79 yr.
mean orbital velocity	5.48 km/s
magnetic field	yes (strong, complex; tilted 50° to planet's axis of rotation)
radiation belts	yes (complex structure)
solar flux at planet (at top of atmosphere)	1.5 W/m^2 (at 30 AU)

Source: NASA.

Dynamic properties and physical data for Pluto

diameter	2,274 km
mass	1.25×10^{22} kg
mean density	2.05 g/cm^3
albedo (visual)	0.3
surface temperature (average)	~50 K
surface gravity	0.4–0.6 m/s^2
escape velocity	1.1 km/s
atmosphere (a transient phenomenon)	nitrogen (N$_2$) and methane (CH$_4$)
period of rotation	6.387 days
inclination of axis (of rotation)	122.46°
orbital period (around Sun)	248 years (90,591 days)
orbit inclination	17.15°
eccentricity of orbit	0.2482
mean orbital velocity	4.75 km/s
distance from Sun	
aphelion	7.38×10^9 km (49.2 AU)
	(409.2 light-min.)
perihelion	4.43×10^9 km (29.5 AU)
	(245.3 light-min.)
Mean distance	5.91×10^9 km (39.5 AU)
	(328.5 light-min.)
solar flux (at 30 AU ~perihelion)	1.5 W/m^2
number of known natural satellites	1

Note: Some of these data are speculative.
Source: NASA.

Physical and dynamic properties of the planet Saturn

diameter (equatorial)	120,540 km
mass	5.68×10^{26} kg
density	0.69 g/cm^3
surface gravity (equatorial)	9.0 m/s^2 (approximate)
escape velocity	35.5 km/s (approximate)
albedo (visual geometric)	0.5
atmosphere	hydrogen (89%); helium (11%); small amounts of methane (CH$_4$), ammonia (NH$_3$), and ethane (C$_2$H$_6$); water ice aerosols
natural satellites	30 (6 major, 24 minor)
rings	complex system (thousands)
period of rotation (a Saturnian day)	0.44 days (approximate)
average distance from the Sun	14.27×10^8 km (9.539 AU) (79.33 light-min.)
eccentricity	0.056
period of revolution around the Sun (a Saturnian year)	29.46 years
mean orbital velocity	9.6 km/s
solar flux at planet (at top of clouds)	15.1 W/m^2 (at 9.54 AU)
magnetosphere	yes (strong)
temperature (blackbody)	77K

Source: NASA.

Selected physical and dynamic properties of the planet Uranus

diameter (equatorial)	51,120 km
mass (estimated)	8.7×10^{25} kg
"surface" gravity	8.69 m/s^2
mean density (estimated)	1.3 g/cm^3
albedo (visual)	0.5
temperature (blackbody)	36 K
magnetic field	yes, intermediate strength (field tilted 60° with respect of axis of rotation)
atmosphere	hydrogen (~83%), helium (~15%), methane (~2%)
"surface" features	Bland and featureless (except for some discrete methane clouds)
escape velocity	21.3 km/s
radiation belts	yes (intensity similar to those at Saturn)
rotation period	17.24 hours
eccentricity	0.047
mean orbital velocity	6.8 km/s
sidereal year (a Uranian year)	84 years
inclination of planet's equator to its orbit around the Sun	97.9°
number of (known) natural satellites	17
rings	yes (11)
average distance from Sun	2.871×10^9 km (19.19 AU) [159.4 light-min.]
solar flux at average distance from Sun	3.7 W/m^2 (approximate)

Source: Adapted by author from NASA data.

Physical and dynamic properties of the Sun

diameter	1.39×10^6 km
mass	1.99×10^{30} kg
distance from Earth (average)	1.496×10^8 km [1 AU] (8.3 light-min.)
luminosity	3.9×10^{26} watts
density (average)	1.41 g/cm^3
equivalent blackbody temperature	5,800 K
central temperature (approximate)	15,000,000 K
rotation period (varies with latitude zones)	27 days (approximate)
radiant energy output per unit surface area	6.4×10^7 W/m^2
solar cycle (total cycle of polarity reversals of Sun's magnetic field)	22 years
sunspot cycle	11 years (approximate)
solar constant (at 1 AU)	$1,371 \pm 5$ W/m^2

Source: NASA.

Physical and dynamic properties of Venus

diameter (equatorial)	12,100 km
mass	4.87×10^{24} kg
density (mean)	5.25 g/cm^3
surface gravity	8.88 m/sec^2
escape velocity	10.4 km/sec
albedo (over visible spectrum)	0.7–0.8
surface temperature (approximate)	750 K (477°C)
atmospheric pressure (on surface)	9600 kPA (~1,400 psi)
atmosphere	
main components	CO_2 (96.4%), N_2 (3.4%)
minor components	sulfur dioxide (150 ppm), argon (70 ppm), water vapor (20 ppm)
surface wind speeds	0.3–1.0 m/s
surface materials	basaltic rock and altered materials
magnetic field	negligible
radiation belts	none
number of natural satellites	none
average distance from Sun	1.082×10^8 km (0.723 AU)
solar flux (at top of atmosphere)	2,620 W/m^2
rotation period (a Venusian "day")	243 days (retrograde)
eccentricity	0.007
mean orbital velocity	35.0 km/s
sidereal year (period of one revolution around Sun)	224.7 days
Earth-to-Venus distances	
maximum	2.59×10^8 km (1.73 AU)
minimum	0.42×10^8 km (0.28 AU)

Source: NASA.

Escape velocity for various objects in the solar system

Celestial body	Escape velocity (v_e) (km/s)
Earth	11.2
Moon	2.4
Mercury	4.3
Venus	10.4
Mars	5.0
Jupiter	~ 61
Saturn	~ 36
Uranus	~ 21
Neptune	~ 24
Pluto	~ 1
Sun	~ 618

Source: Developed by the author from NASA and other sources of astrophysical data.

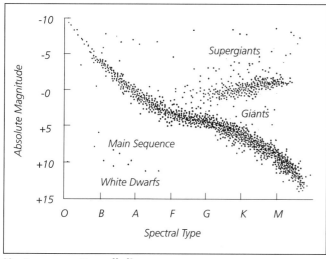

Hertzsprung-Russell diagram

APPENDIX A
Recommended Reading

Angelo, Joseph A., Jr. *The Dictionary of Space Technology.* New York: Facts On File, Inc., 1999.

Angelo, Joseph A., Jr. *Encyclopedia of Space Exploration.* New York: Facts On File, Inc., 2000.

Beatty, J. Kelly, Carolyn Petersen, and Andrew L. Chaikin, eds. *The New Solar System.* 4th ed. New York: Cambridge University Press, 1999.

Chaisson, Eric, and Steve McMillan. *Astronomy Today.* 4th ed. Upper Saddle River, N.J.: Prentice Hall, 2001.

Collins, Michael. *Carrying the Fire.* New York: Cooper Square Publishers, 2001.

Consolmagno, Guy J., et al. *Turn Left at Orion: A Hundred Night Objects To See in a Small Telescope—And How To Find Then.* New York: Cambridge University Press, 2000.

Damon, Thomas D. *Introduction to Space: The Science of Spaceflight.* 3rd ed. Malabar, Fla.: Krieger Publishing Co., 2000.

Dickinson, Terence. *The Universe and Beyond.* 3rd ed. Willowdater, Ont.: Firefly Books Ltd., 1999.

Matloff, Gregory L. *The Urban Astronomer: A Practical Guide for Observers in Cities and Suburbs.* New York: John Wiley & Sons, 1991.

Pasachoff, Jay M. *Peterson First Guide to Astronomy.* New York: Houghton Mifflin Co., 1998.

Seeds, Michael A. *Horizons: Exploring the Universe.* 6th ed. Pacific Grove, Calif.: Brooks/Cole Publishing, 1999.

Spangenburg, Ray, and Diane Moser. *Wernher Von Braun: Space Visionary and Rocket Engineer.* New York: Facts On File, Inc., 1995.

Sutton, George Paul. *Rocket Propulsion Elements.* 7th ed. New York: John Wiley & Sons, 2000.

APPENDIX B
Cyberspace Destinations

Selected Space Organization Home Pages

European Space Agency (ESA)
  http://www.esrin.esa.it/
4th Space Wing (Cape Canaveral/Patrick AFB)
  http://www.pafb.af.mil/
National Aeronautics and Space Administration (NASA)
  http://www.nasa.gov
National Oceanic and Atmospheric Administration (NOAA)
  http://www.noaa.gov
National Reconnaissance Office (NRO)
  http://www.nro.odci.gov/
U.S. Air Force Space Command (AFSPC)
  http://www.spacecom.af.mil/hqafspc/

Selected Space Missions

Cassini mission (Saturn)
  http://www.jpl.nasa.gov/cassini/
Galileo mission (Jupiter)
  http://www.jpl.nasa.gov/galileo/

Other Interesting Space and Astronomy Sites

Lunar and Planetary Institute (LPI)
  http://cass.jsc.nasa.gov/
NASA's Space Science News
  http://science.nasa.gov/
National Air and Space Museum (Smithsonian Institution)
  http://ceps.nasm.edu/NASMpage.html
National Space Science Data Center (NSSDC) (numerous space missions—multinational)
  http://nssdc.gsfc.nasa.gov/planetary/
The Planetary Society
  http://planetary.org

**Nineteenth Century
Empire**

4421

**Eighteenth Century
American Colonial**

**Nineteenth Century
Victorian**

COSTUMES
for you to make

Also by Susan Purdy

BOOKS FOR YOU TO MAKE

JEWISH HOLIDAYS

FESTIVALS FOR YOU TO CELEBRATE

HOLIDAY CARDS FOR YOU TO MAKE

IF YOU HAVE A YELLOW LION

CHRISTMAS DECORATIONS FOR YOU TO MAKE

MY LITTLE CABBAGE

SUSAN PURDY

COSTUMES
for you to make

J. B. Lippincott Company / Philadelphia and New York

ABOUT THE AUTHOR

Susan Purdy was born in New York City and grew up in Connecticut. She attended Vassar College and New York University and took her junior year at the Sorbonne and the Ecole des Beaux Arts in Paris. Mrs. Purdy was a textile designer in New York City until her marriage, and now she and her husband live in a farmhouse in Connecticut. Mrs. Purdy writes and illustrates children's books and, until recently, taught children's private art classes and codirected and taught art at a music-and-art summer day camp. An essential part of the camp art program was creative experimentation with materials and design, and she has drawn on her teaching experiences in adapting many creative art techniques to the projects in her popular activities and craft books.

ISBN-0-397-31317-9

Library of Congress Catalog Card Number 77-151470
Fourth Printing
Typography by Jean Krulis

For Jeanne Vestal
who makes all things possible

Forever to dress up
Never to grow up.

ACKNOWLEDGMENTS

For invaluable help in the selection and preparation of material for this book, I wish to express my indebtedness and gratitude to all who helped, including especially the following: my husband, Geoffrey Purdy, Nancy Gold, Frances J. Gold, Mrs. Bea Joslin, Sara Jane and Michael Chelminski, Little Bobby Hanson, Sophia and Gregg Sterling, Jenny Cappel, Linda Mitchell and all the children who attended the Wilton Music and Art Day Camp, Deborah Sherman, Doris Jason, Mary Haberman, Mary Dwyer, and Ian McDermott; also, the children and and faculties of the Wilton, Connecticut, schools, including Mrs. Gudi Pierson, Miss Mary Elizabeth Jones, Lydia Schneid, Miss Edna Abbott, Angela Donovan, Ruth Peale Vining, Robert Douglass, Mary Bosqui; and Mrs. Esther Franklin, Consultant, Library Services, Sacramento, California; Mary Lee Keath, Director, Department of Library Services, Denver, Colorado, Public Schools; Beverly Petersen, Denver Schools; Elnora Portteus, Directing Supervisor, Cleveland Public School Libraries; Elizabeth M. Stephens, Director, Educational Media Center, Clearwater, Florida; Irene Davis, Houston, Texas, Indepedent School District; and Mildred Nickel, Director of Lansing, Michigan, Public School Libraries.

And for further reference the following books are recommended: Bentham, Frederick, *Stage Lighting*. London: Sir Isaac Pitman & Son, Ltd., 1955; Bradley, Carolyn G., *Western World Costume: An Outline History*. New York: Appleton-Century-Crofts, Inc., 1954; Elicker, Virginia Wilk, *Biblical Costumes*. New York: A. S. Barnes and Company, 1953; Evans, Mary, *Costume Throughout the Ages*. New York: J. B. Lippincott Company, 1950; Grimball, Elizabeth and Wells, Rhea, *Costuming a Play*. New York: The Century Company, 1925; Hewitt, Barnard, *Play Production, Theory and Practice*. New York: J. B. Lippincott, 1952; Köhler, Carl, *A History of Costume*. New York: Dover Publ., 1963; Laver, James, *Costume Through the Ages*. New York: Simon & Schuster, Inc., 1964; Mann, Kathleen, *Peasant Costume in Europe*. London: A. & C. Black, Ltd., 1931; Wilcox, R. Turner, *Five Centuries of American Costume*. New York: Chas. Scribner's Sons, 1963.

TABLE OF CONTENTS

1. INTRODUCTION

HOW TO USE THIS BOOK

It is hoped that this book will guide and inspire you through the process of creating costumes for plays, pageants, pantomimes, Halloween, or costume parties. Many of the ideas are simple enough to be fun for quick dress-up projects. Others are more complex and take longer to complete. Before beginning any costume, it is best to read all the directions through to the end so you will know what is required. All materials are kept as simple as possible, and no previous sewing skills are needed.

If you are looking for costume suggestions, read Where to Find Costume Ideas, *page 11*; also see the picture section beginning on *page 12*. To make the various parts of each of these costumes, you are referred to the appropriate pages throughout the text. These sketches are intended to be only a starting point; while they may guide you, they are by no means the only costumes which can be made with this book. All patterns and costume parts can be adapted to fit nearly any costume need. The more you rearrange, invent, and improvise in designing your costumes, the more fun you will have. The Table of Contents will indicate some, *but not all* of the costumes in this book; other costume ideas appear in various parts of the text. To look up a specific costume or costume part, check the index on *page 119*.

For general information on costume design, color, materials, and makeup techniques, see Section 1. For basic skills such as tracing and transferring patterns, hemming and padding costumes, or for trimmings and accessories which can be applied to any type of costume, see Sections 2 and 3. Section 4 presents basic costume parts, with such all-purpose garments as tunics, aprons, capes, footwear, etc. These are followed by several basic types of hats and headbands in Sections 5 and 6, each with many variations. For short skits or rainy day fun, there are assorted masks and quick disguises in Section 7, and in Section 8, such unusual costumes as a Lion Body Mask and a Two-or-More-Persons Animal. Suggestions for further reading will be found on the Acknowledgments page.

All costumes in this book have been scaled to the average size of a ten-year-old. However, by following the specific instructions, you can fit each costume to your own needs. Throughout the text, the symbol ″ has been used for inches, and ′ for feet. Before working with patterns, read the tracing and transferring directions on *page 28*.

WHERE TO FIND COSTUME IDEAS

Ideas for costumes can come from an infinite variety of sources. You may begin with the library—the young people's as well as the adults' division. For general costume needs, you will find a wealth of ideas in illustrated works of nonfiction and fiction, magazines, fairy tales, children's books, song and dance books, and even some contemporary textbooks.

Pictures of national folk costumes can be found in illustrated travel books, magazines, and histories of a specific country or region. Also look at encyclopedias, record jackets or books of folk dances, and national tourist office travel brochures and posters (often given away free). Inquire at your library, school office, or local newspaper to see if there are people living in your area who are interested in and might have examples of authentic national costumes.

As an accurate guide to historical costumes, look for material on the history of art and the theater, or the history of a particular period or place. Seek out histories of festivals, music, or dance as well as specialized studies of textiles, design, fashion, jewelry, wigs, masks, hats, shoes, furnishings, and architecture. If a class or group is studying an historical period or a geographical area, it will be fun as well as helpful for everyone to investigate a particular aspect of fashion such as color, jewelry, fabric, accessories, general cultural history, etc., and then compile a reference scrapbook which may well become the start of a theatrical reference library.

Explore library picture collections for costume ideas, which are filed under each country and each historical period (by century) as well as under "costumes." Also look at the *National Geographic Magazine*, and magazines dealing with the theater arts, opera, the dance, and all the visual arts.

Museums are a treasure house of costume ideas, particularly if they have costume collections such as the Boston Museum of Fine Arts or the Metropolitan Museum of Art in New York. But do not overlook special museums such as the Museum of Contemporary Crafts in New York, or a museum of natural history, a museum of the American Indian, or a local historical society. Most museums sell photographs or postcards which are valuable reference materials, and many have their own libraries, or special costume libraries, which you may be able to use if you inquire.

COSTUME SUGGESTIONS

The following sketches are intended to suggest various costume ideas. The captions indicate how the costume is made and the pages on which each element can be found. By combining various costume parts, you will be able to create nearly any costume you wish. To find additional ideas, or to look up a specific costume or detail, check the index on *page 118*. The following costumes are illustrated elsewhere in the book on pages noted: Hawaiian boy and girl (*39*), Jester (*43, 79*), Clown (*24, 46*), Ancient Greek (*63*), Ancient Roman (*64*), Little Red Hood (*71*), Cat (*80, 98*), Flower, Fairy, Candle (*100*), Monster (*104*), Pumpkin (*106*), Pig (*107, 109*), Astronaut (*107*), Wolf (*108*), Dog (*108*), Witch (*109*), Lion (*111*), Medieval Knight (*115*), Crusader (*115*), Dragon (*117*). See the endpapers for period costumes from the fifteenth, sixteenth, seventeenth, eighteenth, and nineteeenth centuries.

SCHOOLBUS Three packing cartons (2 large, 1 medium) all about 24″ wide). Glue together as shown (*arrows*). Draw, then cut out windows and armholes. Glue or staple on cardboard wheels and steering wheels, paper or plastic soup-bowl headlights. Cut air holes in "roof" if you want to cover windows with clear plastic wrap. Hold bus up with arms as shown.

TOTEM POLE Large packing carton with look-out hole cut in front and armholes cut into sides. Stiff cardboard shapes cut out and glued to front and back surfaces.

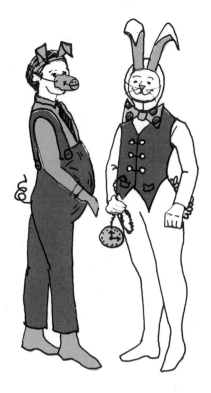

PIG AND RABBIT *Pig*: Ears (*page 98*), snout (*page 107 or 109*), suspenders (*page 50*) on slacks, tail (*page 76*), pink tights, shirt, and socks. *Rabbit*: Ears (*page 99*), hood (*page 79*), bow tie (*page 48*), vest (*page 68*), tail (*page 76*), tights and sweater, watch (painted cardboard circle on paper or brass chain).

BEE Antennae (*page 100*), black hood (*page 77*). Wear yellow sweater and yellow gloves or yellow socks on hands, black tights. Sew black cloth or crepe-paper stripes around sweater as shown. Stuff 2 pairs yellow knee socks with cotton or rags and sew open ends to sides for "legs."

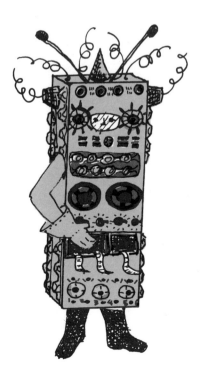

COMPUTER Large packing carton with look-out hole cut in front and armholes cut in sides. Paint box and trim with glued-on egg cartons, dials, discarded foam or plastic food packs, paper cups, ticker tape or adding-machine tape strips, etc. Wire on hardware such as springs, bolts, nuts, and bottle caps. Antennae are wire coat hangers with cork or clay knobs. For special sound effects carry a rattle or noisemaker. To make dials light up, cut holes in box, cover with colored cellophane, wear lighted flashlight tied to waist inside box.

PLAINS INDIANS Both wear slacks with short- or long-sleeved tunics (*page 61*) or jacket (boy) trimmed with fringe (*page 38*), and decorations (*page 30*), and moccasins or loafers with "beaded" design on paper taped or sewn to instep. *Girl*: Headband (*page 100*). *Boy*: War bonnet (*page 101*).

COWBOY Hat (*page 82*), colorful shirt and kerchief, slacks, vest (*page 68*), loose belt (*page 51*) with narrow strip of paper looped as shown and stapled to belt to form cartridge pockets, boots (*page 74*), moustache (*page 110*), holster.

KING AND QUEEN *King*: Crown (*page 102*), moustache and beard (*page 110*), pajamas trimmed with wide diagonal sash of ribbon or fabric, cuffs (*page 47*), cape (*page 71*), with "ermine" (*page 30, #1, (d)*), scepter (30"-long broom handle with foil knob and/or paper trim). *Queen*: Crown (*page 102*) or tiara (*page 98*); long-sleeved blouse and gathered skirt (*page 66*) trimmed with ruffles (*page 37*) and cuffs (*page 47*), fan (accordion-folded paper).

CAPTAIN HOOK AND PIRATE *Captain Hook*: Hat (*page 94*), cravat (*page 48*), moustache (*page 110*), jacket trimmed with lace cuffs (*page 47*), knickers (*page 58*) tied with ribbons, knee socks under boots (*page 74*) with lace cuffs, earrings (*page 32*), wide sash, hook (bent coat hanger covered with tape, held by hand up inside sleeve). *Pirate*: Bandanna, eye patch (paper shape attached to string or elastic), striped tee shirt and solid-colored sash, boots (*page 74*), cutlass or scimitar (*page 57*).

KING AHASUERUS AND QUEEN ESTHER (ANCIENT PERSIA) *King*: Fez (*page 89*) with trim (*page 30*), robe (*page 62*) worn with opening in back, optional collar (*page 42*) of soft fabric which drapes easily, sandals (*page 72*), rope belt (*page 51*), scepter (broom handle with foil knob), beard (*page 110*). *Queen*: Fez (*page 89*) trimmed with chiffon or gauze scarf, fringe (*page 38*) trimmed short-sleeved tunic (*page 61*), toga (*page 64*), and diagonally draped waist sash; jewelry (*page 31*).

THREE KINGS Pajamas or long-sleeved tunic under robe (*page 62*), crown (*page 102*) or tiara (*page 98*), or Shepherd headdress (*below*), belt (*page 72*) or wide sash, sandals (*page 72*), beard (*page 110*).

MARY, JOSEPH, SHEPHERD, ANGEL All wear tunics (*page 61*), sandals (*page 72*), belts (*page 51*); men's beards (*page 110*). *Mary and Joseph*: Toga (*page 64*) or cape (*page 71*). *Shepherd*: Robe (*page 62*), headdress (dish towel tied with twisted wool or rope, also use for Arab man), staff (broom handle with cardboard scroll wired or taped on). *Angel*: Wings (*page 77*), halo (*page 101*).

SAINT NICHOLAS (BISHOP) Long-sleeved tunic (*page 61*), cape (*page 71*), beard (*page 100*), mitre (*page 96*), sandals (*page 72*), staff (broom handle with cardboard scroll wired or taped on).

15

EGYPTIAN KING AND QUEEN Both wear collars (*page 43*), sandals (*page 72*) with curved-wire toes, armlets, anklets, and bracelets (*page 42*). *Queen*: Headdress (*page 87*), short-sleeved tunic (*page 61*). Other women wear a variation of headband (*page 97*) and starched scarf, held in place with bobby pins. *King*: Headdress of Lower Egypt (*page 93*), suspenders (*page 50*) attached to extra-wide bodice which is laced in back (*page 70*), loincloth (*page 59*), beard (*page 110*) trimmed to shape shown. Other crowns are modified forms of fez (*page 89*) with stiff paper trim stapled or glued on.

AZTECS *Girl*: Simple poncho (*page 60*) over long straight, or wraparound skirt or short-sleeved tunic (*page 61*), jewelry (*page 30*), hair in braids curved up to top of forehead. *Boy* (King): Loincloth or breech clout (*page 59*), cape (*page 71*), sandals (*page 72*) and below-knee ribbons both trimmed with triangular paper or felt tabs, beard (*page 110*) trimmed to shape shown, tiara (*page 98*) or headband (*page 97*) with ribbons tied on in back and/or front, club 24″ long, of cardboard.

VIKING WARRIOR Helmet (*page 81*), moustache (*page 110*), rough-textured turtleneck sweater under simple tunic (*page 60*) with side ties, loose trousers or pajama bottoms laced up to knee with strings, sneaker sandals (*page 73*), shield (*page 57*), sword (*page 56*).

16

GREEK AND ROMAN SOLDIERS Both wear armor (*page 114*), sandals (*page 73*), greaves (shin guards, *page 75*), swords (*page 56*), and shields (*page 57*). *Greek soldier*: Helmet (*page 83*), tunic worn under armor (*page 62*). *Roman soldier*: Helmet (*page 85*), tunic worn under armor (*page 64*).

ROBIN HOOD AND MAID MARIAN (ME-DIEVAL BOY AND GIRL) *Boy*: Hat (*page 93*), collar (*page 43*) worn over pajama-top or jacket with notched hem, belt (*page 51*), tights, Medieval slippers (*page 73*). *Girl*: Hat (*page 91*), long-sleeved tunic (*page 61*), bodice (*page 70*).

DRUM MAJOR AND EIGHTEENTH-CEN-TURY SOLDIER Both wear short jackets trimmed with buttons (*page 52*), Mandarin collars, cuffs (*page 47*), and cross straps (*page 50*); boots (*page 74*). *Drum Major*: Hat (*page 90 or 94*), epaulettes (*page 54*). *Eighteenth-Century Soldier*: Tricorne (*page 82*).

GEORGE AND MARTHA WASHINGTON (EIGHTEENTH CENTURY) *Boy*: Tricorne (*page 82*), cravat (*page 48*), jacket trimmed with buttons (*page 52*), cuffs (*page 47*), epaulettes (*page 54*); sword (*page 56*), boots (*page 74*). *Girl*: Tiara (*page 98*), jewelry (*page 30*), blouse trimmed with lace cuffs (*page 47*) and gathered skirt (*page 66*) trimmed with ruffles (*page 37*).

PURITANS Both wear dark blouses or sweaters trimmed with collars (*page 44*) and cuffs (*page 47*), and shoes trimmed with buckles (*page 52*). *Girl*: Cap (*page 95*), apron (*page 65*). *Boy*: Hat (*page 91*), dark knickers (*page 58*), white knee socks, cape (*page 71*).

AUSTRIA (TYROL) *Girl:* Flower chain (*page 40*), blouse trimmed with ruffles (*page 37*), Austrian corselet (*page 69*), gathered skirt (*page 66*), apron (*page 65*). *Boy:* Hat (narrow-brimmed cowboy hat, *page 82*) with brush trim, shorts (lederhosen) with suspenders (*page 50*), shirt and tie, knee socks.

SWEDEN *Girl*: Dutch girl's cap (*page 95*) with decorated cuff, fringed shawl, laced bodice (*page 70*), gathered skirt (*page 66*), apron (*page 65*), white tights. Danish girl wears shawl and Norwegian wears Scandinavian cap (*page 95*). *Boy:* Stocking cap (*page 92*), jacket trimmed with Mandarin collar (*page 45*) and buttons, knickers (*page 58*), knee socks and shoe buckles (*page 52*).

ARGENTINA and BRAZIL *Boy* (*Argentine Gaucho*): Hat is stovepipe (*page 88*) with 2″-tall top, moustache (*page 110*), bolero (*page 68*), wide belt (*page 51*) with foil discs, full knickers (*page 58*), boots (*page 74*), colorful shirt and neck bow. *Brazilian Girl*: Bandanna on head, jewelry (*page 30*), blouse and gathered skirt (*page 66*) trimmed with ruffles (*page 37*).

VIETNAM Both figures wear Southeast Asian cone hat (*page 91*) and sandals (*page 72*) or rubber zoris. *Girl*: Dark slacks, tunic (*page 61*) with long, close-fitting sleeves, side slit in hem, neck opening on diagonal with Mandarin collar (*page 45*). *Boy*: shorts, shirt.

WEST AFRICA *Ghanian Boy*: Fez (*page 89*), tee shirt and slacks, toga (*page 64*), sandals (*page 72*). *Girl*: Dress is fabric 4½' wide by 5' long. One third of width is folded over, fabric is wrapped around body and tucked into itself to hold. Bandanna on head, jewelry (*page 30*).

MEXICO *Girl*: Flower (*page 40*), jewelry (*page 31*), short-sleeved blouse trimmed with ruffles (*page 37*), sash, gathered skirt (*page 66*). *Boy*: Straw hat or modified tricorne (*page 82*) wired around entire brim so it bends up evenly, white shirt and trousers, colorful sash, sandals (*page 72*), serape (striped cloth 18″ wide by 5' long) fringed (*page 37*) at ends.

INDIA *Girl*: Short-sleeved tunic or blouse and straight skirt, slacks underneath (optional), jewelry (*page 31*), sari (same as toga, *page 64*), slippers or sandals (*page 72*). *Boy*: Fez (*page 89*), long white jacket or long-sleeved tunic (*page 61*) trimmed with buttons, red sash, and Mandarin collar (*page 45*), tight white slacks.

NOTES ON COSTUME DESIGN, COLOR, AND MATERIALS

Basic Approach to Costume Design

The most effective costume is the one that has an immediately recognizable silhouette, or basic outline. If you can train your eye to look for the basic shape and style rather than getting lost in small details, you will be on the way to creating successful costumes.

To design a costume for a particular character, first think about his personality (sweet young girl, nasty villain), his location (hot or cold climate), the time in which he lives (is he an historical figure? of what period?) and the action (will he sit still or climb a tree?). Also consider your budget, and try to avoid elaborate and expensive costumes. When working on a large production, consult with the director and the designers of scenery and lighting before carrying out costume designs.

When you have made sketches, review the following points: Does your costume clearly convey its historical period? Do you use color and style to exaggerate your character's personality? A gay young sprite should be in soft lines and cheerful colors, while a villain wears dark tones and sharp angles; a leading character should stand out in a color which contrasts with a surrounding group. Is the costume comfortable to wear and practical for the required activity? Too-long hems can make even the most graceful dancer trip. Consider the theater size if you are giving a play; will the audience be nearby or far from the stage? Bold, bright colors and designs are seen more clearly at great distances than are pastel tones and tiny prints. Will you use stage lighting? If so, try out costume colors under stage lights before the production, as this light can completely change the appearance of some colors.

Elements of Design

When creating costumes for dramatic productions, keep in mind such basic design elements as line, shape, and color. For visual excitement, use a variety of lines, both in costume style and fabric pattern. Verticals make figures seem tall, dignified, and thinner. Horizontal lines make one look broader, fatter. Strong diagonals, spirals, diamonds, and plaids make restless and interesting patterns for the eye to follow. Asymmetrical (or off-balance) designs create movement; symmetrical (balanced) designs are peaceful and still. A group of figures in equally divided black and white tunics look static and boring; for more movement, divide the color areas irregularly and off-center. Vary fabric textures within one production, using rough surfaces like burlap as well as shiny sateens and dull wools. Use repetition of line, shape, and color to give unity within the production.

Color

Color is one of the basic elements used to create the atmosphere or mood of any production. For individual masquerade or Halloween costumes, let your color imagination run wild. For dramatic productions, consider color part of the overall stage design and use it thoughtfully.

Colors used should be appropriate to the dramatic action. Earth tones such as rusts, golds, browns, reds, and oranges are warm in feeling and create a sense of excitement. Shades of blues, purples, and greens are cool and restful and may suggest mystery or evil. A happy, gay musical might use bright yellows, white, gold, orange, scarlet, red-orange, yellow ochre, and lime, with accents of such contrasting colors as purple, lilac, blue, and green. A sad, somber drama might require grays, black, green-gray, blue-gray, beige, tan, rust, purple, brown, and white with magenta, wine, and gold accents. Remember that various historical periods and nationalities favored particular colors for their costumes; for an authentic flavor, research should be thorough. Finally, consider costumes in relation to the surrounding scenery. A white ghost will stand out from a dark sky better than a dark blue ghost.

Materials to Use and Where to Find Them

When selecting costume materials, keep in mind the type of character you are dressing and the time and place in which he lives. Royalty or nobility should wear elegant, heavily draped garments rich in texture and design. Use old brocade, velveteen, satin or sateen, heavy cotton or flannel, corduroy or toweling; old blankets, curtains, or old extra-large bath towels make good capes or tunics. Also use old tablecloths, bedspreads, and pieces of upholstery and drapery fabric. Elegant gowns are made of taffeta, polished cotton, or many synthetics. Easily draped fabrics for togas and shawls include crepe, silk, chiffon, dress lining, and cheesecloth, as well as soft, worn sheets. Peasants wear coarse burlaps, wools, toweling, homespuns, denim, monk's cloth, and unbleached muslin. For metallic-looking uniforms, use gold or silver plastic sheeting (available by the yard) or oilcloth or heavy duty tinfoil. Or, cover shaped cardboard or stiff paper with self-adhesive decorator paper with the desired texture or pattern.

Before buying costume materials, look around in attics and rag bags, textile remnant shops and thrift shops. You can almost always find cheap and intriguing substitutes for store-bought fabrics and accessories.

Crepe paper can be sewn on a machine, or with needle and thread, just like fabric; however, use a slightly wider-spaced stitch on the sewing machine to avoid perforation. Duplex crepe paper usually has different colors on each side, and as it is made of two layers bonded together, is more durable for costumes than regular, single-weight crepe paper. Use either type for costumes which are to be worn just one time; fabric makes the most durable costumes.

Thrift shops are a gold mine of costume materials. You may find usable fringe on old shawls; feathers, flowers, hat crowns and hatbands on old hats; bows and buckles to cut from old shoes and belts; old collars and cuffs which provide lace for ruffles; nightgowns, sheets, pillow cases, and curtains to transform into tunics, capes, and skirts. Long skirts may be hemmed to fit you; cut unwanted tops off evening dresses for skirts or capes; make bustles from scarves; add crepe paper swaths or upholstery fringe to disguise an old umbrella. A scarf or bandanna can be wrapped into an African or East Indian turban; an old dish towel or hand towel can be draped over the head and tied around the forehead with a rope or wool braid making an Arab's headdress. Large old gloves make fine gauntlets for a suit of armor when sprayed silver. If you don't like a garment's color, change it *(see page 32, #7)*. Sometimes your search will turn up old fur pieces and muffs which make lovely trimming. Or, furriers will occasionally sell or give away odd scraps of fur. Sometimes trimming laces can be bought inexpensively from lace mill outlet shops (look in the Yellow Pages).

MAKEUP

Theatrical makeup is used to exaggerate the facial characteristics and make them more visible under stage light and from a distance. It is also used to change or modify the features of a character.

For most school and amateur theatrical productions, makeup should be kept simple; ordinary women's makeup purchased in a five-and-ten-cent store or a drugstore will be adequate. For character makeup (clowns, villains, etc.) more elaborate cosmetics or greasepaints may be used, though they are by no means essential. (If you do use greasepaint, apply it to a base of cold cream rubbed into the face then well blotted. Regular makeup is applied to a clean, dry face.) Theatrical supply stores and theatrical costumers located in most cities sell greasepaint and water-base makeup in sticks or tubes, black tooth wax (non-toxic), nose putty (a claylike substance for molding false features), crepe hair (fake hair sold in a braid which is unravelled, cut, and applied with a special spirit gum), clown white (a water-base foundation color), luminous paints in all colors (which glow in the dark), and fluorescent makeup (which glows when under ultraviolet light). Some of these specialty items may be found in department or toy stores at Halloween.

Note: Generally, makeup will not irritate the skin. However, sensitive skins should use nonallergenic makeup found in drugstores. When applying eye makeup, use extreme caution so eyes are not injured; if possible, have an adult help you.

Basic makeup supplies for a dramatic group should include: cake or liquid foundation—in light, medium, and dark skin tones; flesh-tone stick or cream—for covering blemishes and blanking out eyebrows; face powder—in light, medium, and dark tones, brushed or patted over face to set foundation color and remove face shine; lipstick—light, medium, and dark shades or pink, red, and orange (to change your lip line, cover your own lips with foundation tone and redraw with lipstick); rouge—in cake or stick, used to highlight cheekbones (lipstick applied to fingertip and dabbed on cheek may be substituted); eye liner—brown or black, in soft pencil, liquid, or moistened cake form (the latter two used with a fine sable brush which is also used for drawing wrinkles and character lines); eye shadow—cake or powder or paste form in shades of blue, green, gray, lavender, and white (the latter spread just below the eyebrows and in corners of eyes to exaggerate eye size), as well as gold and other sparkling iridescent colors used for glamorous effects; mascara—brown or black, applied with brush or self-applicator, used to exaggerate eyelashes; false eyelashes—bought in many styles or made from paper (*see page* 38), use instead of, or in addition to mascara; eyebrow pencils—brown, black, and red (soft, dark pencils are used to darken eyebrows or draw wrinkle lines, red pencil or a dot of red lipstick is used at the inside corner of each eye for emphasis; to change eyebrow line, cover natural brows with foundation, then redraw; use pencil to draw beauty mark or "patch" on a girl's cheek); small sable brushes—used for applying eye makeup and red eye dots; cold cream—any type, used with facial tissue for removing makeup or use soap and water; flesh-colored adhesive tape (for dark skins, color tape with felt pens)—make eyes slant upward at the outer corners by pulling the skin out and up between upper lid and brow and fastening with tape.

Character makeup requires a bit of practice, but the technique is basically simple and fun to learn. In general, character lines such as worry wrinkles, jowls, dimples, laugh lines, etc., should follow the natural contours of your face. Assume the facial expression of your character and watch in the mirror to see what happens to your face. For example, to "become" an old man, wrinkle your forehead, suck in your cheeks, tuck in your chin to exaggerate jowls and double chin, turn down your mouth and purse your lips into a thin line. To capture a difficult character, it may help to look at an illustration or cartoon of this individual, as the lines in a drawing exaggerate prominent features.

Model the face with light and dark tones rather than relying entirely upon drawn-in lines. Use light foundation, rouge, or white eye shadow to highlight prominent areas such as cheekbones, jowls, and nose; deepen around eye sockets with gray or brown eye shadow blended in. Or, use a burnt cork for dark smudges. Lightly draw wrinkles and furrowed lines with soft eye-liner pencil, then blend slightly so lines are not too sharp. For boys or men, use lipstick sparingly if at all.

YOUNG GIRL

OLD MAN

CAT

CLOWN

WHITE BATHING CAP

GREEN

WHITE

EGYPTIAN EYE

GREEN

ORDINARY GIRL'S EYE

Moustaches, beards, and sideburns may be drawn on the face with an eyebrow pencil, or made of crepe hair or paper (*see page 110*). Bits of wool or unravelled twine may also be taped to the face for quick effects. To make wigs, nearly any material may be used. Old (but clean) mops may be trimmed to shape, dyed or spray-painted any color, and tied in place with strings. More elaborate wigs may be made by sewing wool, scouring pads (nonsoapy; or stainless steel or colored plastic but *not steel wool, as it sheds*), raffia, twine, etc., to the outside of a hood (*see page 79*) or to the crown of an old hat or skull cap (*page 80*), or to a baseball cap with peak removed. Or, poke holes all over an old bathing cap and tie pieces of wool or raffia into the holes; for a bald-headed man or clown, paint a bathing cap flesh color, and cut out ear holes. To make gray hair, dab on corn-starch or talcum powder (if perfumed, be careful not to get it in your eyes), or brush on white eye liner or clown white. For an old person, remember to gray eyebrows as well as moustache and beard if you are graying the hair.

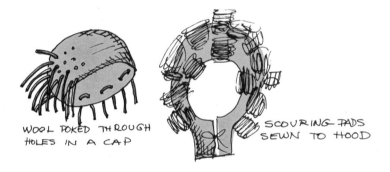

WOOL POKED THROUGH HOLES IN A CAP

SCOURING PADS SEWN TO HOOD

To make fingernails or claws, cut long tapered triangles of paper or foil-covered paper, about ½″ wide at base and 2″ long or more. Place a bit of double-face tape or rubber cement on each fingernail and press on paper. To make fierce, curved talons, tape a piece of thin wire or pipe cleaner to the underside of each triangle, then tape or glue it to fingernail. Bend wire into a curve. Be sure to bend over the end of the wire, or cover it with tape, to avoid scratches.

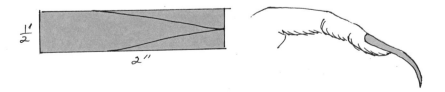

$\frac{1}{2}'$

2″

MATERIALS

The following is a list of all the materials mentioned in this book, although each project will require the use of only a few items at a time. Materials can easily be found in your home, local stationery store, five-and-ten-cent store, art supply or craft shop, fabric or remnant shop, thrift shop, or hardware store. If you have trouble finding materials, look in the Yellow Pages of the telephone book. Use materials suggested or make up your own variations.

ruler, tape measure, yardstick
pencil, colored pencils
crayons
colored felt or nylon-tipped pens (oil and water base)
scissors (regular shears), pinking shears (optional)
stapler
Exacto knife
masking tape, cellophane tape, Mystic cloth tape, florists' tape (optional)
double-face (two sticky sides) masking and cellophane tape
double-face carpet tape ("Carpetak," found in hardware stores)
Velcro tape (double-layered fastening tape with interlocking surfaces, found in
    fabric and department stores and sold by the inch or by the yard)
rubber cement, white glue
brass paper fasteners
notebook ring reinforcements
surgical cotton
self-stick decorative gold and silver braid (found in fabric and dime stores)
raffia, ribbon
wool, bias tape (found in fabric stores, optional)
rope, cotton string, colorless fishing line
darning needle (blunt-nosed and/or regular), fine sewing needle
button thread, regular (#50) thread
chalk (regular blackboard chalk or tailor's chalk found in fabric store, used to
    mark felt or fabric)
straight pins, safety pins
egg cartons (pressed paper or styrofoam)
tempera paint, enamel paints or spray enamel or spray lacquer (use to paint tin-
    foil), textile paints (for permanent color on fabric)
shellac and alcohol (shellac solvent)
paintbrushes

high gloss clear acrylic spray coating
spray starch
iron (optional)
rubber bands or sewing elastic
paper hole punch (optional)
styrofoam balls (1″ or 2″ diameter)
thumbtacks
cotton or other fabric, felt, old sheets or pillow cases, etc.
hula hoop or cardboard barrel top with flat surface removed
half-gallon milk carton (empty and clean)
paper cups (9-oz. size)
8″ paper plates
compass (optional)
extra-large paper (not plastic) bag about 21″ by 30″ (found in discount or
    department stores)
jumbo balloon (about 9″ by 5″ deflated)
1 gallon plastic bleach or cider jug (empty and clean)
heavy-duty tinfoil
foil paper (optional)
construction paper
crepe paper (regular and duplex or double-weight)
cardboard, shirt cardboard or other flexible and lightweight cardboard, 2-ply
    bristol board, oaktag
colored cellophane (optional)
cardboard tubes (from paper towel or wax paper or gift wrap rolls)
wire (#22 spool wire and #18 or other flexible but strong wire)
coat hangers and wire cutters (optional)
decorative, self-adhesive paper (found in hardware and department stores)
miscellaneous trimming (metallic glitter, sequins, cork, bells, buttons, colored
    glass, decorative braid, fringe, etc.)
pipe cleaners (white or colored, found in drugstores or art supply stores)
Rit or Tintex textile dye or powdered aniline dye (found in art supply stores)
foam rubber or polyurethane foam (optional)
paper or plastic doilies
baseball cap or painter's peaked cap (sold cheaply in paint or hardware stores)
cardboard paint mixing bucket (sold cheaply in paint or hardware store)

2. BASIC SKILLS

HOW TO TRACE AND TRANSFER PATTERNS

Always trace patterns, *never* cut book, or you will destroy the directions on the other side of the page, and will not have the patterns to reuse year after year.

1. Place a piece of tracing paper over the whole pattern you want to copy. Pull edges of tracing paper over book and tape them to table to hold drawing steady.

2. With pencil, lightly trace pattern outline onto tracing paper. (If you press too hard, you will tear through tracing and into book.) Also trace dotted lines (which mean *fold*), and all other markings *within* pattern.

3. Lift tracing off book and place it *face down* on scrap paper. With soft, dark pencil, rub all over *back of tracing* (a). To transfer onto dark colored paper, rub with white chalk instead of pencil. *Note*: Instead of scribbling on back of tracing, you can place a sheet of carbon paper, *black side down*, between right-side-up tracing and paper, and draw over it to transfer pattern (b).

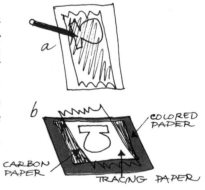

4. Compare transferred pattern to original in book to be sure you have copied all lines. Pattern is now ready to be cut out.

HOW TO CUT CIRCLES

Method I—Use when paper or cardboard may not be folded.

1. On *square* paper, measure and mark off halfway point on each side.

2. To draw circle, sketch arcs connecting each point. Cut out.

Method II—Use when paper may be folded.

1. Fold rectangular paper in half, then in half again as shown.

2. To make a circle 6″ across, mark off points 3″ from bottom right corner (*arrow*). *Note*: The radius, or distance from center to edge of circle, is one half the diameter, or total distance across circle.

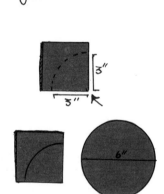

3. Draw a curved line connecting points. Cut along line and unfold circle.

Method III—Easiest method if you have a compass or two pencils and some string.

1. Set compass legs at desired circle radius (3″ for a 6″ circle). Set point of compass in center of paper and swing pencil around, drawing circle. Cut out.

2. The same effect may be achieved by using 2 pencils and some string. Tie string between pencil tops as shown, the distance between pencils is equal to radius of circle (3″). *Keeping both pencils perpendicular to paper*, hold the point of one in paper's center and swing the other around, drawing circle. Cut out.

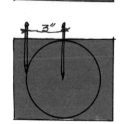

COSTUME DECORATION, FASTENING, AND COLORING

Costumes may be dyed, painted, and decorated in many ways and for many purposes. You may wish to remodel an old costume or transform old clothes into a costume. Or, you may add trimming to brighten an existing costume. Experiment with the techniques below, trying them separately or combining several ideas to create the effect you want.

There is one guiding principle to remember when working with costume decoration: BE BOLD AND SIMPLE. EXAGGERATE. You are creating an illusion, the feeling, not the fact of the real thing. Remember that most costumes will be seen from a distance, and pale or tiny details will be lost.

1. *To make decorative and metallic braid*: (a) Fabric stores sell looped and fringed upholstery which may be bought in small quantities and sewn, stapled, or sometimes glued onto costumes. Colored bias tape and ribbons may be used in the same way. Iron-on colored bias tape adds quick and effective trim for border designs, skirt, apron, and cuff hems. (b) Self-stick metallic braid, found in fabric or five-and-ten-cent stores, can easily be cut with scissors and pressed into place adding instant elegance to crowns, helmets, collars, boot cuffs, etc. Available in gold and silver. (c) To do it yourself, cotton string may be glued into any surface, a belt, a crown, a piece of jewelry or an epaulette, with rubber cement or white glue. Loop or spiral the string as you glue it down. If the entire piece, including background, is to be colored, spray it with gold or silver enamel when glue is dry; if only string is to be colored, paint it (on newspaper) before glueing down. (d) To make "ermine" trim for a royal robe, glue on strips of cotton, dab here and there with black paint dots.

2. *To make fake jewels*: (a) Wad up small balls of tinfoil, glue them down on crown, braclet, sword hilt, etc. Or use rounded shells, pebbles, or bits of colored glass. Leave jewels unpainted or color with enamel or nail polish. (b) To make jewelry settings, press metallic self-stick braid (#1, (b)) around jewel, or glue on an encircling base of wool or painted string to hold jewel in place (#1, (c)). (c) Many jewel effects may be created by cutting and painting styrofoam, foam supermarket packing trays, balsa wood, corrugated cardboard, nutshells, pieces of tree bark and sticks, marbles, colored candy sourballs, corks, pop bottle-tops, pieces of old costume jewelry and old buttons; instead of painting, the above may be covered with pressed-on tinfoil. (d) Mold jewel shapes, and other trim, from self-hardening clay or papier-mâché (made from flour and water paste or

white glue and water with shredded newspaper). For a shining finish, spray completed jewels with high gloss, clear coating or paint with shellac. (e) Brass paper fasteners can be used as the central knob of a jewel while at the same time holding various layers (of cardboard, leather, etc.) together. (f) Crowns, necklaces, breastplates, etc. can be trimmed with dry macaroni, rice, beans, etc., glued on and painted.

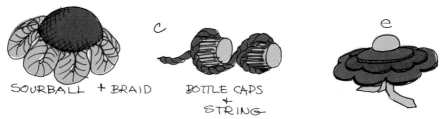

SOURBALL + BRAID   BOTTLE CAPS + STRING

3. *To make sparkling trimming*: Brush rubber cement or white glue over area to be decorated, then sprinkle with metallic glitter or sequins. To make glamorous jewelry trim, glue or pin on glass beads, sequins, or other small metallic decorations with central holes. *Note*: To create the opposite effect, dulling a decoration, spray with matte finish enamel or clear coating. For a dull texture, brush with glue and sprinkle on sand; shake off excess.

4. *To make necklaces, baubles, and bangles*: (a) On clear fishing line or nylon thread, string cut-up drinking straws, nutshells, dry seeds, spools of thread, styrofoam balls, painted dry macaroni (tubular type), bottle caps (with center holes punched with nail and hammer), self-hardening clay or papier-mâché (#2, (d)) beads molded over a nail. (b) Make your own paper beads by cutting colored paper into long tapered triangles about 1½″ at base or larger and about 3½″ to 4″ long. Spread wrong side of triangle with glue and roll tightly over an oiled nail. Roll from base to tip; length and taper of paper determine bead's fatness. Remove bead from nail when glue is dry. (c) Cheap hardware-store chains in brass and chrome make effective jewelry, with simple wire-twist closures. Assorted pieces of hardware (large nuts, drapery rings, etc.) may be strung on chains of string or ribbon by knotting each piece in line. (d) Multicolored electrical wire, found as scrap at construction sites or purchased in hardware or electrical supply stores, may be easily and quickly twisted or braided into bracelets, rings, earrings, anklets, etc. To make belts, *(see page 51)*.

5. *To make rings and earrings*: (a) Sequins, beads, or any central-holed metallic decoration may be pinned into a half styrofoam ball and glued or wired (with #22 or finer wire) to the top of an old costume jewelry ring. (b) To make your own ring, cut one flat end on a small styrofoam ball or ball of crushed tinfoil and push a 3″-piece of #18 wire through ball and loop into a ring with ends twisted flat. (c) Decorated styrofoam balls or pieces of cork, etc. may be made into earrings by wiring them onto the curved area of a regular pair of earrings. Or, push a 6″-pipe cleaner or piece of #18 wire through the ball and bend wire into a loop which fits over the *top* of the ear (rather like bracelets over the ears), with the bauble hanging below earlobe.

6. *To make costume fasteners*: (a) To attach loose-woven fabrics and knits, use brass paper fasteners with masking or adhesive tape placed over sharp ends if against skin. (b) Most fabrics, papers, and foils can be attached "invisibly" using double-face masking, cellophane, or carpet tape (strongest of all) between adjoining surfaces. Quick costumes to be worn one time may be stapled closed when put on. (c) Velcro tape, found in fabric stores, makes a convenient closure for costumes to be taken on and off many times. This tape comes with two strips, whose rough surfaces interlock when pressed together. Sew short strips of Velcro, rough sides out, to facing surfaces of fabric; press to fasten. Safety pins and decorative brooches make good quick fasteners; attach a safety pin on wrong side of fabric to keep it from showing. (d) To make decorative buttons, (*see page 52*).

7. *To color or dye fabrics*: (a) To dye fabrics, the simplest method is to use Rit or Tintex textile dye; powdered aniline dyes, sold in art supply stores, may also be used. In both cases, follow directions on package, mixing dye with warm or hot water. For an even color throughout, wet garment before dyeing. However, costumes appear to have richer textures if colors are uneven. For mottled effect, twist or knot garment before dyeing. Or, dye first in light color, then in darker color. For a 3-dimensional effect, when garment is dry bring out highlights of

design by dry-brushing lighter-colored paint over fabric surface. *Note*: Test a swatch of fabric before dyeing an entire garment. (b) Tie-dyeing, an ancient art form, is easily applied to costumes. Scarves, sashes, skirts, blouses, or trim, anything can be tie-dyed. Areas to be colored are pulled into variously shaped bunches (clusters, cones, accordion-pleated folds, etc.,) and tied with rubber bands or string. Entire garment, or only tied area, is then set in dye as described above. After dyeing, ties are removed, showing pattern. Piece may then be retied in another pattern and/or dyed another color. When dry, iron fabric. Tie-dye any fabric including velvet. (c) Most fabrics can be painted with tempera paint, oil paints, or textile paints (which, along with oils, give most permanent color). Crayons (ironed into fabric between two layers of newspaper) and felt pens can also be used. (d) For quick designs, stencil patterns by cutting paper stencil and pinning to fabric. Be sure area not to be colored is covered with scrap paper. Spray or brush color over stencil. Paper or plastic doilies may be used as a stencil, and sprayed to give lacey effect. Spraying is quickest method, but keep room well ventilated or spray outdoors. Upholstery paints come in spray cans and are very useful for this method. (e) Instead of stenciling, simply pin up fabric or costume and spray color while swirling can around to create spiraling design. (f) Every type of printing technique, such as potato prints, eraser prints, silk screening, etc., can be used for costume design. (*See author's book* HOLIDAY CARDS FOR YOU TO MAKE for printing techniques.) (g) Painted color may be used to change the look of a fabric's texture. Dry-brush cotton or cotton flannel to resemble velvet; dry-brush black or brown over a towel's nap to suggest fur. Black, white, or gold outlines painted around design motifs on fabric make them stand out when seen from a distance. (h) Freehand designs may be painted directly onto fabric; if desired, sketch motifs first with chalk, which later rubs off.

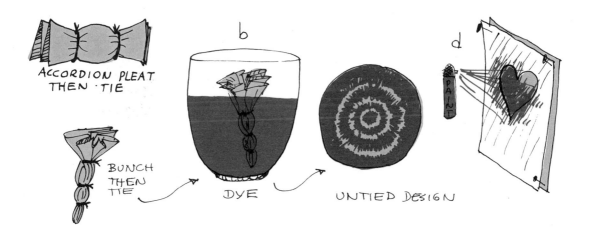

ACCORDION PLEAT THEN ·TIE

BUNCH THEN TIE

b

DYE

UNTIED DESIGN

d

HEMS AND STITCHES

Materials: Fabric, needle and thread, scissors, tape measure or yardstick, chalk or tailor's chalk, straight pins.

Note: Hems are only necessary in fabrics with loosely woven threads which tend to unravel. Felt and paper do not need to be hemmed. Many fabrics may be cut with pinking shears at the edges to prevent unraveling of threads. Hemming takes time, and, for quick costumes, pinking shears or a sewn-on ruffle or fringe can often cover a frayed edge. Selvage, or woven, finished edges do not need hemming.

1. To begin a hem, trim straight any jagged or uneven edges of material. In loosely woven fabric you can pull a single cross thread to find the straight grain.

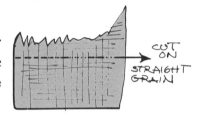

2. Turn edge of fabric up toward the wrong side, folding it the width desired or as specified in directions. Pin to hold. To get width of hem even all around, measure with ruler, and mark with chalk; or, measure and pin up in one spot, then pull this piece over and measure against width of remaining hem (a). Or, put skirt on and have a friend pin an even hem by measuring against a yardstick held on the floor (b). Or, measure the skirt against another one of correct length. To do this, use tape measure to measure distance from bottom edge of waistband to hem of older skirt; then measure and mark this distance around hem of new skirt. Bend fabric under at marked points, pin up hem, sew.

Hem Stitch:

3. While the following hem is not the professional seamstress's technique, it is the quickest and easiest method for the casual costume-maker. Thread needle, knot one end of thread. About ¼" down from edge, stitch from wrong side to right side, so knot stays on wrong side. Take small (¼" or less) stitch on front, bringing needle back to face you. Then take large (1") stitch across side facing you, then small (¼") stitch through to front. Repeat as shown.

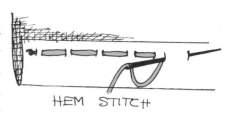

HEM STITCH

4. To end, sew several stitches over each other in same spot, pull needle through stitched thread and loop.

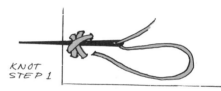

KNOT STEP 1

PASS NEEDLE THROUGH STITCHES TO MAKE LOOP

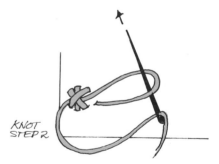

KNOT STEP 2

PASS NEEDLE THROUGH LOOP, PULL THREAD TAUT, CUT THREAD

Running Stitch:

A running stitch is the same as above, except the distance between stitches on front and back is the same, as shown. Running stitches are used for seams and gathers (*see page 65*).

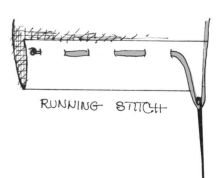

RUNNING STITCH

HOOPS AND PADDING

Hoops

1. The simplest hoop may be made from a hula hoop, which is sewn to the hem of a full skirt or petticoat. Or, strips of 1″-wide cardboard may be sewn or stapled to inside of hem making a hoop which measures about 106″ *around perimeter*.

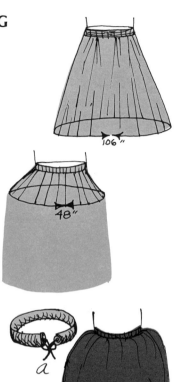

2. The seventeenth-century *farthingale* is a shorter hoop which makes a skirt flare out at the hips, then hang straight to the floor. Use a full petticoat or half slip for a base. Sew a 1″-wide, 48″-long cardboard strip into a hoop about 12″ below waist. Cut off excess length of petticoat.

Padding

1. The sixteenth-century hip roll is worn beneath the skirt to make it flare out at the waist. It also makes one appear fatter. To make a roll (a), cut a string about 50″ long. Over this roll felt, fabric, or newspaper into a tube about 3″ in diameter or more, and about 20″ long. Tape roll, or sew it together to hold layers in place. Arrange roll around waist, tie strings in front. The sixteenth-century woman wore a stiff triangular stomacher (b) on the bodice front. Made of cloth-covered cardboard, the stomacher is decorated and sewn to the bodice with the point about 3″ below the waist, as shown.

2. Stomach padding is most easily effected with a pillow, held in place with a rope or belt (a). More carefully formed padding may be made by gluing together layers of foam rubber or polyurethane foam on a felt base. For a rounded form, use graduated pieces, as shown (b). Foam rubber or polyurethane may be glued with rubber cement, and cut with scissors. Tie padding on with sash, or stuff it inside clothing.

3. TRIMMINGS AND ACCESSORIES

GATHERED RUFFLES

Materials: Strip of soft fabric or crepe paper, needle and thread, scissors, tape measure.

1. Cut material length and width desired, but remember that ruffles are gathered, and original length must be much longer than finished ruffle (as an approximate guide, use about two times finished length for a full ruffle; a 12″-strip gathered, makes a ruffle 6″ long). If using crepe paper, grain runs crosswise on strip.

2. Sew a large running stitch (*see page 35*) along the top long edge of the strip for an Edging Ruffle (a), or down the center of the length of the strip for a Decorative Ruffle (b). *Note*: Edging Ruffles are used for hems and border trims, while Decorative Ruffles are used for sewing into spirals or various designs across a skirt, sleeve, or bodice. Either style may be used as a neck or cuff ruffle. Leave needle in thread at strip end.

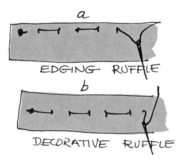

3. Gather, or push, fabric along thread toward knot until it is drawn together as much as you wish. Stitch several times in one place at end of ruffle to hold and knot thread.

EDGING
(SEW DOWN ALONG EDGE)

DECORATIVE
(SEW DOWN BY STITCHING
THROUGH STRIP CENTER)

EDGING RUFFLES ON
BLOUSE

DECORATIVE RUFFLE ON
SKIRT

FRINGE

Materials: Fabric, felt, paper or crepe paper, ruler, scissors.

1. Cut paper or fabric width and length specified in directions. If using crepe paper, be sure grain runs with the direction you want fringe to curl (*step 3*).

2. Measure and mark a base line, A, of width specified in directions. Cut narrow strips of specified width from bottom edge up to base line A. *Note*: For Cowboy or American Indian costume, sew or staple strips of fringe along outside edges of pants, tunic or vest, and cowboy hat.

3. To curl fringe, gently pull each cut strip across scissor blade. Crepe and other soft papers curl very easily; to make soft fabrics curl, spray starch and iron dry before curling.

Variation I: False Eyelashes

Follow basic directions for fringe above. Extra-large eyelashes for a mask might be cut as large as 2″ wide by 4″ long; you need two pieces. Base line A would be ½″ down from top edge, with fringe strips cut about ¼″ wide (a). Lashes for facial makeup are not quite as large, measuring about 1″ wide by 1½″ long for each piece. Make base line A, ¼″ down from top edge with fringe strips ⅛″ wide or less. After cutting fringe strips, trim them into a natural curve, making end lashes shorter, central lashes longer (b), before curling. To wear, attach base line to upper lids with double-face cellophane tape. (*Note*: Test a small bit of tape on eyelid first to be sure it does not irritate your skin.)

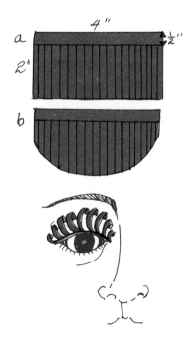

Variation II: Grass Skirt

Follow basic directions for fringe above. To determine waistband size, measure around your waist, then add 2″. In example, 24″ plus 2″ equals 26″ total. Decide length of skirt. In example, 16″. Cut material 26″ by 16″. Draw base line A, 2″ down from top of one 26″ edge. Cut ¼″ fringe strips from bottom edge up to base line. Wrap band around waist, overlap edges and pin to hold. Or, cut a 2″ by 26″ waistband and staple, glue, or sew 16″ lengths of raffia, wool, ribbon, etc., to wrong side of waistband. To wear, turn right-side out, overlap ends of band and pin. *Note*: To complete Hawaiian costume, girl wears swimsuit top, and flower chains (*page 40*) in her hair and around wrists and ankles. Boys wear shorts or Egyptian loincloth (*page 59*) with front panel wrapped around to the side. Both wear Leis (*below*) around their necks and have bare feet.

LEIS

Materials: Regular or duplex crepe paper, scissors, ruler, needle and thread (or darning needle and button thread or wool), stapler, #22 wire, #18 stem wire, florists' tape (optional).

1. Cut strip of crepe paper 2″ wide and 2 to 3 yards long. More may be added as needed. Cut long thread and thread needle; knot one end of thread.

2. Sew large running stitches (*see page 35*) through center of strip length. Holding one end of strip at knot, twist paper around thread while gathering gently toward knot end. Gathered twists resemble flowers. Add more strips for longer chain. Overstitch and knot at end.

TWIST STRIP AROUND THREAD

FLOWER CHAINS

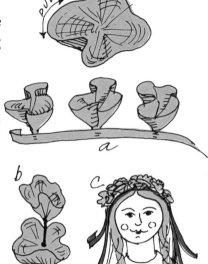

1. Cut many 2″ or 3″ crepe-paper circles (*see page 29*). Flute edges of each circle by pulling gently across grain as shown.

2. Pinch circles together at centers, staple or sew onto long ribbons or ½″ wide band of paper to make chain of desired length (a). Or, stitch through centers of flowers stringing them on a long piece of wool or button thread (b). Add flowers to a headband (*page 97*) for a floral wreath, worn with many national folk costumes. Trim with long ribbons (c).

PAPER FLOWERS

1. Cut several circles (*see page 29*) of graduated sizes (say 3″, 4″, and 5″) and of different shades of one or more colors. Flute all edges and set one circle inside the other from smallest out to largest. Gather at centers and twist to hold.

2. Wrap #22 wire around twisted centers. Wire this to stiff (#18 or heavier) stem wire, and wrap whole length (including underside of centers) with florists' tape or long strip of green crepe glued to hold. Use this flower to trim a clown's hat, for a boutonniere (buttonhole flower), or make several for a bouquet .

FEATHER

Materials: Crepe paper, construction paper or foil, scissors, pencil, ruler, #18 (or stiffer) wire, rubber cement.

1. Cut piece of paper 12″ by 2½″ (or larger or smaller depending on feather size wanted). Divide piece in half lengthwise by folding or drawing a line down center. If using crepe paper, have grain run across feather, side to side.

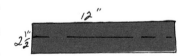

40

2. Sketch, then cut out shape shown, curving sides and tapering to 1"-long stem about ¼" wide. On wrong side, tape a 12"-long wire down center line.

3. Feather may be left as is, and completed as in *step 4*, if you do not mind wire showing on back side of feather. To cover wire, make a second feather shape by drawing around the first and cutting out. Glue second shape wrong side down over first, covering wire.

4. To complete, cut ¼" (or finer) fringe from curved sides in to within ¼" of center wire. Fringe may be curled by gently pulling each piece over scissors blade. Bend center wire into graceful curve.

POMPOM

Materials: Wool, cardboard, scissors.

1. Decide size of pompom needed. Example is 3" in diameter. Cut cardboard rectangle 5" by 3". Cut piece of wool 4 or 5 yards long. *Note*: You want pompom to be full and round; for thick wool use less length, for thin wool use more.

2. Wrap wool around 3" width of cardboard as shown. Cut 6" length of wool (or #22 wire). Place under center of wool-wrapped card (X).

3. Slip scissors blade through top and bottom edges of card (*arrows*) and cut through all loops, taking care not to disturb their order. Carefully slip out card.

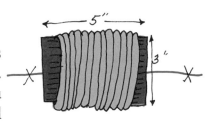

4. Pull up ends of piece X and tie together centers of all wool pieces. Tie tightly and knot. Fluff out wool. If any ends are uneven, trim into ball with scissors. Pin or sew onto costume.

BRACELETS, ANKLETS, AND ARMLETS

Materials: Colored paper, felt, or stiff fabric, ruler, tape measure, pencil, scissors, double-face masking or carpet tape (or Velcro tape for durable articles), felt pens or crayons or other trimming (*see page 30*).

1. Decide length of piece needed. Example is a bracelet which will measure 4″ from wrist up arm. To determine width, measure with tape measure around narrowest part (wrist) and around widest part (4″ up from wrist). Bracelet dimensions might be 6″ at narrowest point to 8″ at widest. Add 1″ to widest dimension for total width (9″). Cut material 4″ by 9″. Fold in half to 4″ by 4½″.

2. Mark point A along fold, ½″ over from left corner. Divide narrowest (wrist) measurement in half (3″). Add ½″ (3½″). Mark point B, 3½″ down from top left corner. Sketch slight curve between A and B. On bottom edge, mark point C, ½″ in from right corner. Draw straight line between B and C, and curve between C and top right corner D.

3. Cut out lines A-B-C-D. Do not cut across fold. Open flat, decorate (*see page 33*). To wear, overlap edges until they fit comfortably, press on double-face tape, or sew on Velcro tape (*see #6, page 32*). *Note*: Wide bracelets, anklets, and armlets are a characteristic part of Egyptian costumes.

EGYPTIAN COLLAR

Materials: Felt, lace-patterned cloth, or any fairly stiff material or crepe paper, ruler, scissors, chalk or tailor's chalk, pins or Velcro tape or brooch, tape measure, small bells, trimming.

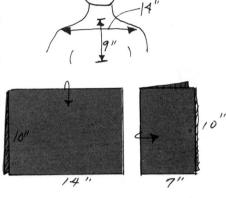

1. To determine width of collar, measure across top of your chest from shoulder to shoulder (in example, 14″). Decide depth of collar by measuring from base of neck to bottom collar edge (9″); add 1″ for shoulder fold-over, making total depth 10″. Multiply this by two (20″) for total pattern size of 20″ by 14″. Cut material this size. Fold in half to 10″ by 14″, then in half again to 10″ by 7″, with folds at top and left.

2. Measure and mark points 3″ out from corner X between top and left side folds; draw curve connecting points. Cut out curve, making neck hole.

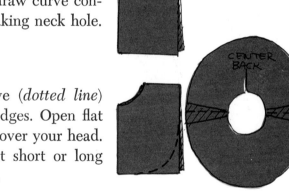

3. Sketch, then cut bottom curve (*dotted line*) between bottom and right side edges. Open flat and try gently to slip center hole over your head. Enlarge hole if necessary, or cut short or long slit through center back of collar.

4. While collar is flat, decorate with trimming (*see page 33*), felt pens, crayons, etc. Cut hem into scallops, triangles, etc. For harlequin, or jester, sew bells on point tips. For Victorian collar, trim with lace doilies, or *see #7,(d), page 33*. For Egyptian collar, use geometric design shown. To wear, fasten with pin or brooch or tape (*see #6, page 32*). *Note*: To make paper collar lie flat, tuck and glue wedges at shaded area on shoulder lines (*step 3*).

JESTER

EGYPTIAN

VICTORIAN

PURITAN COLLAR

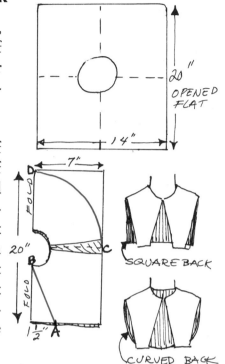

1. Collar should be white, worn on a gray, black, or blue top. Use *Materials* and steps 1 and 2 of Egyptian Collar above, making *rectangular* collar with neck hole. Open piece flat, then fold lengthwise, wrong side out, to 7″ by 20″.

2. On bottom edge, mark point A, 1½″ to right of fold. Draw, then cut, along line A-B (to edge of neck curve). Collar can be worn as is, with all squared corners; put on with slanted edges center front, pin at neck to hold. Or, cut curved back edge by sketching, then cutting line C-D from outside of shoulder fold (*dotted line*) to top left corner. *Note*: To make paper collar lie flat, tuck and glue wedges at shaded area on dotted shoulder line. For complete Puritan costume, *see page 18*.

SEVENTEENTH- AND EIGHTEENTH-CENTURY GIRL'S COLLAR OR SAILOR'S COLLAR

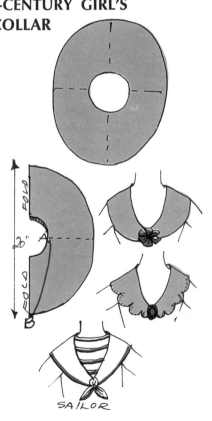

1. Use *Materials* and *steps 1, 2, and 3* of Egyptian Collar, making *oval* collar with neck hole. Open piece flat, then fold lengthwise, wrong side out, with fold at left 20″ long.

2. Draw, then cut line from point A (center of neck edge at dotted shoulder line) to bottom left corner B. While collar is flat, decorate right side with doilies or lace, or other trim (*page 33*), or cut bottom edge into scallop or points. Wear with pointed ends center front; overlap or tie tips, hold with brooch or artificial flower (*see page 40*). This is a good collar for a flower girl or milkmaid.

3. For a sailor's collar, follow *step 1* of Puritan Collar; then follow *step 2* immediately above, making pointed ends. Add a decorative stripe across the square back and tie ends in front.

CHINESE MANDARIN COLLAR

Materials: Any fairly stiff fabric, felt, or paper, tape measure, pencil or chalk, scissors, pins, needle and thread.

1. This collar can be added to any blouse, shirt, or pajama top which has a round neck. Measure around garment neckline (in example, 12″) and add 1″ all around if you plan to hem collar (*see page 34*). Cut strip of material 12″ by 3″ wide.

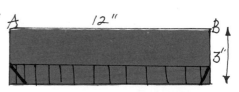

2. Measure, mark, and draw line across strip 1″ up from bottom long ledge. Cut 1″-wide fringe (*see page 38*) from bottom edge up to line. Cut both end pieces of fringe on slant as shown. Cut both top corners (A and B) rounded as shown.

3. Existing collar on garment may be folded under and sewn down on wrong side, or cut off; or it may be left in place and worn along with mandarin collar. To attach collar, open neck of garment flat; *wrong* side facing up. With *wrong* side of collar *strip* facing you, pin its fringe tabs to inside base of old collar (slanted fringe ends are at collar edges). Sew fringe down with running stitch (*see page 35*). To wear, put garment on, stand collar up as shown, with rounded edges in front of neck. *Note*: To make an Elizabethan woman's high lace collar, follow directions above but use stiff, or starched, lacey fabric cut 6″ to 8″ wide.

ELIZABETHAN NECK RUFF

Materials: Duplex crepe paper or other pliable paper or fabric with body, needle and thread or small stapler, tape measure, scissors, safety pin.

1. Cut neck band 2″ wide by 24″ long.

2. Determine ruff width desired (in example, 3″). Cut material 3″ wide by 3 yards long. Or, use several shorter strips combined. (Double width, and fold in half lengthwise to 3″ if using regular crepe paper. Also be sure grain is running crosswise on strip.) Depth, or thickness, of this ruff will be about 3″. *Note*: Depth varies with size of loops; larger loops make a deeper ruff and require a longer strip of material.

3. Gather strip as shown, into roughly 14 "s" loops; at dots, fasten sides of loops with needle and thread or stapler. Loops are fastened *only* on the side which will touch the neck; fronts of loops are not fastened.

4. To attach ruff to neckband, place band flat over fastened ruff edge, as shown (a). Fold over and sew about 2″ of each end of band to each end of ruff (b). Band rests flat atop loop ends. Twice along length or ruff, or about every 5 loops, pinch band down between a pair of loops and fasten in this position (c).

5. To wear ruff, place neckband around neck, safety pin ends together. If band is too loose, pinch band down between more loops, as in *step 4*, (c). Use for clown as well as Elizabethan man or woman.

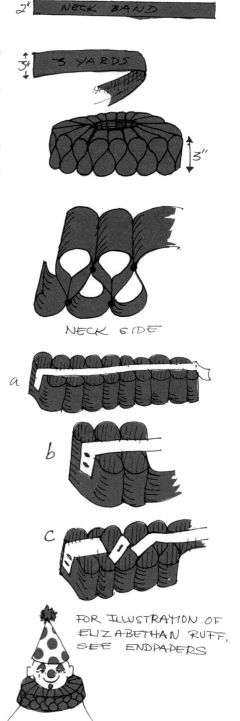

46

CUFFS

Materials: Any flexible fabric or felt or crepe paper, scissors, straight pins, tape measure, needle and thread or stapler.

Note: These cuffs can be added to any sleeves, or to gloves (for gauntlets) or boot top (for Cavalier) or pants or knickers hem (for seventeenth-century courtier.) If made to fit a neckline, this style cuff can be turned into a ruffled collar (national folk costumes).

1. Use tape measure to determine distance around hem of sleeve (or other garment). In example, 10″; add 4″ for total width of cuff pattern (14″). Decide length of finished cuff (3″) and add 1″ for turnover (4″). Cut material in strip 14″ by 4″.

2. With right sides together, fold strip in half to 7″ by 4″. With piece arranged as shown, measure and mark point A, 1″ in from top right corner. Draw, then cut line from A to B at bottom right corner.

3. With right sides together, sew *(see page 35)* or staple seam about 1″ wide along edge A-B.

4. With right sides still together, fit narrow end of cuff up inside sleeve opening. Working with one hand up inside sleeve and the other on outside, pin, then sew narrow end of cuff to sleeve about 1″ in from edge. You may find it easier to hold cuff if it is doubled in half or gathered slightly.

5. Turn cuff back over outside sleeve edge. If using crepe paper, flute edges for ruffled effect by pulling across grain. Or glue, sew, or staple paper doilies onto cuff before attaching to sleeve. *Note*: Cuff may be made entirely from 10″ or 12″ plastic or paper doily; cut center hole in doily and sew or staple to sleeve.

BOW TIE AND LONG NECKTIE

Materials: Paper or stiff felt, ruler, scissors, felt pens or crayons, stapler or needle and thread or glue, string or elastic cord, old man's necktie.

Bow Tie

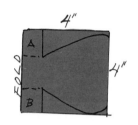

1. Bow tie can be made any size. For clown or humorous costume, use oversize tie such as example. Cut material 8″ by 4″. Fold in half to 4″ by 4″. Trace and transfer (*see page 28*) pattern from *page 49*. Cut out. For a gentleman's formal tie or an eighteenth-century shoe trimming, use material 5″ by 2″.

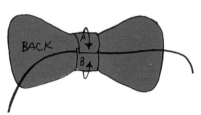

2. Open material, decorate right side with felt pens or crayons. On dotted lines, fold tabs A and B over onto back side. Glue or staple tab A to tab B, making a loop between tabs and tie back. Cut a 25″-string or elastic and pass it through this loop. To wear, hold bow at collar front, pull string ends under collar and tie in back.

Long Necktie

Use any old or discarded man's tie. Spray or paint (*see page 32, #7*) desired color. To widen tie, open seams and iron flat. To make paper tie, trace around ready-made tie, attach to shirt front with pin or double-face carpet tape.

SEVENTEENTH-CENTURY MAN'S CRAVAT

Materials: Soft fabric, lace or doily, needle and thread, tape measure.

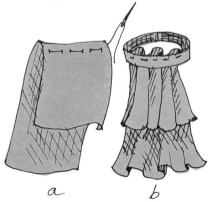

Cut material about 5″ by 16″. Fold one end over the other as shown, (a), covering about ⅔ the length. Sew a running stitch across the top of folded edge as for Edging Ruffle, *page 37, step 2,* (a). Gather material on thread, knot to hold. Cut a band 1″ by 18″; sew center of band over top of gathers (b). To wear, center ruffle at neck front, pin or tie ends of band behind collar. Can be used for Captain Hook, Cavalier, or George Washington.

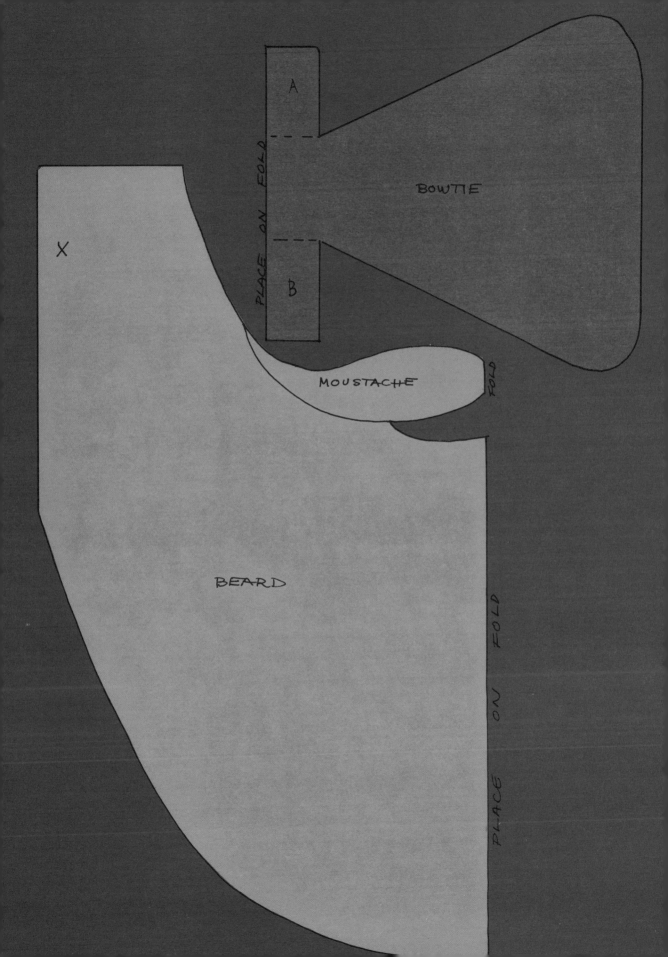

A

B

PLACE ON FOLD

BOWTIE

X

MOUSTACHE

FOLD

BEARD

PLACE ON FOLD

SUSPENDERS

Materials: Any fairly strong fabric or paper, tape measure, scissors, safety pins or brass paper fasteners or needle and thread.

1. Use tape measure to determine distance from waist in back over your shoulder to waist in front (30″ in example); add 2″ for total (32″). Cut 2 strips of material 32″ by 1½″ (or preferred width). To wear, pin suspender ends inside your waistband. Or, in very loosely woven fabrics, attach bands with paper fasteners, but put masking tape over sharp ends on inside surface.

2. To make Tyrolean lederhosen suspenders, cut a paper or fabric oval or rectangle about 3½″ high and 8″ wide. Decorate as shown, with flower motifs, and sew, glue, or staple short ends to suspenders so panel will be at chest center. For an authentic Tyrolean look, make another decorative panel for center front of your waistband or belt, and wear with shorts.

3. To make soldiers' cross straps, cut two strips of material 43″ by 1½″. To wear, cross straps in front and back, tuck excess length inside pants and attach to waistband as in *step 1*, above. For a military look, wear a wide belt with buckle (*page 52*), and add a Chinese Mandarin collar (*page 45*) to jacket. For Greek or Roman Armor straps, (*see page 114*).

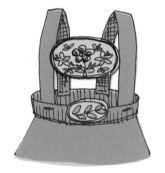

CUFFS ON PAGE 47

4. To make overalls bib, cut fabric or paper panel 8″ by 10″. Sew, staple, or glue the 8″ sides to suspenders as shown. Add 4 decorative buttons and use for many boys' national folk costumes, Hansel, Pinocchio, etc. Add to waistband of apron (*page 66*) for milkmaid's or servant girl's costume.

BASIC BELTS

Materials: Any fabric, felt, paper or duplex crepe paper, scissors, safety pin, double-face carpet or masking tape or Velcro tape, measuring tape.

1. To determine belt length, measure around your waist (23″) and add 3″ for overlap; total is 26″. Cut belt and width desired (in example, 2″). Cut strip 26″ by 2″.

Note: If you plan to hem fabric (*see page 34*), allow 1″ extra all around.

2. Cut one end of strip into curve or taper. If buckle is desired, make as below and glue, staple, or sew it to tapered end, as shown.

3. To wear, wrap belt around waist, overlap tapered end over square end, and fasten (*see #6, page 32*) with pin, double-face tape or sewn-on Velcro tape (especially good for belts which must be taken on and off many times).

TWISTED WOOL

4. Belts can also be made from rope or wool (plain or braided), or bias tape, ribbon, raffia, etc. Many of the necklace techniques in #4, *page 31*, can be used for belts. Tie strings through painted and cut-out paper-cup bottoms, discs or loops of tinfoil, or brass curtain rings, or add rings or trim to an old ready-made belt.

BRAIDED WOOL

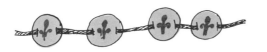

DISCS AND RIBBON

RINGS AND STRING

BUCKLE

Materials: Any paper, foil-covered cardboard, or stiff felt, ruler, scissors, double-face carpet or masking tape or paper clips, stapler (optional).

1. Buckle may be made any size you like. Example, 2″ by 3″, is suitable for a Puritan's hat or shoe, Santa Claus's belt, or a witch's hat. Cut rectangle of material 2″ by 3″ (larger or smaller if you prefer).

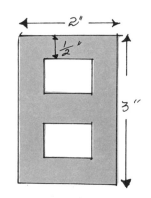

2. Mark shape shown, making buckle even ½″ all around edges and in mid-section. Cut away white sections.

3. Attach buckle to hatband or belt with glue, staples, or tape. To attach it to a shoe, use double-face tape or a paper clip which fastens center buckle strip to top of instep.

DECORATIVE BUTTONS

Use same materials as above. Cut buttons any size or shape you want, using cardboard, construction paper, felt, or paper covered with glued-on fabric, self-adhesive decorator paper, or pressed-on tinfoil. Buttons may be painted or decorated with felt pens or crayons. Attach buttons to costume with double-face tape, or pin, or sew. Brass paper fasteners may be used when attaching to paper or very loosely woven fabrics. *Note*: Fasteners make holes in close-grained cloth.

PAINTED PAPER

PAPER FASTENER

CARDBOARD

FOIL ON CARDBOARD

PAINTED FLOWER

STRING AROUND CORK ON DISC

MILITARY MEDALS

Materials: Colored construction paper, felt, or striped ribbon about 1½″ wide, flexible cardboard, tinfoil, compass (optional), scissors, safety pins, rubber cement, stapler, ruler, felt pens or crayons, pencil.

1. Cut circle (*see page 29*) about 1½″ across of cardboard or construction paper covered with tinfoil. A 1½″ circle can be covered with a 2″ square of foil, excess folded onto back. Medals may also be in shape of stars, flowers, etc.

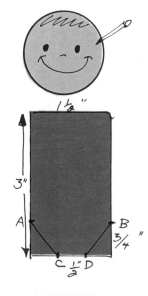

2. Decorate front of medal with paints, felt pens, or crayons, or draw with pencil, making an embossed design in foil.

3. Cut paper or fabric ribbon 1½″ wide by 3″ long. Measure and mark points A and B, ¾″ up from bottom edge. Measure and mark points C and D each ½″ in along bottom edge. Draw, then cut taper along A-C and B-D.

4. Decorate paper "ribbon" with stripes or other designs. Staple or glue medal, front side out, onto tapered end of ribbon. Tape or pin onto costume. *Note*: Several medals may be made and glued to a 1″-wide strip of paper which is then pinned to costume.

SOLDIER'S EPAULETTES

Materials: Felt (stiffened with starch) or construction paper, heavy foil paper or tinfoil-covered cardboard, self-stick gold or silver decorative braid or cotton string, glue, gilt paint, ruler, scissors, double-face carpet or masking tape or brass paper fasteners.

1. Measure from base of neck to outside edge of shoulder. In example, 5″. Add 1″ for total length (6″). Cut two pieces of material 6″ long by 6″ wide. *Note*: 6″ width is the same whatever the length.

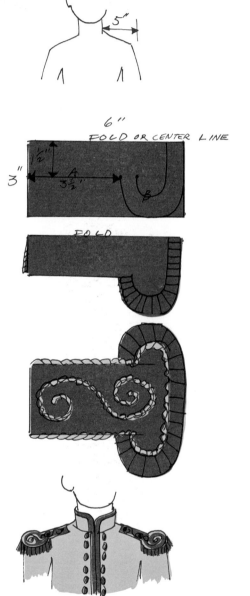

2. If using flexible material, fold each piece in half, wrong side out, to 3″ by 6″, and arrange with fold at top as shown. *If using stiff material,* leave flat, wrong side up, and draw a line lengthwise down the center. Then draw pattern shown on either side of center line. To complete pattern, make a point 1½″ down from top left corner (*center line*). Draw line A, 3½″ over from marked point. Then draw curved line down to bottom right edge and up side as shown. Draw line B, 1″ inside curved end.

3. Cut around line A. Cut ¼″ fringe (*see page 38*) around curve up to line B. Open piece flat. Draw around it to make second epaulette. To decorate (*on right sides*), use metallic braid or gilded string as described in #1, *page 30*. Pieces of wool or other trim may be added to fringe. Bend fringe over onto wrong side, making it stand at right angles to decorated flat top. To wear, stick to shoulder tops with double-face tape, pins, or paper fasteners (*see #6, page 32*).

COLONIAL FLINTLOCK RIFLE

Materials: Brown construction paper, straight stick or sawed-off broom handle 44″ long, ruler, scissors, stapler, felt pens.

1. Cut paper 8″ by 36″ (or tape 2 pieces of paper together to complete length). Fold in half to 8″ by 18″. Sketch basic form with dimensions shown. *Note*: This basic form can be used for other types of rifles besides the flintlock.

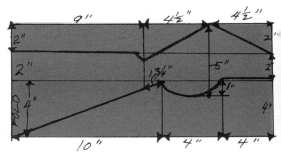

2. Trace (*see page 28*) lock pattern *below,* and transfer it to both sides of folded paper as shown, lining up pattern edge X with narrow end of paper. Turn pattern wrong side up to transfer it to back side of rifle. With felt pens, draw lock and trigger guard details as shown.

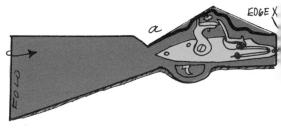

3. Cut around heavy black outlines as shown (a), but do not cut through fold. Staple bottom edge together about ¼″ up from bottom. Then insert stick or broom handle (*dotted line*) all the way in to fold at end of stock (b). While stick is inside, staple top edges closed over it.

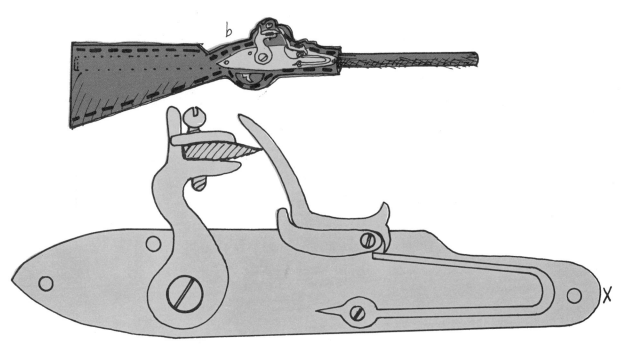

SWORDS

Materials: Construction paper, flexible cardboard or 2-ply bristol board, foil paper or heavy-duty tinfoil, ruler, scissors, cardboard tube (from paper towel, gift wrap, etc.), thumbtacks or cellophane tape, paint or spray paint (optional), gold or silver-colored Mystic tape (optional).

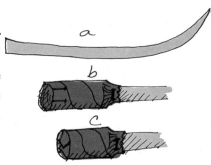

1. To make sword blade, cut cardboard whatever length and shape desired. In example, blade is 2″ wide, 36″ long, and curved at one end (a). Handle is a 6″ piece of cardboard tube. Ends of tube are squeezed closed over top of sword blade and stapled together (b). Top of handle is cut into fringe 1″ deep and ½″ wide which is overlapped and taped down making flat top (c).

2. Sword guard is cut from paper or flexible card-board 6″ by 12″. Piece may be painted gold or silver or wrapped in tinfoil. Place piece as shown and divide in half lengthwise with line A across center (3″ in from edge). On top and bottom, mark halfway points B and C, 6″ in from ends. To left of line drawn between B and C, sketch an oval as shown, with a knob at center of left end. In center of oval, draw cross lines in a circle as shown, about 1½″ across. To right of line B-C, draw a 2″-wide strip centered 1″ above and 1″ below line A. Round end of strip.

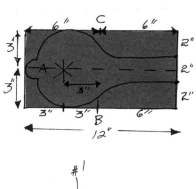

3. Cut out around outline of shape. Cut into cross lines in center of oval. Put guard onto handle as shown, oval piece at bottom with center cross cuts fitting over handle top. Enlarge cuts if necessary. Fold long strip up and over onto handle top and tack or tape down (*arrow #1*). Tack or glue or tape triangles of cross cuts to bottom of handle (*arrow #2*).

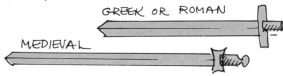

56

SPEARS, BATTLE-AXES, TOMAHAWKS

Materials: Same as Swords.

Cut cardboard or paper blades whatever style desired, using basic size about 12″ high by 6″ wide or as indicated. Staple, tape, or wire blades to broom handles or sticks of indicated length. Blades may be colored, painted, or spray painted.

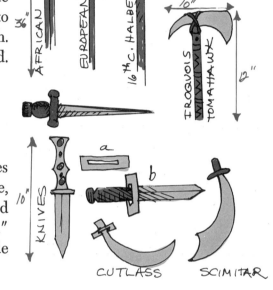

KNIVES

Materials: Same as Swords.

Cut cardboard or stiff paper blades and handles in one piece. Handle width is basically 1½″ wide, overall length about 10″. If desired, blade guard may be made as a separate piece (a) 3½″ long, 1″ wide, with 1½″ slit in center. Fit slit up over blade and tape in place at handle base (b).

SHIELDS

Materials: Cardboard, 2″-wide sewing elastic or flexible cardboard, stapler or masking tape or brass paper fasteners, plastic or metal garbage pail lids (optional).

Cut shield from cardboard in shape desired. Spray paint or brush on over-all tone, then add decorative details. To make hand hold, cut strip of wide elastic or flexible cardboard 8″ by 2″. Cut two slits 2″ long, 2″ apart in shield center. Thread strip through slits so both ends face back side; staple or sew strip ends together in back (a). To use, grip handle and hold shield right side facing out. Or, decorate insides of plastic or metal garbage pail lids (use enamel or spray paints) and hold them by the handle. The latter are especially suited to the round Greek or Roman shields (b).

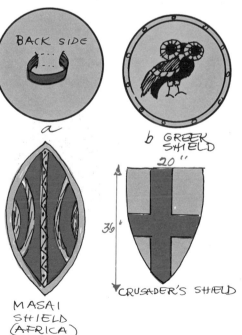

57

4. BASIC COSTUME PARTS

INTRODUCTION

In order to avoid the complexities of advanced dressmaking, it is assumed for the purposes of this book that the amateur costume maker has available ready-made, and well-worn, garments such as shorts, trousers or jeans, tee shirt or turtleneck jersey, tights, pajamas, bathrobe, blouse, and shoes. These, together with the basic costume parts which follow, will enable you to make nearly any costume you wish.

To make knickers, which are used in many national folk and historical costumes, use an old pair of trousers or jeans. Tuck the bottoms of the pants into high socks or boot tops; or if you can, cut them off about 3″ below the knee, roll the edges into 1″ cuffs, and tie with strings or ribbons.

To make bloomers, cut the legs short on an old pair of wide pajama bottoms. Gather the hem of each leg as you would for a ruffle (*page 37*), using thread, or elastic in place of the thread. Or a hem may be turned up over a piece of sewing elastic which is pulled snug to gather hem. If you do not have wide pajamas, make a very short gathered skirt (*page 66*), put it on, and pin front and back together between your legs. *Remove skirt*, cut on pin line, then sew each half together making wide pants. Gather and hem as for bloomers. *Note*: Elastic in gathered hems makes garments easier to put on and take off.

KNICKERS   PINNED SKIRT   BLOOMERS

INDIAN BREECH CLOUT

Materials: Fabric (long scarf will do), belt or waist-length rope, scissors, tape measure.

Cut fabric about 8″ wide by about 48″ long (or more for longer draped panels). Wear swimming trunks underneath if desired; wear loose belt or tie loose rope around waist. Fit fabric strip into belt as shown, pulling fabric between your legs and over belt in back and front to hang down in even panels. *Note*: Border designs may be added to cloth before draping if desired. For Indian designs (*see page 101*).

EGYPTIAN LOINCLOTH

Materials: Stiff fabric or crepe paper, scissors, tape measure, safety pin.

Decide length of loincloth (it should be length of shorts); in example, 14″. Cut fabric or paper 14″ by 54″. Wear swimming trunks or shorts beneath loincloth. To wear, wrap one end around waist as shown (a), then pull other end around and pin at top where they overlap, (*arrow*, (b)). Fold back loose end and press along fold to hold in triangular panel. Pin panel back if necessary to hold.

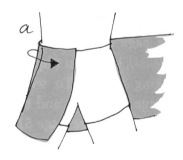

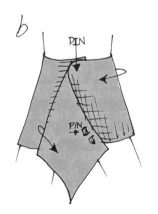

4. To make short-sleeved, straight-sided all-purpose tunic, follow directions (*p. 61*), *but make following changes in pattern*: Mark point I, 4″ past point C; draw dotted line straight down from I to line E-G. Cut off sleeve along this dotted line. (For longer or shorter sleeve, vary position of this dotted line.) Cut from dotted line over to E, then cut straight sides from E down to D. Open piece.

5. To complete either style tunic, try it on with wrong sides out. Pin seams along open edges until tunic fits comfortably. Seams may vary from 1″ to any width necessary for good, but slightly loose, fit. Tunic is put on over head, so neck opening must be large enough. Slit neck as in (a) or (b) for greater ease in dressing. Sew (*see Running Stitch, page 35*) along pinned seams. Iron seams flat if you wish, then turn tunic right side out. Bottom edge and sleeve cuffs may be hemmed (*see page 34*) or cut with pinking shears or trimmed with decorative braid. *Note*: To make a long robe, as for a king, cut tunic down center front to hem. Trim edges as on *page 30, #1, (d)*.

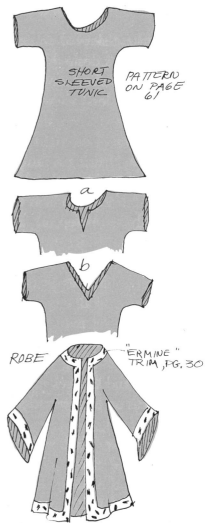

GREEK TUNIC (Doric Chiton)

Materials: Soft, easily draped fabric, or 2 large bath towels or old sheet, needle and thread, straight pins, tape measure, belt or rope, chalk, safety pins.

1. This tunic, or *chiton*, is similar to Simple Tunic, except it has the characteristically Greek draping. To determine length, measure from base of neck to bottom hem (in example, 36″). Greek boys' tunics were generally worn above the knee; girls the same or ¾ length. Add about 12″ to length for draping (total length is 48″). Cut 2 panels each 48″ long and 36″ wide. With right sides together,

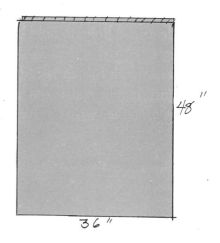

sew panels (*see Running Stitch, page 35*) all the way up long sides, making 1″ seams (a). Or, fold a sheet in half, fold at side, and trim it to correct size. Sew up edge opposite fold (b). Turn right side out.

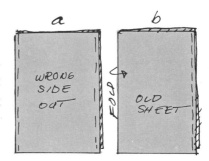

2. On right side, decorate edges with border designs as shown. Greek tunics were many colors besides white; to dye fabric, (*see page 32, #7*); or use colored cloth. If desired, hem edges (*see page 34*).

3. Measure and mark center of top edge of front and back panels (18″ in from sides). Mark points A and B, 4″ out to each side of center. Neck opening is between these points; armholes are to each side, from A-C to B-D.

4. Slip tunic on over head, bringing top edge down just below armpits (a). At points A and B, pull front and back top edges up onto shoulders and safety pin them together. Pin again about 2″ outside A and B (b). Tie belt or rope around waist; blouse fabric up over belt until desired hem length is reached. *Note*: Greek boys often wore a short cape (*see page 71*), called a *chlamys*, draped over their left shoulder and pinned together at the right shoulder. The chlamys is worn (c) over the tunic. Girls do not wear a cape. With this costume, legs are bare, feet bare or covered with sandals. (*see page 73*). Greek men and women may wear an outer garment called a *himation,* which is about 4′ by 8′ and wraps around the body like a toga (*page 64*).

ROMAN TUNIC AND TOGA

Materials: Same as Greek Tunic.

Note: Roman boys and girls wore a knee-length tunic, or *tunica*, similar to the Greek, *page 62*. Over this, they often wore a toga, or wraparound robe of white trimmed with colored border designs. A girl's toga is called a *palla*. With this costume, legs and feet are bare, or sandals (*see page 73*) may be worn. The toga may also be worn as part of a West African costume or as an Indian sari.

Tunic

1. Make Greek Tunic, *page 62*. This is suitable for girls and boys (a). *Or*, cut two panels desired length (36") by 28" width. Place panels right sides together and sew (*See Running Stitch, page 35*) 1" seams to within 12" of top edge (b). Turn right side out.

2. Slip tunic on over head, so top of side seams is at armpits. Gather up and pin (or sew) about 8" of fabric's top edge along *right* shoulder. Fold remaining top edge toward inside, so fabric lies flat, and arrange in diagonal from top of right shoulder to below left armpit. Pin fabric under left arm to hold firmly, if desired. Wear with or without a belt.

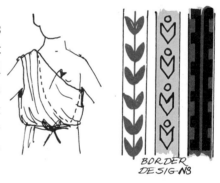

BORDER DESIGNS

Toga or Palla

Toga or *palla* may be made from a long piece of fabric, roughly 1 yard wide by 3 yards long. Trim borders with colorful Roman designs as shown. Boys may wear togas alone or over a tunic, wrapping it as shown (a) with inside corner X tucked into pants or shorts. Girls wear tunic and drape *palla* over head (b); they may also wear a tiara, *step 3, page 98*.

SIMPLE, UNGATHERED APRON

Materials: Any type of cloth or crepe paper, needle and thread or stapler, ruler, scissors, tape measure.

1. Measure across waist from center of one side to center of other. In example, 12″. Add 3″ to this for total apron width—15″. Decide length of finished apron; example is 14″. If you plan to hem apron (*see page 34*) add 2″ to length.

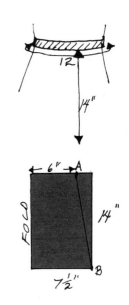

2. If using crepe paper, always be sure grain runs vertically from waist to hem. Cut material 15″ x 14″, or to your own dimensions. Fold in half to 7½″ x 14″. Take half of *step 1* waist measure (12″), or 6″. Measure this distance out from fold on top of fabric and mark point A. Draw, then cut line from A to B in bottom right corner. To complete, *see step 2, below*. Use for many national folk costumes or servant girl's costume.

GATHERED APRON

1. Measure waist from side to side as in *step 1* above (12″). Add 6″ to this for total apron width (18″). Decide apron length as in *step 1* above (14″). Cut material 18″ x 14″. Arrange open flat, with short ends at sides. If using crepe paper be sure grain runs vertically. Thread needle and sew running stitch (*page 35*) about 1″ down from top edge. Gather (*page 37*) along top edge until width is about 12″, or your side to side waist measure. Stitch and knot at end to hold.

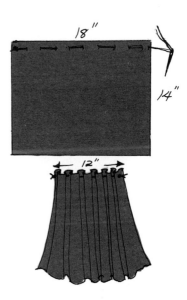

2. To make waistband, measure all way around waist (24″). For a waistband that has a tablike closure held with a pin in back, add 2″, for total band length of 26″. For a waistband that ties

with a bow in back, add at least 36″, making total 60″. Cut waistband 4″ wide and preferred length. Fold waistband in half lengthwise. Insert top edge of apron between open edges of band. Center apron so even amount of band extends out beyond each side. Staple paper or sew fabric along dotted line, fastening front edge of waistband to back edge and sandwiching top in between.

3. To complete, hem (*page 34*) or cut bottom edge with pinking shears or in scallops or leave plain. Trim with decorative borders using paints, etc., or embroider (a). *Note:* Overalls bib (*page 50, step 4*) with over-shoulder straps may be attached to apron top (b). Use for historical or national folk costumes, or for milkmaid, Cinderella, etc.

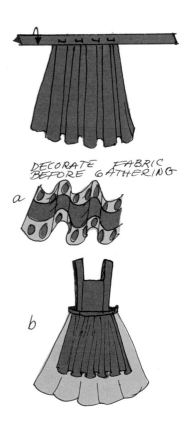

DECORATE FABRIC BEFORE GATHERING

a

b

GATHERED SKIRT

Materials: Soft, pliable fabric (such as cotton percale) or crepe paper, needle and thread or stapler, ruler, scissors, tape measure, safety pin or Velcro tape.

1. Read directions for Gathered Apron. To begin, decide skirt length (in example, finished length is 18″ from waist to hem) then add 2″ for hem (*see page 34*), and about 1″ for waistband (total is 21″). Approximate width needed for a full gathered skirt is 108″, or about 3 yards. If using old curtain or sheet or other very wide fabric, cut it 108″ by 21″. Cut 36″-wide fabric 3 yards long. For crepe paper, have grain running from waist to hem. *Note:* If you do not have one piece 3 yds. wide, use two pieces each 1½ yd. and make 2 seams as in *step 2*; one seam goes full length, the other stops 6″ below top.

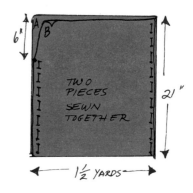

6″

A
B

TWO PIECES SEWN TOGETHER

21″

1½ YARDS

2. F material... e

seam...

half with wrong...

ends, staple...

stitch (...

Stop se...

shown, ...

3. Turn i...

seams ma...

out. While...

paints, etc.

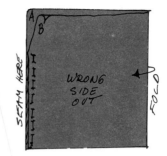

4. After deco... ...are dry, top may be gathered. About 1″ down from top edge, gather (*see page 37*) entire width of material from corner A to corner B. Pull fullness along thread until distance between A and B measures your waist size (24″). Knot and over-stitch gathering thread to hold.

5. To complete skirt, make waistband and hem as in Gathered Apron, *steps 2 and 3, page 65*. To wear, place open-topped seam (*step 2*) in center back or on side, tie or pin waistband closed. Or, sew 1½″ strips of Velcro (rough-sides-out) to facing surfaces of waistband ends; press rough surfaces together to fasten. For a Queen's or Princess's gown, use this skirt trimmed with sewn-on Decorative Ruffles (*page 37*) as shown. Also use for girls' national folk costumes, with full petticoats underneath.

VEST OR BOLERO

Materials: Felt, stiff fabric, or duplex crepe paper, tape measure, pencil or chalk, scissors, stapler or needle and thread.

1. For vest length, use tape measure to determine distance from top of shoulder to waist. (In example, 15″). Add 1″ for shoulder seam (total, 16″). For vest width, measure loosely around chest just under armpits (34″). Cut material 34″ by 16″. Bring short ends together, folding piece in half to 16″ by 17″; at halfway point, mark on right side with chalk (*arrow*).

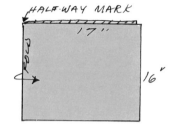

2. Arrange piece flat open, right side up. Fold short ends over, lining them up with center mark. Width of folded piece now should equal half of chest measurement (half of 34″ is 17″), and each folded panel should be half of that (half of 17″ is 8½″).

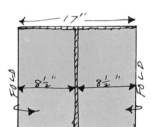

3. On top edge, measure and mark points A and B, 2″ in from each folded end. Mark points C and D, 2″ in from A and B. On each folded end, measure and mark points 8″ down from top. Draw curves on each side, from these points to points A and B. Cut along each curve. Mark point E, 1½″ down on curve from A; mark point F, 1½″ down on curve from B. Draw and cut on lines E-C and F-D.

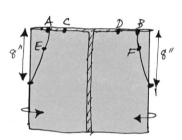

4. Refold piece so short ends are together as shown. Cut open edges in curve as shown, or cut in any other shape (see sketches), making front of vest.

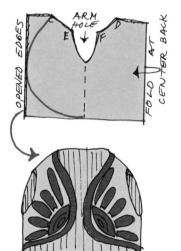

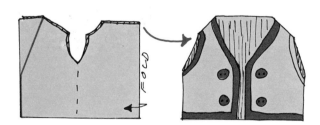

5. Fold as in *step 3*, with wrong sides out. Sew or staple ¾"-wide seams along slanted edges E-C and F-D. Turn piece right side out, open shoulder seams and press them flat. Decorate with trimming (*page 30*), or crayons or felt pens. For Cowboy or Cowgirl, trim front and bottom edges of vest with 1" deep, ¼"-wide fringe (*see page 38*) cut directly into vest material or stapled or sewn on in strips. For national folk costumes, trim with decorative buttons. Front halves may be closed with Velcro tape (*page 32, #6*) laces, or pins.

Note: For Swiss, Austrian, Scandinavian, Italian, and many other national folk costumes, as well as such storybook characters as Heidi and Gretel, the Austrian Corselet or the Simple Laced Bodice is worn with a long- or short-sleeved blouse, a gathered skirt (*page 66*), an apron (*page 65*), and colored tights or knee socks.

AUSTRIAN CORSELET

Make vest as described on *page 68*, but in *step 4*, cut the open edges as shown, beginning cut at point 6" above bottom and curving in about 4" from left side and up to points C and D. To complete, see Simple Laced Bodice, *steps 2 or 3*.

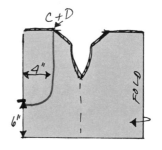

CURVED ENDS

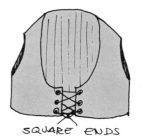

SQUARE ENDS

SIMPLE LACED BODICE

1. With tape measure, measure around waist (in example, 24″) and add 2″, for total length needed (26″). For straight-ended bodice, width is 4″ (or as much as 6″). Cut material 26″ by 4″. For curve-ended bodice, cut 6″ wide, *see step 3*.

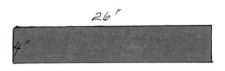

2. To complete straight-ended bodice: Fold over and glue or sew down a 1″ flap on wrong side of each short end (a); ½″ in from each folded end, poke 3 evenly spaced holes through folded layers as shown (b). For fabric, make holes with scissors, for paper use pencil point. On paper bodice, glue notebook ring reinforcements around front and back sides of each hole. Rings may be colored with felt pens. To wear, wrap bodice around waist, right side out, and lace with 36″ string, wool, or ¼″-wide ribbon, etc. To lace, begin by threading string from inside to outside of bodice. Center string between bodice ends, then cross lace and tie strings at bottom (c).

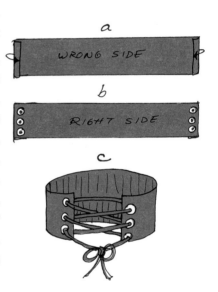

3. To make curve-ended bodice: Cut material 26″ by 6″. Place flat, wrong side up, and draw lines A-B and C-D across each short end, 1″ in from edge. Measure, mark, and draw dotted line across length of piece, 1″ down from top edge. Mark points E and F on top line and G and H on bottom line 2″ out from cross points of dotted line with A-B and C-D, (a). Fold over end flaps along A-B and C-D and glue or sew down. Draw, then cut out curves between points A and E, F-C, G-B, and H-D (b). Complete bodice as in *step 2*, (b) and (c), but make 4 holes instead of 3 (c). Suspenders (*see page 50*) may be attached to top of bodice for a simple dirndl or peasant dress top (d).

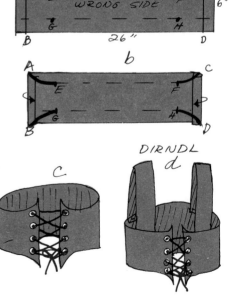

ROYAL CAPE

Materials: Any felt, wool, toweling, or other fabric (or substitute an old blanket or sheet or curtain, etc.), tape measure, scissors, ribbon, string, brooch, chalk.

1. Decide cape length by measuring from base of neck to desired hemline. Add 2″ if you plan to hem (*see page 34*). In example, cape length is 36″ with no hem. Fabric width needed is twice this distance, or 72″. Cut material 36″ by 72″. Fold in half to 36″ by 36″.

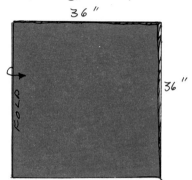

2. With chalk, sketch then cut neck curve 5″ down and 5″ to right of top left corner A. Sketch, then cut hem curve from bottom left corner B to top right corner C as shown. For circular cape, leave as is. For less wide cape, cut on dotted line. Hem if desired. Decorate with border or all-over designs (*see page 33, #7*). To fasten, pin with brooch or tie string through holes in neck corners as shown.

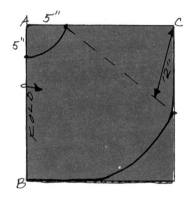

3. A simple collar may be made by folding back a border around neck opening (a), or by sewing on a decorative collar (b) as described on *pages 43-45*.

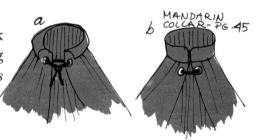

4. To make a hooded cape, sew on hood from *page 79*, for monk or Little Red Riding Hood. To trim royal cape with "ermine," *see page 30, #1(d)*.

71

ALL-PURPOSE SANDALS

Materials: Paper or felt, or plastic or leather cloth, ruler, scissors, string or leather thong, pencil or chalk.

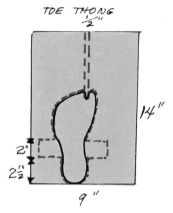

1. Cut 2 pieces of sandal material 9″ by 14″. Center heel of foot on bottom edge of each piece as shown and draw around foot. Also draw line against inside edge of each big toe from tip end of its nail to base.

2. On each piece, draw ½″-wide strip centered on big toe line, from toe base to top edge of material.

3. On each side of foot, 2½″ up from heel, draw instep tabs 2″ wide and 2″ high. Repeat for second foot.

4. On each piece, cut out around dotted lines, following foot outline *plus* two side instep tabs and toe thong. Try on as shown, pulling up narrow strip between big and second toes and pulling up instep tabs on each side. Cut off big toe thong at point where it meets side tabs. Remove sandals.

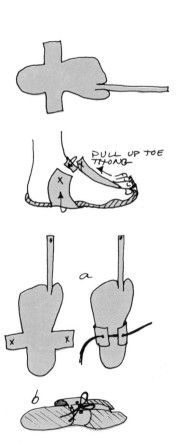

5. With pencil point or scissors, poke holes (X) in ends of big toe thong and side tabs. If sandal is paper, glue notebook ring reinforcements on wrong sides of all holes. Also, reinforce inside surfaces with pressed-on Mystic or masking tape for extra strength. Cut two 16″ strings or leather thongs. For each sandal, put foot flat on sole, pull up side tabs, and thread string through both from underneath, as shown (a). Then pull big toe thong up between toes and thread *both* ends of string through hole in this strip *from under side*. Pull strings gently until fit is snug; tie in bow (b).

Note: To make sandals more durable, draw around soles of feet on cardboard, cut out and glue to bottoms of sandal soles. For oversize clown feet, glue or staple sandals to extra-large cardboard soles (c).

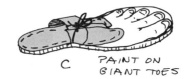

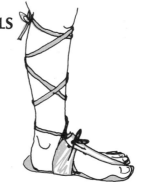

C PAINT ON GIANT TOES

GREEK AND ROMAN SANDALS

Make Sandals *above*. Then cut two long, (48" or more) strings or thongs and pull them under your instep (before putting on sandals). Put on sandals as described above, then crisscross strings up leg and tie on calf.

SNEAKER SANDALS

An old pair of sneakers or other discarded canvas shoes may be cut away with scissors or Exacto knife as shown, leaving laces and ankle seam intact.

TURKISH OR MEDIEVAL SLIPPERS

Materials: Extra large pair of socks, chalk, needle and thread, cotton (surgical type), wool or felt for trim and sole (optional).

1. To make first slipper, turn a sock inside out. Put foot in sock and make chalk mark where your toes end. Remove sock and set flat as shown. Sew a running stitch (*page 35*) from top and bottom of instep in lines curving up to tip of toe.

2. Turn sock right side out. Pull toe point all the way out, coaxing fabric with a pin if necessary. Stuff toe with cotton, making toe into firm up-curved shape. Sew a bell or pompom (*page 41*) on tip.

3. Complete cuff as you like. Top may be rolled down into a rolled cuff (a), trimmed with a felt cuff (b) from *page 47*. For a sturdy sole, draw around your foot on a piece of felt, cut out and glue or sew to bottom of sock. Repeat for second slipper.

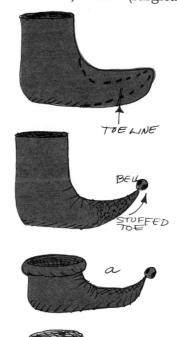

TOE LINE

BELL

STUFFED TOE

a

b

BOOTS

Materials: Regular pair of shoes, felt or paper (any paper about as stiff as construction paper), or plastic or imitation leather, tape measure, scissors, pencil or chalk, plastic or paper doilies or lace (optional), masking tape.

1. Boots in example are 12″ tall, with 4″ folded-over cuff and they fit over your regular shoes. To make boots fit your legs, take following measurements:

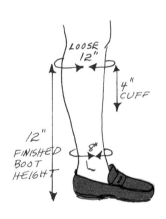

   a. Measure distance from floor to boot top (12″). Decide width of turn-over cuff (4″) and add them together (16″). Add 2″ more for under-foot tabs, making total height (18″).
   b. Use tape measure to measure around widest part of calf of your leg (12″). Add 8″ for total width (20″).
   c. Measure loosely around ankle (8″). Divide this in half (4″) and add 2″ for total (6″).

2. Cut material for one boot the height of measurement in *step 1*, (a), (18″), and width of *step 1*, (b) (20″). With right sides together, fold in half to 18″ by 10″ and arrange as shown.

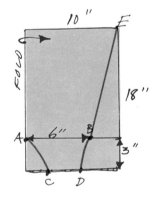

3. Measure, mark, and draw a line across material at point A, 3″ up from bottom. Measure and mark point B, 6″ (*step 1*, (c)) over along line A. On bottom edge, mark point C, 1½″ over from bottom left corner, and point D, 2″ past C. Draw, then cut curved lines A-C, D-B, and straight line B-E (*top right corner*).

4. Staple or sew seam about ¾″ wide along edge B-E. If using fabric, sew seam with wrong sides out, then turn right side out. If desired, cut scallops or notches in top cuff edge. Fold down cuff. Repeat for second boot.

5. Trim cuffs with doilies, lace, etc., or leave cut edge as in *step 4*. To wear, remove shoe, slide boot top up leg with seam in back; *put shoe back on*, then pull flaps C-D under bottom of shoe instep and tape flaps together under shoe to hold.

ROMAN AND GREEK SOLDIERS' GREAVES (SHIN GUARDS)

Materials: Colored paper, stiff felt or flexible cardboard (covered with tinfoil if desired), tape measure, ruler, scissors, pencil, notebook ring reinforcements, string, ribbon or leather thong, trimming material (*see page 30*), silver or gold spray paint, paper hole punch (optional).

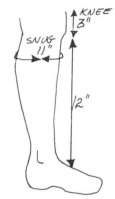

1. Part of a suit of armor, greaves protect the leg from the top of the instep to the top of the knee-cap. First, measure around widest part of calf with tape measure (in example, 11″); subtract 2″ for total width of pattern (9″). For height, measure from top of instep to bottom of knee (12″), then add 3″ to cover kneecap; total height is 15″. Cut two pieces of material 9″ by 15″.

2. Place material as shown, short ends top and bottom, wrong side up. Measure, mark, and draw line dividing piece in half lengthwise (5½″ in from each side, dotted line). Measure around your ankle (7″), subtract 2″ for total (5″) bottom width. Draw lines A and B, 5″ apart, centered on dotted line (A and B are 2½″ to each side of dotted line). Draw line C across top of piece at instep-to-knee height (12″). Mark point D, 1″ up from bottom edge on dotted line.

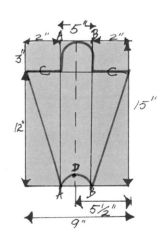

3. Draw, then cut curve A-D-B. Draw, then cut out straight lines from A to left edge of line C, and from B to right edge of line C. Cut in along both sides of line C *just* until you cross lines A and B; sketch, then cut curve from cross points around top center of material. Shape of this curve may vary as you wish.

4. With paper punch or pencil point, poke about 5 holes ¾″ in from edges A-C and B-C as shown. *Note*: Space holes evenly and try to get them opposite each other; if material is folded in half, both sides can be punched at once. Glue notebook ring reinforcements around each hole. Cut two 28″ strings or thongs. Tie one string in top hole of each side. Center on leg right side out and crisscross laces up back; tie at bottom to hold. Repeat *steps 2, 3, and 4* for second shin guard.

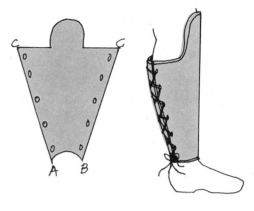

ANIMAL TAILS

Materials: #18 or stiffer flower stem wire or coat hanger and wire cutters, flexible cardboard, masking tape, scissors, tail covering material (wool, felt, raffia, etc.), glue.

1. Decide length of tail needed. Whatever this length, use wire twice as long. In example, 8″ poodle tail uses wire 16″ long. Bend wire in half.

2. Cut 4″ circle (*see page 29*) of cardboard. Poke 2 holes about 1″ apart in circle center. Poke ends of wire through holes and bend about 1½″ of each end over at right angles taping it firmly to bottom of cardboard disc.

3. To attach tail to costume:
   a. Tied-on tail:
   At opposite sides of disc (*made as above*), poke a hole 1″ in from edge. Tie a 30″ string in each hole, place tail in position on costume, and tie string around hips, knotting in front. String can be outside or inside costume.
   b. Permanently attached tail:
   Cut small hole in costume rump where tail will be. Sew or safety pin disc to *inside* seat of costume, making tail stick out hole.

4. To complete tail, curve wires into desired shape, and cover with glued-on wrapping of wool, rope, felt strips, raffia, cloth, fake fur, crepe paper, etc. For poodle, trim tail with pompom (*page 41*), sewn on end.

WINGS

Materials: Stiff paper or cardboard, stiff but flexible wire, string or elastic sewing cord, ruler, scissors, pencil, stapler, masking tape, cheesecloth or net or organdy, glue or needle and thread.

1. To make paper or cardboard angel's wings, cut material 16″ by 36″. *If using flexible material*: fold in half, fold left, to 16″ by 18″. To begin, make center square: Along center fold, mark point A, 6″ down from top, and point B, 1½″ to its right. Mark point C, 3″ below A, and D, 1½″ to right of C. Draw rectangle connecting points (*dotted area*). *If using stiff material*: leave flat, 16″ by 36″, and draw line dividing it in half crosswise (18″ from end). Copy pattern shown on both sides of center line; center square will be 3″ by 3″. Wings may be drawn any shape you like, (those shown are for angels, fairies, birds, etc.), but *must* start at edges of center square, which is for fastening wings.

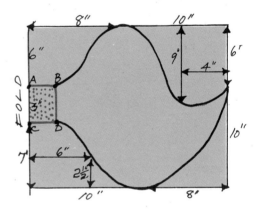

2. Cut wings out; open folded wings flat. To stiffen wings, cut wire 52″ long. Tape down as shown (*dotted line*, (a)); be sure wire crosses center square twice. At this point, if you want to cover wire, a second pair of wings can be cut by drawing around the first pair. If you do this, add cardboard square from *step 3* below, then glue layers together, covering wires (b). Or, tape or glue a 3½″-yard wire around wing outline, 1″ in from edge. Glue wing edge over onto wire (c).

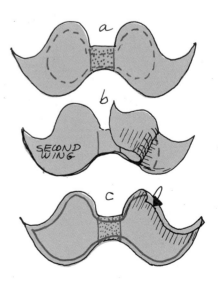

3. Center square may be stiffened by glueing a 3″ square of stiff cardboard over wires on top of original center square.

4. To fasten wings, poke two holes about 2″ apart in center square between wings. Cut a 60″ piece of string and thread it through holes (a). Put wings on your back, tie string in front of chest. Hold knot in place with pin or catch it on button. Or, poke 2 sets of holes about 2″ apart in center square as shown (b). In each, thread a 24″ piece of ⅛″ wide elastic cord; tie elastic ends, making two loops. To wear, place wings flat against back, pull loops over shoulders. Retie loops if necessary for snug fit. To decorate wings, paint, crayon, spray, or *see page 31, #3*.

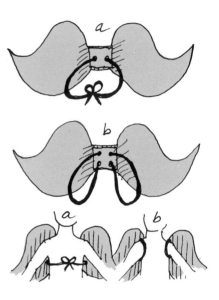

5. To make transparent insect or fairy wings, cut one pair of paper wings (*p. 77, step 1*) as a pattern. Cut a wire about 3½ *yards* long (or whatever fits outline of wings) and bend it around the pattern outline. Twist the ends to hold; then cut a 12″ wire and cross it several times over center (mid-wing) area to hold outline wire in shape (a). Discard pattern, Cut a 3″ square of stiff cardboard and tape it firmly to crossed central wires. Cover outline wire with cheesecloth, net, organdy or other delicate and transparent cloth (b) sewn or stapled over wire. To trim, *see page 31, #3*. Use for butterfly, insect, bird, etc.

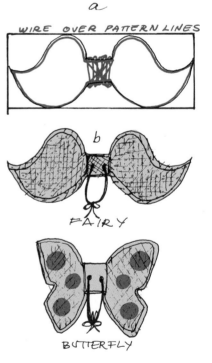

5. HATS

HOOD

Materials: Fabric or felt, tape measure, chalk, scissors, needle and thread, straight pins.

1. Cut fabric or felt 28″ long by 12″ wide for close-fitting hood (*pattern a*). For peaked hood, cut material 18″ wide (*pattern b*). Bring short ends together, folding piece in half crosswise, with right sides together, fold at top.

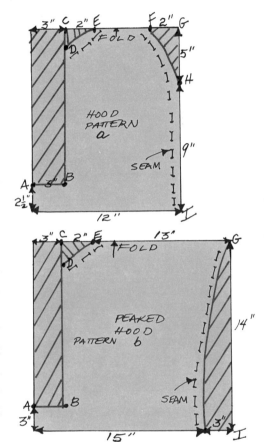

2. On left edge, mark point A, 2½″ up from bottom left corner. Mark point B, 3″ over to right of A. Mark point C, 3″ over from top left corner. Draw line A-B-C; cut on this line and discard striped area. Mark point D, 1″ below point C, and point E, 2″ to right of C along fold. With chalk, draw curve between D and E; pin, then sew (*see Running Stitch, page 35*), along this line. Mark point F, 2″ over from top right corner G, along fold. Mark point H, 5″ below G on right side. Draw, then cut along curve between F and H; discard striped area. Pin, then sew 1″-wide seam between points F and I, at bottom right corner.

3. Turn hood right side out, so seams are hidden. Seams may be ironed to make stiff fabric lie flat. Edging of line around face (A-B-D) may be hemmed (*see page 35*), cut with pinking shears, or left as is. Or, it may be covered with decorative braid, etc. (*See page 30, #1*). To wear hood, fasten overlapped neck tabs below chin using safety pin, snaps, etc. (*See page 32, #6*). To make medieval peaked hood, follow directions above, but use dimensions shown in *pattern b*.

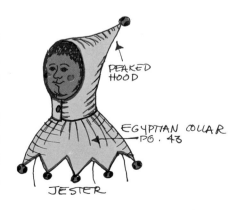

Hood may be used for animal characters with ears added as shown (felt or cardboard triangles sewn or stapled on). To cover shoulders, as for a jester, make an Egyptian Collar, *page 43*, of same fabric as hood, and wear them together.

BASIC HATBAND

Materials: Construction paper or flexible cardboard or stiff fabric (such as buckram), tape measure, scissors, stapler or needle and thread or glue.

Note: This hatband will sometimes be used as part of a hat and other times be only a pattern for the head size of hat brim.

1. Measure with tape measure around head just above the ears. Add 2″ for overlap fastening to determine total length (in example, 21″ plus 2″, or 23″).

2. Cut strip of material 1½″ wide by 23″ long. To fasten, wrap around head, overlap for snug fit, and staple or sew overlapped ends together.

BASIC CAP

Materials: Same as above, plus heavy-duty tinfoil and felt or fabric.

1. Make Basic Hatband as *above*. In addition, cut 2 strips of material ½″ wide by 12″ to 14″ or more long. (Strip length depends upon how deep a cap you want; a close-fitting skull cap may be 12″, a high-crowned cowboy hat 16″.) Staple or sew ends of bands to hatband, making a cross as shown. Try cap on; it should sit comfortably, well down on head.

CAP FRAME

2. Cut two 16″ (or larger) squares of heavy-duty foil. Set cap upside down on the two layers of foil and pull foil up and cover bands with it. Press foil over cross strips while supporting cap inside and out with hands. Try not to crush in or flatten cap. Pinch foil edges flat and round out cap.

FOIL COVERED

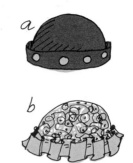

3. If desired, basic cap may now be covered with 16″ or larger circle of felt or fabric, stapled or sewn to hatband with hem folded under toward inside. To make a Gothic cap (a), staple or sew on a 1½″ by 23″ cuff trimmed with buttons or paper disc. For an eighteenth-century girl's cap (b) cover foil with cotton or lace and trim with 1″- or 2″-wide decorative ruffles around base (*see page 37*).

VIKING HELMET

Materials: Same as for Basic Cap, *page 80*, plus construction paper, masking tape, and enamel paint or spray (optional).

1. Make Basic Cap, *page 80.*

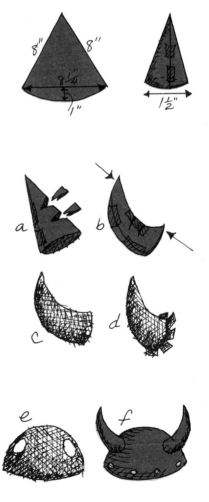

2. Draw, then cut 2 wedges of construction paper with sides 8″ long, bottom width 8¼″, and curve 1″ deep. Roll each wedge up into a cone and tape so base opening is about 1½″ across.

3. In each cone: Pinch sides together around the middle and cut out two small wedges about ⅜″ at top, stopping about halfway from other side of cone (a). Gently push ends of cone together (*arrows* (b)), curving cone up and closing wedge-shaped openings in side. Tape cone together in this position making horn. Keep horn rounded by pushing out from the inside. Cover horn with 6″ by 9″ piece of heavy-duty foil wrapped around (c). Repeat for second horn. In base of each horn, cut fringe ¾″ deep and ¾″ wide. Fold fringe upward (d). Cut two holes each about 1½″ across in cap, above ears, as shown (e). Poke horns through holes from inside out; tape fringe to inside of cap (f). Paint helmet whatever color you wish (such as brown, with horns left silver) or leave it unpainted. 1″-construction-paper discs or foil-covered buttons may be glued around base of helmet for trim. For complete Viking costume, *see page 16.*

COWBOY HAT

Materials: Same as Viking Helmet, *page 81*, plus lightweight cardboard.

1. Make Basic Cap, *page 80*. For brim, cut a 15"-cardboard circle (*see page 29*). Cover both sides of circle with heavy-duty tinfoil pressed on. Overlap and flatten foil on edges.

2. Set foil-covered brim flat on table. Center basic cap over center of circle; holding cap steady with one hand, lightly draw around its outline with the other hand as shown (a). Remove cap from brim. Measure and mark a second ring 1" *inside* drawn hatband ring (b). Cut out *inner* circle. Cut 1" deep, 1"-wide fringe all around inner circle up to original hatband ring as shown (c). Fold fringe over onto brim, so it stands up as shown (d).

3. Set cap down over fringe; reach inside and tape fringe to inside of cap hatband. Crease cap in center of top. Spray cap brown or other color. Ribbon or paper band may be added around base of cap, and fringe around brim.

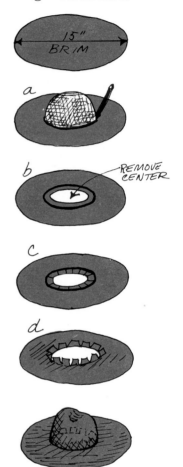

TRICORNE

Materials: Same as Viking Helmet plus stiff but flexible wire (such as #18, or coat hanger), and rubber cement.

1. Make Basic Cap and brim as in Cowboy Hat *above, except make two brims*, both of construction paper or other flexible material instead of cardboard. Foil cap may be covered with paint or sewn on fabric. Set aside second brim, and complete Cowboy Hat.

2. Cut 45″-long piece of wire. Tape wire firmly to top side of hat brim as shown (a), making triangle with points equally spaced, about 15″ apart, all around brim. Brush glue over top of wired brim, then press on second brim, sandwiching wire beneath it (b). Curve the three sides of brim up as shown (c), folding on dotted lines so each point of the wire triangle becomes the center of a curved side (d). Use tricorne for eighteenth-century man such as George Washington. To make a pirate hat, use this design or *see page 94.*

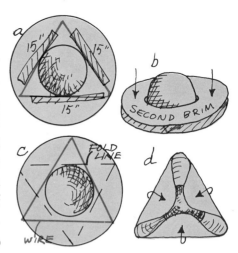

GREEK HELMET

Materials: Same as Cowboy Hat, *page 82*, plus silver-colored cloth tape (Mystic), and ruler.

1. Make foil-covered Basic Cap, *page 80.* Put cap on head. To measure width of brim, measure from bottom of hatband to top of your nose (in example, 3½″). Add 1″ for total width (4½″). To make brim, cut a piece of heavy-duty tinfoil 18″ by 27″. Fold foil across its width making a 4-layered piece 4½″ by 27″.

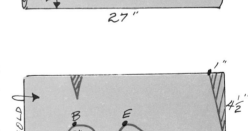

2. Fold foil piece in half to 4½″ by 13½″ and set flat, fold at left. Measure and mark a point (*arrow*) 1″ in from top right corner. Cut off diagonal from this point to bottom right corner G. Draw bottom edge with shape and dimensions shown. To begin, set ruler along bottom edge. Starting at fold, mark off points at 1″ (A), 3½″, 4½″ (C), 5½″ (D), 6″, and 9″ (F). Mark points B and E 2″ above the 3½″ and 6″ points. In the top edge, above point B, draw, then cut out a wedge 1″ wide and 1½″ deep. Draw curved bottom edge as shown, connecting points A, B, C, D, E, F, and G at bottom right corner. Cut out curved edge.

3. Open half-folded strip (being sure to retain its many layers) as shown (a). Overlap and tape wedge (*see step 2*) edges closed, making brim flare out. Below wedge, press a flat dart. Fit top edge of brim to inside cap headband and tape in place. Staple or tape (with silver-colored tape) ends of brim together in back (b).

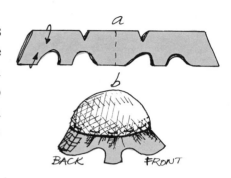

BACK   FRONT

4. To make helmet decoration, cut a cardboard semicircle 18″ in diameter (cut an 18″ circle in half, *see page 29*). Mark circle center, 9″ from edge. With compass or string, draw semicircle A, 4″ out from center point. Draw semicircle B, 5″ out from center. Mark point C, 1½″ in from bottom left corner, and point D, 7″ in from same corner. Draw straight line up from D to semicircle A; this marks point E. Measure 2″ out to left of E and mark point F. Draw line between E and F. Draw curve connecting C to F. Cut out along heavy black lines. Cover with foil if desired. Decorate with designs by inscribing foil with pencil or coloring with felt pens.

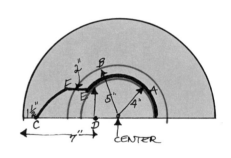

5. In border area between curves A and B, cut a 1″-wide fringe. Fold each piece of fringe in the opposite direction, one pointing to one side, one to the other (a). Place front end of shape about 3″ above front of cap base as shown (b) and fasten with silver tape down center of cap—pointed "tail" toward back.

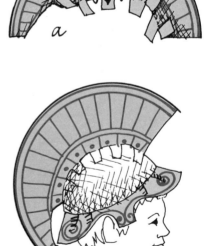

84

ROMAN HELMET

Materials: Same as Greek Helmet plus double-face tape.

1. Make foil-covered Basic Cap, *page 80*.

2. To make neck guard, use construction paper covered on both sides with heavy-duty tinfoil or use 4 layers or more of foil alone. Cut paper or foil layers into a piece 20″ by 14″; fold in half to 20″ by 7″, then in half again to 10″ by 7″. Trace and transfer (*see page 28*) pattern with dotted outline from *page 86*. Cut shape out.

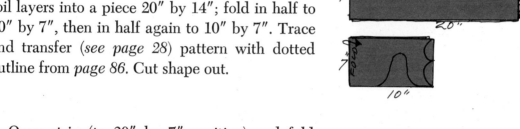

3. Open strip (to 20″ by 7″ position) and fold over several small tucks in top edge between A and B; press tucks flat to gather edge slightly and make brim flare out. Tape top edge to inside back of cap, so A and B are just over your ears and flaps 1 and 2 rest on cheeks. Trim cheek pieces if they are too large for you. Fold under any rough foil edges around brim.

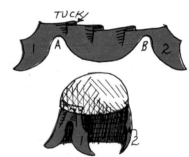

4. To make front brim, cut foil-covered construction paper (or several layers of foil) 12″ by 2½″. Fold in half to 6″ by 2½″. Draw curved shape shown, making small curl at right end about 1″ wide by 1½″ high. Cut out and unfold. Attach with double-face tape to outside front of cap as shown, *step 5*.

5. To make plume, use helmet plume, *page 112, step 2*. Or, trace and transfer (*see page 28*) orange pattern from *page 86* to foil-covered or plain cardboard. Cut shape out. Cut, bend, and attach fringe to cap top as in *step 5*, Greek Helmet, *page 84*.

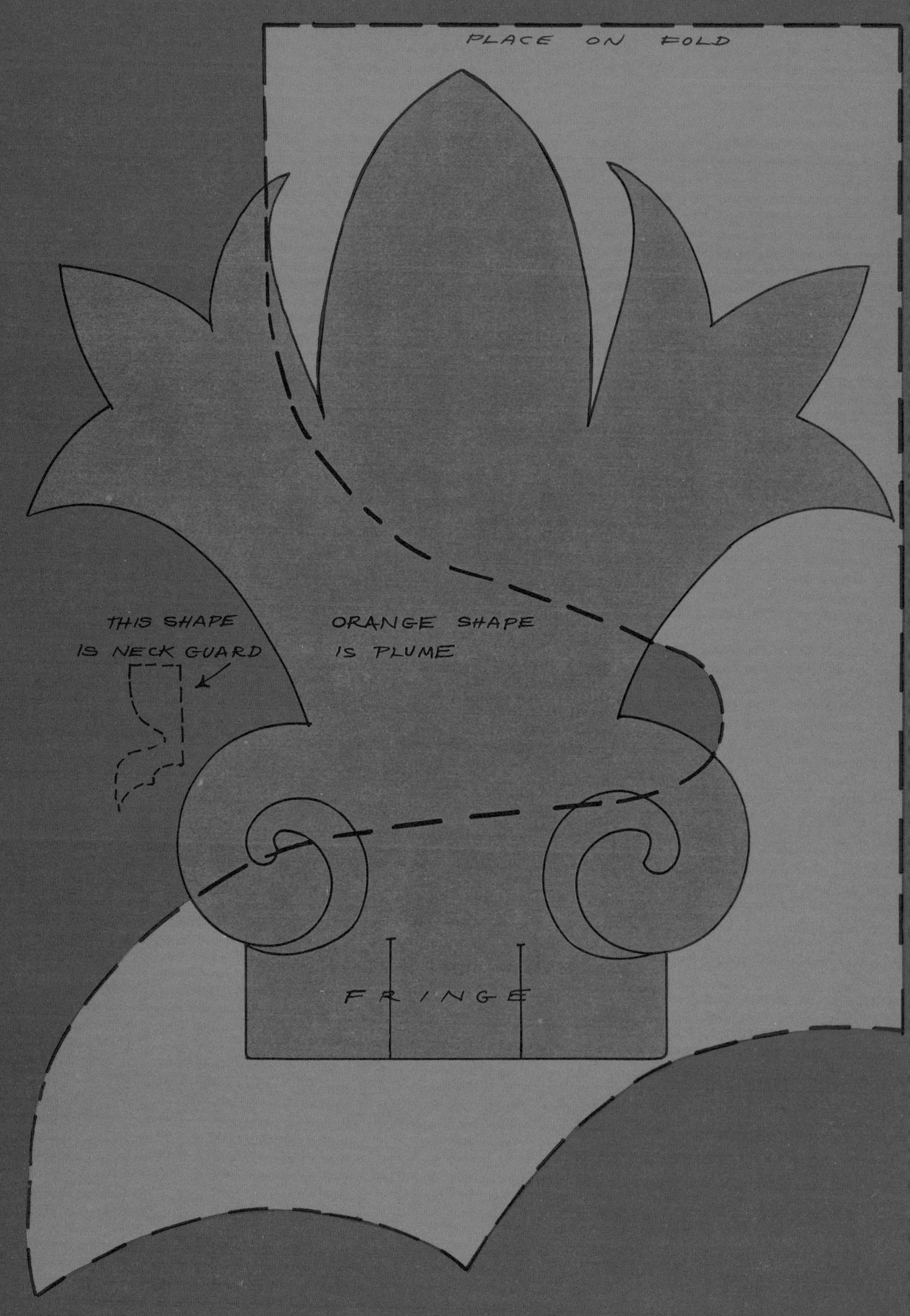

PLACE ON FOLD

THIS SHAPE IS NECK GUARD

ORANGE SHAPE IS PLUME

FRINGE

EGYPTIAN QUEEN'S HEADDRESS

Materials: Same as Basic Cap, *page 80*, plus stapler, felt or fabric, and needle and thread.

1. Make Basic Cap as on *page 80*. Paint cap in all-over feather design. *Note*: To paint tinfoil, use enamels or oil-base felt pens.

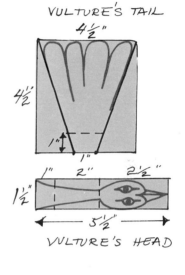

2. Cut vulture head and tail from construction paper. Tail is basically a flat-ended triangle 4½" at top, 1" at bottom end, and about 4½" tall. Draw a dotted line across, 1" up from bottom end. To make head, cut a piece 1½" by 5½" Draw dotted lines 1" up from bottom and 2" above that as shown. Shape rounded head and pointed beak in remaining 2½" as shown. Cut out, tapering width of neck down to about 1¼".

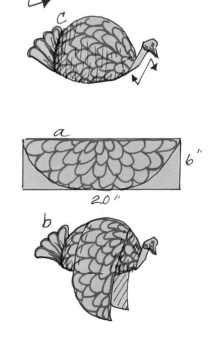

3. Decorate front and back of tail in feather design. Then fold under bottom 1" as shown (a); fit center back of cap into fold and staple or sew down (b). Draw eyes on top of vulture's head. Bend head down on first dotted line, and bend bottom of neck backward on second dotted line as shown. Fit center front of cap into *bottom* neck fold and staple or sew in place (c).

4. To make wings, cut felt, foil, or fabric strip 6" by 20". Cut the short ends in a curve as shown. Decorate with feather designs (a), then sew down as shown (b) over top of cap so wings are centered and extend below cap an even distance on both sides. You may have to tuck or gather edges of strip slightly while sewing, to make piece lie smoothly against cap.

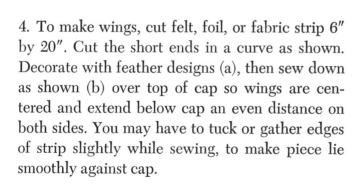

STOVEPIPE HAT

Materials: Construction paper or bristol board (2-ply) or flexible cardboard, ruler, scissors, pencil, stapler, rubber cement or masking tape.

1. Make Basic Hatband, *page 80*, and staple it into a ring, correct size. Make brim of hat as in Cowboy Hat, *page 82, steps 1 and 2, but* use hatband instead of cap for measuring hole in center of brim. It is not necessary to cover brim with tinfoil. Save circle removed from brim center for later use as top of hat.

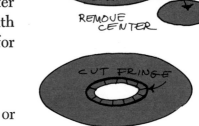

2. If you wish to cover hat brim with fabric or paper, trace twice around brim on covering material, making top and bottom brim covers and glue them on. Or, if brim is left plain, entire hat may later be painted.

3. To make "stovepipe," decide how tall you want finished hat to be (in example, 10″). Add 1″ to this for total unfinished height (11″). Cut piece of hat-colored construction paper (or cardboard to which colored fabric has been glued, or plain cardboard which may later be painted) 11″ by 23″. Measure, mark, and draw a line across 1″ down from top edge. In this border (a), cut 1″-wide fringe down to drawn line. Turn material wrong side up and bend top fringe down (b).

4. Pull short ends of piece around (with right side out, fringe pointing in to center) to make a tube same size as hole in brim; tape or glue tube sides together (a). Tape or glue brim fringe to *inside* of tube's bottom (unfringed) edge (b).

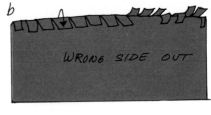

5. Set hat brim down. Spread glue on tops of top fringe (*arrow*, a) and also around edges of brim

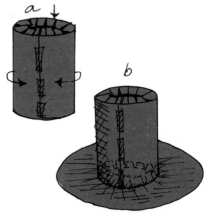

circle saved from *step 1*. Press glued surfaces together, fitting circle onto tube top with one hand inside for support. Paint or decorate hat, trim with ribbon or paper hatband. Use for Mad Hatter from *Alice in Wonderland*, or *Doctor Dolittle*, or nineteenth-century man such as Abraham Lincoln. For a clown's hat, make stovepipe 5″ tall and cut brim down to 2″ wide.

DANIEL BOONE CAP, PILLBOX, OR FEZ

Materials: Same as Stovepipe Hat plus crayons, paints, or scraps of fake fur.

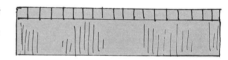

1. Make Basic Hatband, *page 80, but* cut strip 3½″ wide by correct length. Do not staple strip into ring at this time. Measure, mark, and draw line across one long side 1″ in from edge. Cut 1″-wide fringe in this border.

2. Decorate strip below fringe, coloring gray, brown, and black like raccoon fur, or glue on strips of fake fur. Fold fringe over onto wrong side, then pull short ends of piece around into a ring, with fringe end up and fringe pointing inward. Fit ring on head as for hatband, and overlap, mark, and staple ends of band together.

3. To make top of hat, set hatband on another piece of similarly colored material. Draw around outline of ring as shown (a). Cut out circle, glue (b) to fringed top of hatband (*see Stovepipe, step 5*).

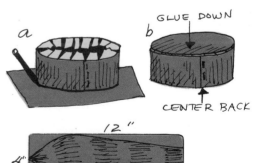

4. Cut out tail about 4″ wide at widest point and about 12″ long, tapering to points at each end. Color like hat and staple or glue tail to center back of hat. *Note*: To complete Daniel Boone's costume, decorate pajama jacket and trousers with fringe (*page 38*) and make rifle (*page 55*).

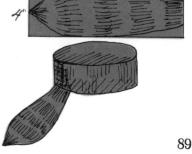

5. To make pillbox, follow *steps 1, 2, and 3* above. Omit tail, and color or trim as you wish. Use for national folk costumes such as a Greek woman's hat or a Yugoslavian shepherd's cap.

YUGOSLAVIAN SHEPHERD'S CAP

GREEK PILLBOX (EPIRUS)

10"

6. To make Middle Eastern, African, or East Indian fez, make 6″ or 7″ tall cap as in *steps 1, 2, and 3 above*. Omit tail. Or, make Puritan Man's Hat, *page 91, without brim.*

WEST AFRICA   MOROCCO TURKEY   INDIA INDONESIA NIGERIA

TOY SOLDIER'S OR DRUM MAJOR'S HAT OR CIVIL WAR CAP

Materials: Same as Stovepipe, *or* lightweight baseball cap or painter's peaked cap and cardboard paint-mixing bucket, 5¾″ tall with top diameter 7″ (*see Materials List*), stapler, paint.

1. *Method I*: Make a 9″ tall hat *without a brim*, as *on page 89, steps 1, 2, and 3*. To make peak, cut a half-oval as shown (a), with 8″ base and 5″ at center. Draw curve at base line and cut 1″ fringe into it as shown. Fold up fringe; glue, or tape it to inside front of hat (b).

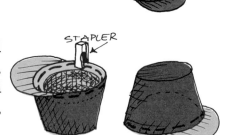

a 8"
5"
9"
b

2. *Method II*: Place cardboard bucket opened end up on table. Fit cap upside down inside it (a), and staple edge of hat to top edge of bucket and peak. Trim either style hat with braid (*page 30, #1*), pompom (*page 41*), or feather (*page 40*).

STAPLER

3. To make Civil War Cap, make basic hat as *above, Method I or II, but* cut top of hat on a slant (a) before making cover for it. Paint navy blue or gray and trim with paper or ribbon strap which rests above peak and is fastened by 2 small brass (yellow) discs.

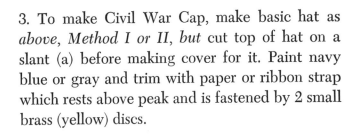

a

PURITAN MAN'S HAT

Materials: Same as Stovepipe.

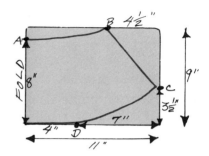

1. Make brim as in Stovepipe, *step 1*. Make brim covers if you wish, *step 2*. To make top of hat, cut paper 9″ wide by 22″ long (or whatever your hatband length is). Fold (without pressing too hard along fold) strip in half to 9″ by 11″. Set as shown, fold at left. Measure and mark point A, 1″ down from top left corner; mark point B, 4½″ over from top right corner. Mark point C, 3″ up from bottom right corner, and point D, 7″ to left of bottom right corner. Draw curves A-B, and C-D; draw line B-C. Cut on lines and open flat.

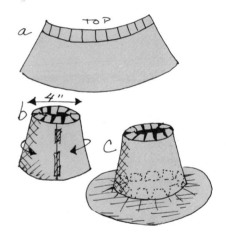

2. Measure, mark, and draw a line across 1″ down from top curved edge; cut 1″-wide fringe in this border (a). Fold fringe over into wrong side of strip, then pull short ends around (with fringe on top and pointing to inside) making a "cone" with top opening approximately 4″ across. Bottom opening should be big enough to fit inside hat brim. Tape or glue sides of hat together (b). Tape or glue brim fringe to inside of cone bottom (c).

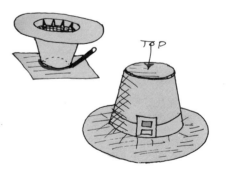

3. To make Puritan Hat top, turn "cone" upside down on paper and draw around top edge. Cut this circle out and glue it to top fringe as in Stovepipe, *step 5, page 88*. If you have not used colored paper, complete hat by painting black or navy blue and trimming with gray or silver hatband cut 21″ by 1½″. Add buckle (*see page 52*) to hatband in center front. This may also be used as witch's hat.

CONE HAT

Materials: Construction paper or flexible cardboard, tape measure, scissors, rubber cement or stapler, pencil, masking tape.

1. To make cone, cut paper 14″ wide by 24″ long. Mark center (*arrow*) of one long edge (12″

NAPOLEON'S OR PIRATE'S HAT

Materials: Construction paper, starched felt, or stiff paper, ruler, scissors, stapler, pencil, felt pens or crayons.

1. Make Basic Hatband, *page 80*. Cut two pieces of material 21″ by 12″. Place one piece wrong side up and draw shape shown. To do this, measure, mark, and draw line across 3″ up from bottom edge. Mark points 6″ in from each side along this line. Mark center of top edge (10½″ from end). Draw curved line as shown, between marked points. Cut on dotted line. Trace around this shape on second piece of paper, then cut out. *Note*: If you use a paper 42″ by 12″ or 21″ by 24″ folded in half, you can cut both halves at once.

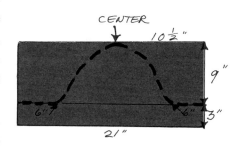

2. On wrong side of each hat half, mark center point of bottom edge; staple these points to two places directly opposite each other on hatband.

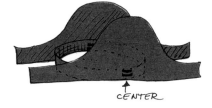

3. Staple or glue narrow ends of hat sides together (a). Trim with cockade, made of 3″ circle colored red, white, and blue, with 2 ribbon ties (b). Glue cockade to lower left front of hat for Napoleon. For pirate, draw or paste on cut-out design shown (c). For soldier or drum major, trim as shown (d). *Note*: If hat is too tall to feel comfortable, cut down top edge.

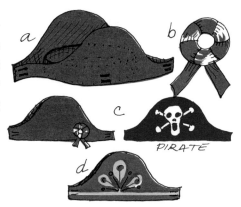

SUNBONNET

Materials: Construction paper, duplex crepe paper, felt, or fairly stiff fabric, scissors, ribbon or string, stapler or needle and thread, tape measure or ruler, pencil or chalk.

1. Cut material 21″ by 11″. With wrong side out, fold in half to 10½″ by 11″. With fold to left, draw

94

outline as shown. Measure, mark, and draw a dotted line across piece 2¼″ up from bottom edge. Mark point A on bottom edge 2″ out from fold; then cut a straight line from point A up to dotted line. Mark point B, 2½″ in from bottom right corner, and point C, 5″ up from this corner. Draw brim curve B-C-D as shown. Cut out curve.

2. Place piece flat, right side up and decorate with crayons, paints, etc. Then place flat, wrong side up, and fold up along dotted line (a). Fold up center flap #1, then pull flaps #2 and #3 up and towards the center, overlapping them and covering #1 to make a closed square end (b). Staple, sew, or glue all flaps together.

3. Cut 2 ribbons or strings each 20″ long; for an old-fashioned bonnet, use ribbon 2″ wide. Staple or sew ends of ribbon to wrong side of bonnet on each side just before brim curve as shown. Tie ribbons under chin.

4. Brim may be cut into scallops (*in step 1*) or folded back into a cuff, or trimmed with a ruffle (*page 37*). Use this bonnet for a milkmaid, a Puritan woman, or a nineteenth-century woman.

DUTCH GIRL'S CAP

Make Sunbonnet as above. Fold back brim into 2½″ wide cuff. Wire may be taped inside cuff for stiffness. Fold brim ends up into triangles as shown.

SCANDINAVIAN CAP

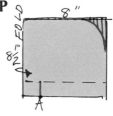

Cut material 16″ by 8½″; fold in half to 8″ by 8½″. Draw dotted line and point A as in Sunbonnet *step 1*. Cut on line A. Decorate right side as shown, then fold up and fasten as in *step 2*.

BISHOP'S MITRE

Materials: Construction paper or other stiff paper, ruler, scissors, paints, pencil, tape measure, glue or stapler.

1. Measure around your head as if for a hatband (*see page 80*) (in example, 21"). Add 2" for overlap fastening, making total (23"). Cut material 23" long by 17" high. *Note*: To use this pattern for a Ukranian girl's hat, cut it 12" high. Set material flat with short ends at sides. Measure and mark point A on bottom edge 2" over from right corner. Fold left short end over to line up with point A, leaving a 2"-wide strip at the right in single layer.

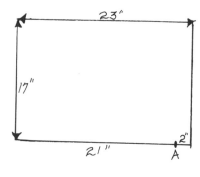

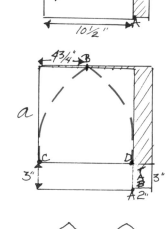

2. Measure, mark and draw line across 3" above bottom edge, from fold at left to right edge of single layer. Cut out shaded area on single layer leaving 3" by 2" tab on bottom as shown. Along top edge, measure and mark center point B of folded paper, (4¾" over from left fold). Sketch, then cut out curves (dotted lines) from points C and D, above 3" border line to top center B. Opened up, hat looks like this (b).

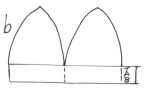

3. Decorate front and back panels. To complete, fold 2" by 3" tab over on dotted line so it faces wrong side. Glue or staple tab to inside edge of opposite side, fastening hatband. Use for Saint Nicholas's hat.

BISHOP

UKRANIAN GIRL

6. HEADBANDS

BASIC ADJUSTABLE HEADBAND

Note: There are two types of headbands, one made just like the Basic Hatband, *page 80*, whose size is fixed to fit your head exactly and the other an adjustable band (*described below*) which can be worn by people with different head sizes. The latter is best for hats which will be shared. A fabric headband may be made from a 2″ by 24″ ribbon or cloth pinned at back of head, or a 2″ by 19″ cloth sewn at both ends to a 2″-long piece of elastic.

Materials: Construction paper or fabric, tape measure, scissors, stapler or glue or needle and thread, pencil or chalk, string or rubber bands, notebook ring reinforcements.

1. To determine length of band, use tape measure to measure around head across forehead (in example, 21″). Cut strip width specified in directions (usually 1½″) by this length (21″). At each end of strip, fold back and staple, sew, or glue down a 1″ tab. In center of each folded tab, poke a hole, and (for paper only) glue a ring reinforcement around hole.

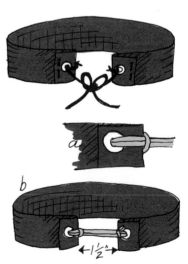

2. For under-chin or back of head ties, cut two strings or ribbons about 14″ long and tie one through each hole. For elastic fastening, use a sturdy rubber band (not less than 1/16″ wide and about 3″ long). To attach rubber band, pull one end of it through a hole, then back under its own end loop (a). Tie other end of rubber band in other tab hole and knot it as if it were string (b). Unstretched space between ends is roughly 1½″ after knotting. *Note*: A 3″ to 4″ piece of sewing elastic about ½″ to 1″ wide may be sewn or stapled to tabs for similar effect.

3. To wear, elastic goes in back of head or at nape of neck, under hair for girls. Strings or ribbons

are tied under chin or in back of head. Two of the simplest variations of this headband are the Tiara (a 4″-wide band cut with curved center and tapered sides (a)), and the Laurel Wreath (a 1½″-wide band trimmed with 24 tinfoil leaves each about 2½″ long, stapled on with leaves facing center front, (b)).

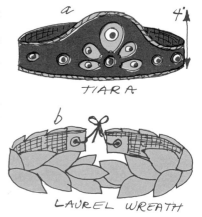

TIARA

LAUREL WREATH

CAT, DOG, MOUSE, MONKEY HEADBAND

Materials: Same as Basic Adjustable Headband.

1. Make 1½″-wide Adjustable Headband, *page 97*, with string ties. Make ears as shown; for most purposes, ears can be about 4″ high, tapered or rounded to 2″-wide base. Cut two pieces of material about 4″ by 4″ (or larger for larger ears). Draw ear shape on one piece, then hold pieces together and cut both shapes (a) out together. Fold up ½″ border on bottom of each ear (b).

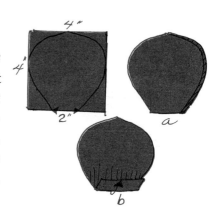

2. Divide headband in half and pinch or mark lightly at center point. Attach ears at equal distances from each side of center (a). To attach ears, slip headband inside the fold of each ear as shown (b), so ears fold down over top of headband. About ¼″ in from fold, staple ears to headband (c). Fold ears back along staple line (d), so they stand straight up. Tie strings under chin. *Note*: For additional cat costume, *see page 24*.

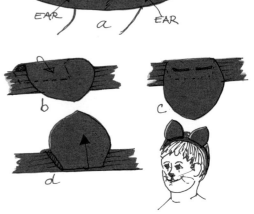

RABBIT OR FLOWER HEADBAND

Materials: Same as Basic Adjustable Headband plus flexible but stiff wire (#18); felt or fabric may be used instead of paper.

1. For Rabbit, make 3″-wide Basic Adjustable Headband, *page 97*, with string tie. Fold in half lengthwise for double-layer band 1½″ wide. To make ears, cut 2 pieces of material 10″ by 6″. Fold in half to 10″ by 3″. Mark X in center of top edge as shown (a), 1½″ from side. Draw curved top from X to bottom corners, then draw bottom edge in 2″ taper. Cut out each ear (b).

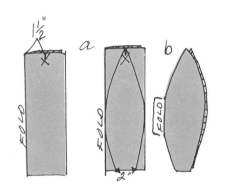

2. Cut 2 wires each 10″ long. Open each ear and tape a wire down center of inside. Spread both inside ear surfaces with glue and close, sandwiching wire inside. Bend up ¾″ at bottom of each ear.

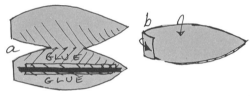

3. Divide headband in half and mark lightly at center point. At equal points from center, insert bottom edge of each ear between open edges of headband. Staple paper band closed over ear, or sew felt or fabric band. If stapling, be sure to place one staple over each ear wire. To wear band, bend ears straight up from band and curve them forward slightly at the tips. Tie strings under chin.

CENTER
RABBIT

4. For flower, first determine length of headband by measuring around head and under chin. Add 2″ for foldover of ends (in example 24″). Cut strip 1½″ by 24″. Complete ends as for Basic Adjustable Headbands, *page 97*, with string ties.

FLOWER

5. Cut 8 or 9 pieces of paper each 10″ by 6″. Cut, fold, and wire pieces into petals as for Rabbit Ears, *steps 1 and 2*. Attach petals side by side along entire length of headband as in *step* 3. Bend

petals over at right angles to headband and curl over the tips. To wear, tie strings under chin. Face is flower center and should be gaily made up (*see page 23*). *Note*: To complete flower costume, wear green turtleneck sweater and green tights; cut out large paper leaves and pin them to your sleeves.

CANDLE, INDIAN, INSECT, OR FAIRY HEADBAND

Materials: Same as Basic Adjustable Headband, plus #18 wire or chenille pipe cleaners (for insect or fairy).

1. Make Basic Adjustable Headband, *page 97*, 1½″ wide with string ties. For candle or feather, cut a stiff paper 5″ by 11″. Fold lengthwise to 2½″ by 11″. Sketch shapes shown (a); candle has "flame" about 2″ high and feather tapers to point at top and 2″-wide base at bottom. If desired for stiffness, cut 11″ piece of wire and tape to inside of candle or feather as in Rabbit Ears, *page 99, step 2*. For insect or fairy antennae, cut 2 pieces of wire or pipe cleaners 10″ long (b). (Twist 2 pipe cleaners together if necessary to make up length.) Curl down about 3″ of each wire into a spiral. Total length after spiral is roughly 7″.

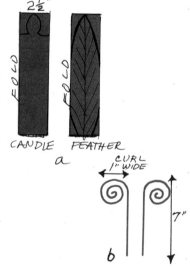

2. Tape or staple base of candle or antennae to inside center front of headband (a and b). Tape over back of staples so they are not rough. Wear as shown, fastened in back of head. Tape or staple feather (c) near tab at end of headband, so when worn feather is in center of back. *Note*: Indian motifs may be drawn on Indian brave's headband. Fairy headdress may be decorated by brushing wires and band with glue and sprinkling on glitter or sequins.

INDIAN WAR BONNET

1. Make Basic Adjustable Headband, *page 97*, 1½" by 42". Decorate front of band with Indian motifs as shown. Cut 17 feathers as for Indian *p. 100, step 1, (a)*. Staple or glue bases of feathers side by side along inside of entire headband. To stiffen feathers, they may be wired in the center as for Rabbit Ears, *page 99, step 2*.

2. Fold headband in half, right side out. Place its center in middle of forehead, wrap ends around head until band fits snugly; staple bands together where they meet (*arrow*). Staple or glue together remaining band ends, making single feathered tail hanging down back. Side discs 2" across and fur tails may be added to front of headband as shown.

HALO

Materials: Same as Basic Adjustable Headband, plus #18 wire or pipe cleaners, tinfoil, rubber cement or white glue, metallic glitter.

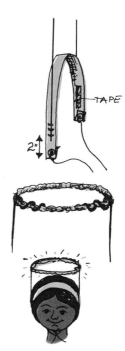

1. Make 1"-wide Basic Adjustable Headband, *page 97*, with string ties. Cut 2 wires each 7" long. Tape one end of each wire 2" up from string hole on each end of headband as shown. Staple over wire ends 2 or 3 times for added strength, then tape over backs of staples to cover sharp edges.

2. To make halo, cut 24" wire or several pipe cleaners twisted together, and cover with 24" by 1" strip of tinfoil pressed on. Brush foil with glue and sprinkle on glitter. To attach halo, bend the end of each straight wire around opposite sides of halo. To wear, place band over top of head, tie strings under chin.

CROWN

Materials: Same as Basic Adjustable Headband plus #18 wire, rubber cement, and trimming.

1. Cut paper 24″ long, 12″ wide. With right sides together, fold in half lengthwise to 24″ by 6″. With fold at bottom, wrap strip around head and mark where ends overlap. Remove and set flat as shown, marked edge up and fold at bottom.

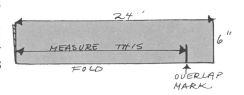

2. Measure distance from overlap mark to end farthest way (*dotted line*, in example, 22″). Divide this distance by 4 (5½″). At top and bottom edges, measure and mark sections 5½″ apart from overlap mark to farthest end. Draw vertical lines between top and bottom section marks. Draw line A across band 2″ up from folded edge. Above line A, draw scallops between each 5½″ section as shown. Bottom edges of scallops sit on line A, and vertical lines mark centers of each point.

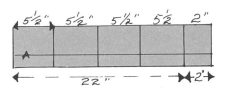

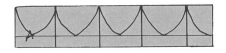

3. Cut out scallops. With marked side up, open piece flat. Cut 4 wires each 6″ long and tape a wire down center of each peak (over the vertical section lines). Brush glue over entire surface facing you, then fold bottom half up, joining edges evenly and sandwiching wire between the two halves. Trim with decorative braid or fake jewels (*page 30, #1 and #2*). Pull ends of band around and staple at overlap mark, making 4-pointed crown. Bend points over into graceful curves. Use for King or Queen.

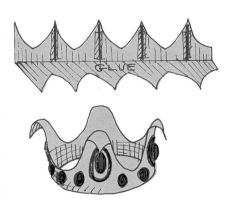

7. MASKS AND QUICK DISGUISES

CAT, OWL, AND FAIRY EYE MASKS

Materials: Construction paper, pencil, scissors, notebook ring reinforcements, string, crayons or paints.

1. For each mask, cut construction paper 18″ by 5½″; fold in half to 9″ by 5½″. Trace (*see page 28*) and transfer cat, owl, or fairy pattern from *page 105*. Set pattern's center line on paper's fold. At pattern points A and B, draw lines as shown, extending across width of paper to make side straps. Cut out.

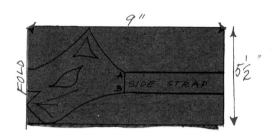

2. Open mask flat and decorate as follows:
   Cat: Draw face as shown, or make up your own designs. Whiskers may be drawn or made from glued-on straw. *Note*: To complete cat costume, wear hood (*page 79*), pajamas, tail (*page 76*), and claws (*page 25*).

COMPLETE AS IN STEP 3

   Owl: To make beak point outward, fold each side of mask over toward the front along dotted lines of beak (a). Then pinch sides of beak together. Decorate as shown (b), or make up your own designs. Feathers may be glued on for trim. *Note*: To complete owl costume, make full-length cape (*page 71*), and cover it with sewn-on 6″-long felt or crepe-paper feathers. Wear with sweater and tights of same color.

   Fairy: Add sparkle to mask and antennae by brushing on glue and sprinkling on metallic glitter or real feathers. *Note*: To complete fairy costume, make transparent wings (*page 78*), and very short gathered skirt (*page 66*) of net or organdy. Wear with leotard or body

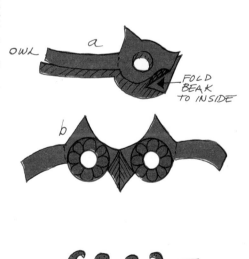

tights. To make wand, shape wire star, cover
it with foil and metallic glitter, tape to 12″
silver stick.

3. To complete masks, poke a hole ½″ in from
end of each side strap (*step 1*); glue notebook
ring reinforcements around holes. Cut 20″ strings,
tie one in each hole. To wear, tie strings behind
head.

MONSTER MASK

Materials: Plastic, 1-gallon bleach or cider jug (thoroughly washed and empty),
scissors, crayons or oil-base felt pens, string.

1. Wash out jug, then screw on the top. Place jug
as shown, and sketch two eye holes in flattened
area, or "face," beneath handle. Handle becomes
nose. About 3″ down from handle, draw a ring
entirely around lower half of jug. With scissors,
poke into this line, then cut around it removing
base, which is discarded.

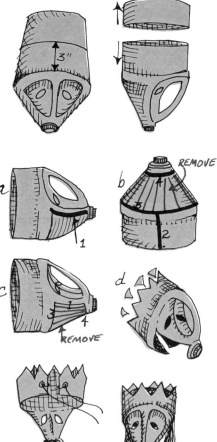

2. Place jug as shown, handle up. On *each side*
of flattened face area, draw line 1 as shown (a).
Then stand jug on end, cap up, and directly opposite handle, draw line 2, up the 3″-wide base.
Then draw rings 3 and 4 as shown (b) around base
of the cap and base of jug neck, *stopping at both
lines 1* (face sides, c). Cut up line 2, around 3, 1,
4, and 1 to remove striped area. To make a crown
headband, cut triangles into jug bottom as shown
(d). Cut out eye holes.

3. Color outside of mask as desired. Poke a hole
in each end of headband (jug bottom), cut two
15″ strings and tie a string in each hole. To wear
mask, tie strings behind head.

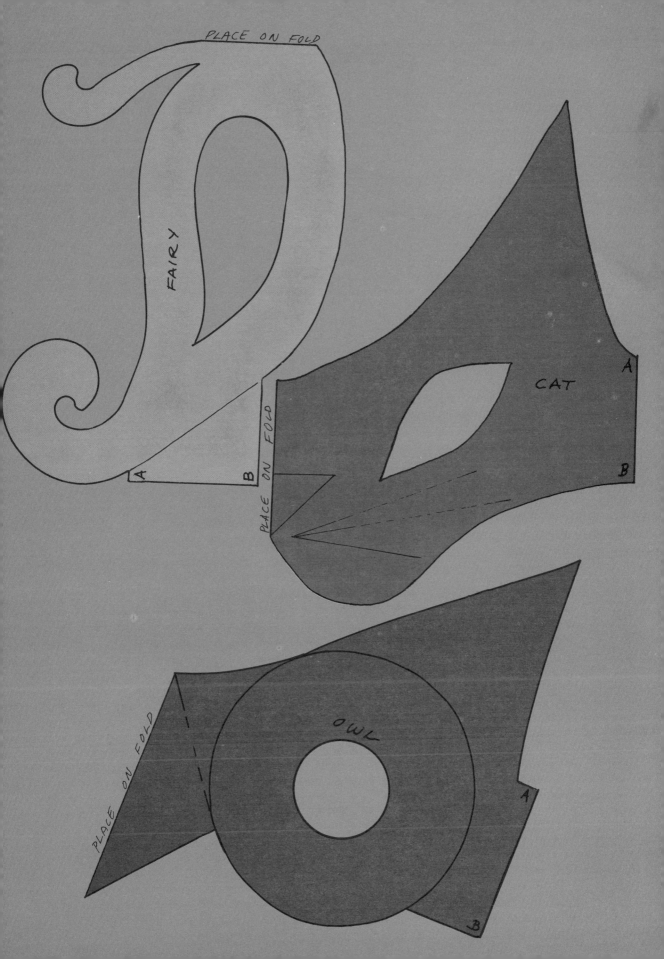

PUMPKIN HEAD MASK

Materials: Jumbo balloon (about 9″ by 5″ when deflated), heavy-duty tinfoil (at least 3 lengths 36″ by 18″), scissors, felt pens, spray enamel or spray lacquer, needle, paper cup or box, rubber cement and construction paper (optional).

1. While the example given is a pumpkin, this basic mask may be made for any character. Note that mask will cover your whole head. To begin, blow up balloon as large as you wish mask to be (minimum size should be slightly larger than your head); knot end. Completely cover balloon with at least 3 layers of heavy-duty foil pressed on. Ears may be added at top by pinching up foil points.

2. With needle, puncture balloon. In bottom of foil ball, cut a hole big enough to fit over your head. Fold under all cut foil edges and pinch them flat.

3. Put mask over your head, and gently press in eye positions. *Then remove from head, and cut out eye holes while supporting mask with one hand inside.* Cut mouth and nose holes if desired. Fold under and pinch flat all cut foil edges. Try mask on for fit, then remove and enlarge any holes that need it. Cut air holes in mask sides if needed.

4. Shape foil into final desired form, working with one hand inside and the other outside mask. Spray paint desired color. Trim as desired. You can make paper eyelashes and claws for an animal (*see pages 25, 38*), or use felt for pumpkin stem and leaves, or animal ears, tongue, nostrils, etc. Straw whiskers for a cat can be poked through the foil and glued in place. Glue or wool or raffia mane for a lion or horse. A black felt pen outline will exaggerate features.

PIG OR DRAGON HEAD MASK

Make basic mask as described above with the following additions: In *step 1*, between second and third layers of foil, set a paper cup (for a pig) or an empty oatmeal box or long tinfoil box or egg carton (for a dragon, horse, etc.) against the balloon to make a snout. The pig's cup should be taped in place bottom out (a); the dragon's snout-box should be taped on closed end out (b). When snout is well taped, add third layer of foil to cover, molding it well around the cup or box. Spray the pig pink or magenta and add round cut paper nostrils to end of cup and draw mouth curve with felt pen, (c). Spray dragon green (or purple or red, etc.), paint or glue on cut paper teeth. Trim as in *step 4, p. 106* (d). Cut additional air holes in mask sides if needed.

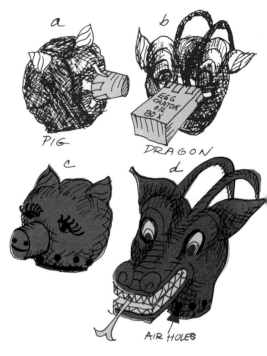

ASTRONAUT'S HELMET

Make Pumpkin Head Mask of tinfoil, and cut out a central face panel as shown. Bend under all sharp foil edges. Tape clear plastic or cellophane over face panel. Be sure to cut air holes around mask and in bottom of face panel. *Note*: To complete Astronaut's costume, use pajamas for suit, large metallic-colored boots (*page 74*), gauntlets (*page 47*), and various decorative dials, name tags, pockets, zippers, etc. sewn on. Back pack is a cardboard box attached like wings, (*page 78, step 4*(b), with dials and tubes of rubber or discarded vacuum-cleaner hoses.

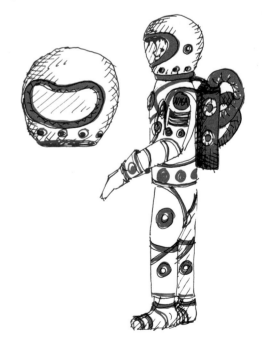

107

WOLF MUZZLE OR DOG SNOUT

Materials: Empty, ½-gallon milk carton, styrofoam ball (1″, 2″, or larger in diameter), rubber cement, paint or felt pens, string, scissors, Exacto knife (optional), masking tape, paper cup (9-oz. size).

1. To make Wolf Muzzle, tape top of empty ½-gallon carton closed (as it was before ever being opened). With knife or scissors point, cut air holes in side pockets of top fold as shown (*arrow*). Cut bottom end off carton, making muzzle as long as you wish. Wolf muzzle in example is 7″ long. Poke a hole (Y) in each side of carton, about 1″ up from bottom edge. Cut two 20″ strings, tie one in each hole.

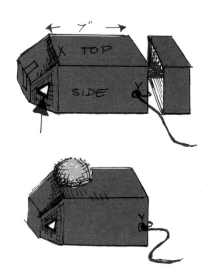

2. Cut styrofoam ball in half and glue one half onto carton at point X on top front.

3. Paint carton with temperas or spray enamel or decorate with felt pens. For wolf, make jagged teeth all around carton sides. When dry, pinch sides of carton together as shown (a); make crease along center of top (*arrow*). To wear, fit creased top over bridge of your nose, bottom of carton at or below your chin, and tie strings behind head (b). *Note*: If you need more air, remove mask and cut additional air holes in carton sides.

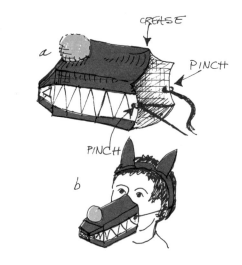

4. To make Dog Snout, cut 2 "bites" out of sides of paper cup as shown, one about 2″ across and 1″ deep (to fit over your mouth), the other 1½″ wide and ¾″ deep (to fit over your nose). In each side between cutouts, poke a hole (*arrows*). Cut two 20″ strings, tie one in each hole.

5. To make dog's nose, cut a curved piece off one side of a styrofoam ball (a) and glue ball to bottom of cup's side, in direct line with nose cutout (b). With pencil, poke air holes in bottom end of cup. Draw nostrils on styrofoam ball with felt pens. Draw mouth, teeth, etc. on cup sides. To wear, set up over your nose, with nose and mouth cutouts in place, and tie strings behind head (c). *Note*: For a pig's snout (d), leave off styrofoam ball, draw nostrils and mouth on flat cup bottom.

PINOCCHIO NOSE OR POINTED BEAK

Materials: Construction paper, scissors, pencil, string or colorless fishing line, felt pen, stapler or rubber cement, notebook ring reinforcements.

For a 3″-long nose or beak, cut in half a 6″ paper circle (*see page 29*); for a longer nose, use a larger circle, cut in half. To complete, pull half-circle around into a cone with a base about 1½″ across (a). Overlap side edges and staple or glue to hold. Poke string holes in opposite sides of cone about ½″ up from open end. Glue ring reinforcements around holes on insides. Cut two 23″ strings, and complete by tying one into each hole (b). Poke or cut air holes around cone base. To wear, set cone over your nose, tie strings behind head (c). This nose may also be used for a witch.

EYEGLASSES

Materials: 8″-long colored chenille pipe cleaners or regular pipe cleaners, colored cellophane (optional), cellophane tape, scissors, ruler.

1. Cut two 8″-long pieces of pipe cleaner. Bend each into a ring about 2″ across. Twist ends over each other to hold.

2. Cut one 3″ piece of pipe cleaner. Twist 1″ of it onto the inside of each ring, leaving a nose bridge about 1″ wide.

3. Twist one 8″ pipe cleaner onto outside of each ring as shown, making stems. Curve ends to fit over ears. Try glasses on; if stems are too long, cut off and reshape. Curve nose bridge out slightly. To hold glasses firmly to head, you may need bobby pins over the stems at your temples. *Note*: Wear glasses with Pinocchio Nose *page 109*, or, tape nose to bridge of glasses (*step 2*) for a one-piece disguise. Colored cellophane "lenses" may be glued onto glasses.

MOUSTACHE AND BEARD

Materials: Construction paper, ruler, tracing paper, scissors, string, double-face masking tape, felt pens or wool (optional), rubber cement (optional), notebook ring reinforcements.

1. Cut paper 8″ by 10″; fold in half to 8″ by 5″. Trace (*see page 28*) and transfer pattern from *page 49*. Cut out, but only cut through fold where indicated. Draw on hair, or glue on wool or raffia.

2. Poke a hole in each side of beard at point X. Glue a ring reinforcement on back side of each hole. Tie one 23″ string into each hole; to wear, arrange beard, then tie strings behind head. *Note*: To make moustache without beard, use only top part of pattern. Hold moustache on upper lip with double-face masking tape.

8. MISCELLANEOUS COSTUMES

LION BODY MASK

Materials: Extra large paper (*not plastic*) bag (usually found in discount or department stores) wide enough to reach easily across width of your shoulders and long enough to reach from head to knees (roughly 21″ by 30″), scissors, felt pens or crayons or paints, stapler or rubber cement, construction paper.

1. To begin, ask a friend to help you. With your arms at your sides, the bag should be slipped over your head. Have your friend pinch in, then *gently and lightly* mark with a crayon where your eyes, nose, and shoulders (points A and B) hit the bag. Remove bag. A and B will be about 11″ down from the top of bag (its closed end).

2. To make armholes, measure and mark points C and D on each side of bag, 7″ below shoulder marks A and B. Cut out curves between A-C and B-D.

3. To draw face, such as lion shown, first sketch a masklike shape around both eye marks as shown. Then lightly draw a wide oval or circle for the face; this can reach half or ⅔ the bag length. Draw a vertical line dividing the oval in half, for nose placement. Complete lion face as shown, or make up your own design. The lion's nose can be at least 4″ long, the space between nose and chin also 4″. Around the face draw or glue on wool for a mane. Draw on whiskers or glue on drinking straws or broom straws. Through front layer of bag only, cut out eye holes in mask area. If desired, cut out small nose hole.

4. To complete, cut 2 construction paper ears (about 5″ tall, 2″ wide at base) and staple or glue them onto top corners of bag. Paint mask color

desired; if you don't want brown bag, paint front and back a color before completing *step 3* (face). Be sure to color bright accent lines around sides of each eye hole. To add eyelashes, *see page 38.* For claws, *see page 25.* Wear with colored tights, and make tail as shown (drawn on back) or as on *page 76.*

SUITS OF ARMOR

Materials: Construction paper or 2-ply bristol board, heavy-duty tinfoil, ruler, scissors, notebook ring reinforcements, rubber cement, string or silver-colored ribbon, silver-colored cloth tape (Mystic tape), brass paper fasteners, masking tape, feathers (optional), cardboard tube (about 1½″ diameter), old tee shirt (of size that fits armor wearer), felt pen, stapler, Exacto knife (optional), old magazine, double-face carpet tape.

Helmet

1. For Roman and Greek helmets, *see pages 85 and 83.* For medieval helmet, make Basic Cap of foil, *page 80.* Then complete as described below.

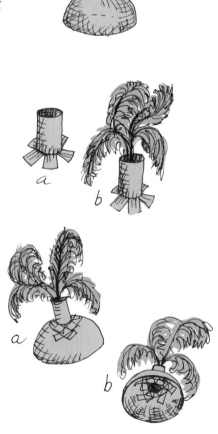

2. To make helmet plume decoration: Use Roman Helmet plume, *page 85, step 5.* Or, cut piece of cardboard tube about 3″ long and cover it with foil. Cut about 1″-deep fringe on bottom end. Bend fringe up (a). Make colored paper feathers (*page 40*), or use real feathers; poke feathers down into tube and tape their bases to inside of tube bottom, just above fringe (b).

3. To attach plume to cap, tape fringe firmly to top of cap as shown (a), using silver Mystic tape. Or, cut a hole diameter of tube in cap top; insert tube in hole from cap's wrong side, so fringe hooks on inside and only tube and feathers project out. Tape fringe to inside of cap (b). *Note:* If you follow *step b,* you may prefer to add feathers to tube after attaching tube to cap.

4. To make visor: Cut paper or flexible bristol board 6″ by 15″. Fold in half crosswise to 6″ by 7½″. Measure and mark point A, on open edge, 2½″ down from top; mark point B, 2½″ up from bottom, making area, between A and B, 1″ wide. Draw, then cut lines from A to corner C, and B to corner D. Open piece; wrap it completely with 12″ by 16″ piece of foil. Press foil to fit, pinch all edges flat (a). Place visor flat on magazine and, using knife or scissors, cut out 5 slits, (b) each about ½″ wide, across front of visor (b). Around sides of slits, pinch flat the jagged edges of foil. To attach visor to cap, use a brass paper fastener at each tapered visor and (*arrows* (c)); push fastener through sides of cap and tape over fastener tips on inside, so there are no sharp edges near your head.

Note: A Roman Helmet neck guard (*page 85, steps 2-3*) may be added to basic medieval helmet. Or, helmet may be worn on top of a gray or silver hood, *page 79*.

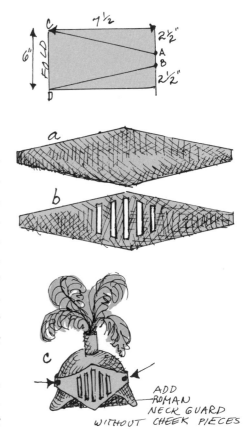

Medieval Breastplate

1. Set one of your tee shirts on a piece of cardboard or bristol board about 2″ larger than shirt all around (fold shirt arms in as shown). Draw outline about 2″ larger than shirt, and above each shoulder, draw 1″-wide tab (X). Cut out around outline, including shoulder tabs. Make second piece by tracing and cutting around first.

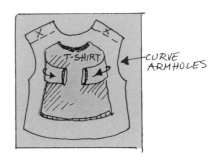

2. Cover both panels with tinfoil or paint or spray silver or gray. Poke about 4 holes ½″ in from each side of breastplate's front and back. Glue ring reinforcements (painted silver) on front and back of each hole, if it needs strengthening. To wear, cut two 30″ silver strings or ribbons and lace or thread them through holes in breastplate sides. Attach shoulders with staples or double-face tape or Mystic tape, overlapping X tabs and fastening. Make skirt as described on *page 114*.

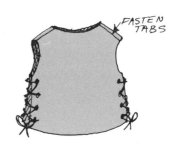

Greek or Roman Breastplate

1. Make Simple Laced Bodice, *page 70, step 2*, but measure width from your waist to armpit (about 7½″). Bodice has straight ends and will lace at one side or in center back. Use gray or silver material (*see page 21*), foil over paper, or painted paper.

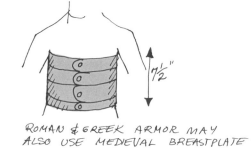

ROMAN & GREEK ARMOR MAY ALSO USE MEDIEVAL BREASTPLATE

2. For shoulder straps, make Suspenders, *page 50, step 1*, with the following changes: Cut width 2″; measure length over shoulder to 1″ below top of bodice in front and back (about 22″). For Roman straps, cut ends in scallops as shown (a). Leave Greek straps with straight ends. Suspenders slightly overlap outside surface of bodice top when attached. To fasten use staples, double-face carpet tape, or paper fasteners (ends taped on inside). Decorate Greek and Roman armor as shown (b) and (c), using felt pen or glued-on string (*see page 30, #1, (c)*). "Skirt" on Greek armor is described below.

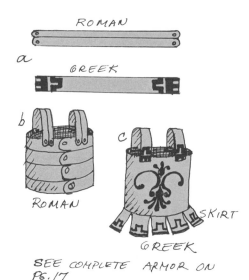

SEE COMPLETE ARMOR ON PG. 17

Medieval or Greek Skirt Armor

1. Cut 2 strips lightweight cardboard or paper, 20″ by 6″. Cover with foil. For medieval skirt, draw on metal bands across each piece as shown (a). Staple top of each strip to inside bottom edge of each half of breastplate, overlapping and folding tucks as you go, to make strip fit. Bottom of skirt flares out slightly (b).

2. For Greek skirt, cut fringe about 4½″ deep and 2″ wide in each foil-covered strip. Pinch cut foil edges flat. Staple uncut top border (1½″ wide) to bottom edge of bodice as shown, overlapping and folding tucks as you go, to make strip fit. Skirt flares slightly.

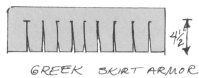

GREEK SKIRT ARMOR

Arm and Leg Armor

Greeks and Romans wear greaves (shin guards), made on *page 75*; they do not wear arm armor. Medieval armor requires tubelike guards from wrist to elbow, and from elbow to armpit, as well as from ankle to knee, and from knee to top of thigh. Make these pieces following Armlets, etc., *page 42*. Or, simply wear gray or silver long-sleeved sweater and tights to cover arms and legs.

Shoulder, Elbow, and Knee Armor

Needed only for medieval armor. Make six more visors as described on *page 113, step 4 but do not cut slits*. Shape each visor into "cup" to fit over shoulder, bent elbow or knee. Attach to arm and leg guards (or fasten ends of visor to each other) using double-face carpet tape or paper fasteners (ends taped inside).

Medieval Foot Guards

Cut two pieces of paper about 8″ by 7″. Mark center point of one 7″ end on each piece, then draw and cut out curved lines as shown (a), from straight corners to marked point. Cover each piece with foil; draw crosswise metal bands with black felt pen. Shape into half-cones to fit over top of foot (b). Attach to shoe tops with double-face carpet tape. Wear as shown.

Chain Mail

Worn by the Crusaders in the Early Middle Ages, chain mail may be made from an old, worn-out long-sleeved sweater of black or gray. Use turtle-necked sweater if possible. Dry brush silver or gray paint over knit texture for chain mail effect. Make gray knit hood (*page 79*) and paint as above. Black or gray tights or narrow trousers may also be painted with chain texture. A simple tunic (*page 60*) is worn over chain mail, along with Basic Cap of foil (*page 80*), with top shaped into slight point.

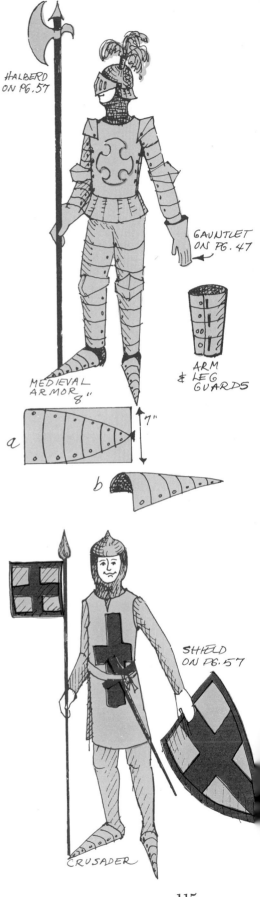

HALBERD ON PG. 57

GAUNTLET ON PG. 47

ARM & LEG GUARDS

MEDIEVAL ARMOR 8″

a

b

7″

SHIELD ON PG. 57

CRUSADER

115

TWO-OR-MORE-PERSONS ANIMAL

Materials: Fabric (old sheet, blanket, etc.), tape measure, stapler, scissors, needle and thread, wool or raffia, chalk, straight pins, mask materials (*see* Pumpkin Head Mask, *page 106*).

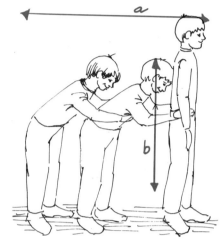

1. Line up two or more people to make up the animal. Stand together as shown, with first figure standing straight, the others bent over, each with arms around waist of figure in front. Have another friend take the following measurements of the group: (a) Distance from head of first person to tail, or back, of last person (in example 60″). Add 6″ for total (66″). Distance can be anything from 5′ to 10′ or more. (b) Distance from just above bent heads to just below knees (30″). Or, if you wants legs hidden, measure to ankles.

2. Cut fabric twice measurement (b) (2 x 30″ is 60″ total) by total length (a) (66″). Bring long edges of fabric together, folding it in half with right sides together. Along fold, measure and mark with chalk point W, 12″ in from left top corner. Pin, then sew (Running Stitch, *page 35*) curve from W to X at bottom left corner. *Note*: Seam W-X will rest on bent back of last person in line. Pin and test seam for comfort before sewing it. After sewing seam, leave a 2″ border *above* seam and cut off excess (striped area) triangle of fabric. On right side, sew 1″-wide seam from top corner Y to bottom corner Z.

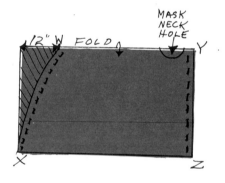

3. Make Dragon Head Mask, *page 107*, shaping head of Dragon or horse or elephant or other creature, but add a stiff foil neck coming out of mask base. Make neck long enough to reach your shoulders, where it will sit when worn. Cardboard tube can be taped to inside of neck to stiffen it, and the bottom edge bound with tape for greater comfort. If your animal has an extra-long neck, cut a hole in it for you to look through. Hole may be covered with dyed cheesecloth or left open.

4. Spray or paint mask to match color of fabric body. To make mask hole in fabric, cut a 2″ hole (*arrow, step 2*) in top of fabric fold beginning about 1″ behind seam Y-Z. Turn fabric right side out to hide seams. Enlarge mask hole until it is *just* large enough for first figure to poke his head through. Edges of hole can be turned under and hemmed, or cut with pinking shears. *Note*: If you are troubled by having such a large neck opening, have the first person wear a turtleneck sweater or Egyptian Collar (*page 43*) the same color as fabric animal body. To complete the beast, set mask on first person's head, resting foil neck on his shoulders. Other figures stand as in *step 1*, and all wear pants or tights of same color as animal body.

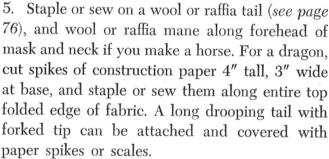

5. Staple or sew on a wool or raffia tail (*see page 76*), and wool or raffia mane along forehead of mask and neck if you make a horse. For a dragon, cut spikes of construction paper 4″ tall, 3″ wide at base, and staple or sew them along entire top folded edge of fabric. A long drooping tail with forked tip can be attached and covered with paper spikes or scales.

PAINT ON
SCALES

ARM HOLE
MAY BE
CUT INTO
SIDE

CLAWS ON
PG. 25

INDEX

Note: Costume suggestions appear in boldface type.

Fifteenth Century

**Seventeenth Century
France and England**

**Sixteenth Century
Elizabethan**